Mapping Wildfire
Hazards and Risks

Mapping Wildfire Hazards and Risks has been co-published simultaneously as *Journal of Sustainable Forestry*, Volume 11, Numbers 1/2 2000.

The *Journal of Sustainable Forestry* Monographic "Separates"

Below is a list of "separates," which in serials librarianship means a special issue simultaneously published as a special journal issue or double-issue *and* as a "separate" hardbound monograph. (This is a format which we also call a "DocuSerial.")

"Separates" are published because specialized libraries or professionals may wish to purchase a specific thematic issue by itself in a format which can be separately cataloged and shelved, as opposed to purchasing the journal on an on-going basis. Faculty members may also more easily consider a "separate" for classroom adoption.

"Separates" are carefully classified separately with the major book jobbers so that the journal tie-in can be noted on new book order slips to avoid duplicate purchasing.

You may wish to visit Haworth's website at . . .

http://www.haworthpressinc.com

. . . to search our online catalog for complete tables of contents of these separates and related publications.

You may also call 1-800-HAWORTH (outside US/Canada: 607-722-5857), or Fax 1-800-895-0582 (outside US/Canada: 607-771-0012), or e-mail at:

getinfo@haworthpressinc.com

Mapping Wildfire Hazards and Risks, edited by R. Neil Sampson, R. Dwight Atkinson, and Joe W. Lewis (Vol. 11, No. 1/2, 2000). *Based on the October 1996 workshop at Pingree Park in Colorado,* ***Mapping Wildfire Hazards and Risks*** *is a compilation of the ideas of federal and state agencies, universities, and non-governmental organizations on how to rank and prioritize forested watershed areas that are in need of prescribed fire. This book explains the vital importance of fire for the health and sustainability of a watershed forest and how the past acceptance of fire suspension has consequently led to increased fuel loadings in these landscapes that may lead to more severe future wildfires. Complete with geographic maps, charts, diagrams, and a list of locations where there is the greatest risk of future wildfires,* ***Mapping Wildfire Hazards and Risks*** *will assist you in deciding how to set priorities for land treatment that might reduce the risk of land damage.*

Frontiers of Forest Biology: Proceedings of the 1998 Joint Meeting of the North American Forest Biology Workshop and the Western Forest Genetics Association, edited by Alan K. Mitchell, Pasi Puttonen, Michael Stoehr, and Barbara J. Hawkins (Vol. 10, No. 1/2 & 3/4, 2000). *Based on the 1998 Joint Meeting of the North American Forest Biology Workshop and the Western Forest Genetics Association, Frontiers of Forest Biology addresses changing priorities in forest resource management. You will explore how the emphasis of forest research has shifted from productivity-based goals to goals related to sustainable development of forest resources. This important book contains fascinating research studies, complete with tables and diagrams, on topics such as biodiversity research and the productivity of commercial species that seek criteria and indicators of ecological integrity.*

> *"There is clear emphasis on the genetics, genecology, and physiology of trees, particularly temperate trees. . . . These proceedings are also testimony to what does or should distinguish forest biology from other sciences: a focus on intra- and inter-specific interactions between forest organisms and their environment, over scales of both time and place." (Robert D. Guy, PhD, Associate Professor, Department of Forest Sciences, University of British Columbia, Vancouver, Canada)*

Contested Issues of Ecosystem Management, edited by Piermaria Corona and Boris Zeide (Vol. 9, No. 1/2, 1999). *Provides park rangers, forestry students and personnel with a unique discussion of the premise, goals, and concepts of ecosystem management. You will discover the need for you to maintain and enhance the quality of the environment on a global scale while meeting the current and future needs of an increasing human population. This unique book includes ways to tackle the fundamental causes of environmental degradation so you will be able to respond to the problem and not merely the symptoms.*

Protecting Watershed Areas: Case of the Panama Canal, edited by Mark S. Ashton, Jennifer L. O'Hara, and Robert D. Hauff (Vol. 8, No. 3/4, 1999). *"This book makes a valuable contribution to the literature on conservation and development in the neo-tropics. . . . These writings provide a fresh yet realistic account of the Panama landscape." (Raymond P. Guries, Professor of Forestry, Department of Forestry, University of Wisconsin at Madison, Wisconsin)*

Sustainable Forests: Global Challenges and Local Solutions, edited by O. Thomas Bouman and David G. Brand (Vol. 4, No. 3/4 & Vol. 5, No. 1/2, 1997). *"Presents visions and hopes and the challenges and frustrations in utilization of our forests to meet the economical and social needs of communities, without irreversibly damaging the renewal capacities of the world's forests."* (*Dvoralai Wulfsohn, PhD, PEng, Associate Professor, Department of Agricultural and Bioresource Engineering, University of Saskatchewan*)

Assessing Forest Ecosystem Health in the Inland West, edited by R. Neil Sampson and David L. Adams (Vol. 2, No. 1/2/3/4, 1994). *"A compendium of research findings on a variety of forest issues. Useful for both scientists and policymakers since it represents the combined knowledge of both."* (*Abstracts of Public Administration, Development, and Environment*)

Mapping Wildfire Hazards and Risks

R. Neil Sampson
R. Dwight Atkinson
Joe W. Lewis
Editors

Mapping Wildfire Hazards and Risks has been co-published simultaneously as *Journal of Sustainable Forestry*, Volume 11, Numbers 1/2 2000.

Food Products Press
An Imprint of
The Haworth Press, Inc.
New York • London • Oxford

Published by

Food Products Press®, 10 Alice Street, Binghamton, NY 13904-1580 USA

Food Products Press® is an imprint of The Haworth Press, Inc., 10 Alice Street, Binghamton, NY 13904-1580 USA.

Mapping Wildfire Hazards and Risks has been co-published simultaneously as *Journal of Sustainable Forestry,* Volume 11, Numbers 1/2 2000.

The development, preparation, and publication of this work has been undertaken with great care. However, the publisher, employees, editors, and agents of The Haworth Press and all imprints of The Haworth Press, Inc., including The Haworth Medical Press® and Pharmaceutical Products Press®, are not responsible for any errors contained herein or for consequences that may ensue from use of materials or information contained in this work. Opinions expressed by the author(s) are not necessarily those of The Haworth Press, Inc.

Cover design by Thomas J. Mayshock Jr.

Library of Congress Cataloging-in-Publication Data

Mapping wildfire hazards and risks / R. Neil Sampson, R. Dwight Atkinson, Joe W. Lewis, editors.
 p. cm.
 "Co-published simultaneously as Journal of sustainable forestry, volume 11, numbers 1/2 2000."
 Includes bibliographical references.
 ISBN 1-56022-071-6 (alk. paper)–ISBN 1-56022-073-2 (pbk.: alk. paper)
 1. Wildfires. 2. Fire ecology. 3. Fire risk assessment. I. Sampson, R. Neil II. Atkinson, R. Dwight (Randy Dwight), 1954-III. Lewis, Joe W.

SD421.M353 2000
363.37'9–dc21
                       00-039406

INDEXING & ABSTRACTING

Contributions to this publication are selectively indexed or abstracted in print, electronic, online, or CD-ROM version(s) of the reference tools and information services listed below. This list is current as of the copyright date of this publication. See the end of this section for additional notes.

- *Abstract Bulletin of the Institute of Paper Science and Technology*
- *Abstracts in Anthropology*
- *Abstracts on Rural Development in the Tropics (RURAL)*
- *AGRICOLA Database*
- *Biostatistica*
- *BUBL Information Service, an Internet-based Information Service for the UK higher education community. <URL: http://bubl.ac.uk/>*
- *CNPIEC Reference Guide: Chinese National Directory of Foreign Periodicals*
- *Engineering Information (PAGE ONE)*
- *Environment Abstracts. Available in print–CD-ROM–on Magnetic Tape. For more information check: www.cispubs.com*
- *Environmental Periodicals Bibliography (EPB)*
- *FINDEX (www.publist.com)*
- *Forestry Abstracts; Forest Products Abstracts (CAB Abstracts) <www.cabi.org/>*
- *Human Resources Abstracts (HRA)*
- *Journal of Planning Literature/Incorporating the CPL Bibliographics*
- *Referativnyi Zhurnal (Abstracts Journal of the All-Russian Institute of Scientific and Technical Information)*
- *Sage Public Administration Abstracts (SPAA)*
- *Sage Urban Studies Abstracts (SUSA)*
- *Wildlife Review*

(continued)

Special Bibliographic Notes related to special journal issues
(separates) and indexing/abstracting:

- indexing/abstracting services in this list will also cover material in any "separate" that is co-published simultaneously with Haworth's special thematic journal issue or DocuSerial. Indexing/abstracting usually covers material at the article/chapter level.

- monographic co-editions are intended for either non-subscribers or libraries which intend to purchase a second copy for their circulating collections.

- monographic co-editions are reported to all jobbers/wholesalers/approval plans. The source journal is listed as the "series" to assist the prevention of duplicate purchasing in the same manner utilized for books-in-series.

- to facilitate user/access services all indexing/abstracting services are encouraged to utilize the co-indexing entry note indicated at the bottom of the first page of each article/chapter/contribution.

- this is intended to assist a library user of any reference tool (whether print, electronic, online, or CD-ROM) to locate the monographic version if the library has purchased this version but not a subscription to the source journal.

- individual articles/chapters in any Haworth publication are also available through the Haworth Document Delivery Service (HDDS).

Mapping Wildfire Hazards and Risks

CONTENTS

ABOUT THE EDITORS

R. Neil Sampson is President of The Sampson Group, Inc., of Alexandria, VA, and Senior Fellow with American Forests in Washington, DC. He is a career conservationist who has held Executive Vice President positions with American Forests and the National Association of Conservation Districts, as well as 16 years of service with USDA's Soil Conservation Service (now Natural Resource Conservation Service). He is Affiliate Professor in the Department of Forest Resources, University of Idaho, and Adjunct Professor of Resource Policy at Virginia Tech, Blacksburg, VA. He holds a BS in Agronomy from the University of Idaho and an MPA from Harvard University.

R. Dwight Atkinson directs atmosphere-to-water modeling efforts for the Environmental Protection Agency's Water Office in Washington, DC. Since 1988, he has served in a variety of positions with the EPA in Washington, including Chief of its Air Policy Branch and Director of the Multi-Media and Strategic Analysis Division. He holds a PhD in civil engineering from Virginia Tech. Dwight and his wife Beth reside in northern Virginia and have two children.

Joe W. Lewis is an economist with the USDA, Forest Service. He has been with the Forest Service for 22 years, most recently with the Forest Health Protection unit in Washington, DC, where he has provided leadership in a national forest health risk mapping exercise. Joe has bachelors degrees in economics and political science from the University of Minnesota, and a masters degree in economics from the University of Wisconsin.

Participant List as of April 21, 1999
Indexing Resource Data Workshop
Pingree Park, CO, Sept. 29-Oct. 4, 1996

Dwight Atkinson
U.S. Environmental Protection Agency
499 South Capitol St. SW, Room 811
Fairchild Building (mail code 4504-F)
Washington, DC 20003
202/260-2771 fax 202/260-0512
atkinson.dwight@epamail.epa.gov

Bob Averill
740 Simms St.
Golden, CO 80401
303/275-5061 fax 303/275-5075
hotel6@aolcom

Brian Banks
USDA Forest Service
740 Simms St.
Lakewood, CO 80255
303/275-5133

Paul Beier
Box 15018
Northern Arizona University
Flagstaff, Arizona
86011-5018
520/523-9341 fax 520/523-1080
paul.beier@nau.edu

Jock Blackard
Department of Forest Sciences
Colorado State University
113 Forestry
Fort Collins, Colorado 80523
970/491-7531
jockb@cnr.colostate.edu

Don Brady
U.S. Environmental Protection Agency
401 M Street, S.W. 4503 F
Washington, DC 20460
202/260-5368 fax 202/260-7024
Brady.Donald@epamail.epa.gov

Eric Butler
725 University Ave.
Boulder, CO 80302
303/473-1991

Coleen Campbell
Colorado Air Pollution
Control Division APCD-TS-B1
4300 Cherry Creek Dr. South
Denver, CO 80246
303/692-3224 fax 30/782-5493
coleen.campbell@co.state.us

Pamela Case
USDA Forest Service
200 E. Broadway
Missoula, MT 59807
406/399-3638 fax 406/329-3359
sprucefir@gnn.com

Kermit Cromack, Jr.
020 Forestry Sciences Laboratory
Oregon State University
Corvallis, Oregon 97331-7501
541/737-6590 fax 541/737-1393
cromack@fsl.orst.edu

Denis Dean
Department of Forest Sciences, CSU
113 Forestry
Fort Collins, CO 80523
970/491-7736 fax 491-6754

Don Despain
National Biological Service
Box 227
Yellowstone National Park, WY 82190
307/344-2230 fax 307/344-2211
don_despain@nps.gov

Bruce Durtsche
Bureau of Land Management
Building 50 Denver Federal Center
Denver, CO 80255
303/236-6310 fax 303/236-6450

Maia Enzer
American Forests
P.O. Box 2000
Washington, DC 20013
202/667-3300 ext. 237 fax 667-7751
email: menzer@amfor.org

Richard Everett
Wenatchee Forest Sciences Lab., USFS
1133 N. Western Avenue
Wenatchee, WA 98801
509/662-4315 fax 664-2742
(home)
2504 Number 1 Canyon Road
Wenatchee, WA 98801
(509) 663-6685

Christian Giardina
University of Hawaii
461 W. Lanikaula St.
Hilo, HI 96720
808/934-9512 fax 808/974-4110

Ron Gosnell
CO State Forest Service
457 Old St. Vrain Rd.
Lyons, CO 80540
303/823-5245 fax 303/823-6114

Colin Hardy
USDA Forest Service Fire Lab
P.O. Box 8089
Missoula, MT 59807
406/329-4978 fax 329-5179
chardy/rmrs_missoula@fs.fed.us

Jihn-Fa (Andy Jan)
Department of Forest Sciences
Colorado State University
113 Forestry
Fort Collins, CO 80523
janjf@cnr.colostate.edu

Jim Hubbard
CO State Forest Service
203 Forestry Building
Fort Collins, CO 80523
970/491-6303 fax 491-7736

Leah Juarros
Boise National Forest
1750 Front Street
Boise, Idaho 83702
208/364-4235

Pete Lahm
USDA Forest Service
c/o ADEQ
3033 N. Central
Phoenix, AZ 85012
602/207-2356 fax 602/207-2366

Joe Lewis
USDA Forest Service
Auditors Bldg 2-SC
201 14th St. SW
Washington, DC 20250
202/205-1597 fax 205-1139

Alicia Lizarraga
Department of Forest Sciences
Colorado State University
113 Forestry
Fort Collins, CO 80523
alicia@cnr.colostate.edu

Debby Martin
US Geological Survey
3215 Marine Street
Boulder, Colorado
80303-1066
303/541-3024 fax 303/447-2505
damartin@usgs.gov

Roy Mask
USDA Forest Service
216 N. Colorado
Gunnison, CO 81230
970/641-0471 fax 641-1928

Jim Menakis
USDA Forest Service
Rocky Mountain Research Station
5775 W. Highway 10
Missoula, MT 59802
406/329-4958 fax 329-4877
jmenakis/rmrs_missoula@fs.fed.us

Melanie Miller
Bureau of Land Management
National Office of Fire and Aviation
3833 Development Avenue
Boise, Idaho
208.387.5165 fax: 208.387.5179
mmiller@nifc.blm.gov
DG: M. T. Miller: R04F02a

Leon Neuenschwander
College of Forestry, Wildlife & Range
University of Idaho
Phinney Hall, Bldg 1133, Room B14
Moscow, ID 83844-1205
208/885-2101 fax 208/885-6226
leonn@uidaho.edu

Dave Neufeld
ESRI 4875 Pearl Street East Circle #200
Boulder, Colorado 80301

Bruce Polkowsky
US Environmental Protection Agency
MD-15, USEPA
Research Triangle Park, NC 27711
polkowsky.bruce@epamail.epa.gov

Helen Rigg
Idaho Division of Environmental Quality
1410 N. Hilton
Boise, ID 83706
208/373-0502 fax 208/373-0576

Neil Sampson
American Forests
5209 York Road
Alexandria, VA 22310
703/924-0773 fax 703/924-0588
nsampson@compuserve.com

Rob Sampson
Natural Resources Conservation Services
9173 W. Barnes Dr., Suite C
Boise, ID 83709
203/378-5700 fax 208/378-5735
rsampson@id.usda.gov

Rusty Scott
College of Natural Resources
Colorado State University
Fort Collins, Colorado 80523

Rose Smyrski
c/o John Youngquist
4046 Weger Rd.
Verona, WI 53593
608/831-1970

Tom Stephens
Colorado Natural Heritage Program
254 General Services Building
Fort Collins, CO 80523
970/407-1702 fax same
Stephens@meeker.cnr.colostate.edu

Roger Stocker
WESTAR
1001 S.W. 5th Avenue, #1100
Portland, OR 97204
503/220-1660 fax 503/220-1651

Cathy Tate
US Geological Survey
Box 25046, MS-415
Denver, Colorado 80202
303/236-4882 ext. 287
fax 303/236-4912
cmtate@usgs.gov

Tracey Woodruff
U.S. Environmental Protection Agency
401 M Street, S.W. (3202 Mall)
Washington, DC 20460
202/260-6669 fax 202/260-0512
woodruff@epamail.epa.gov

Ron Zeleny
920 E. Lake St.
Fort Collins, CO 80524
970/493-9249

Preface

That fire is vitally important to the health and sustainability of many forested watersheds is a concept whose acceptance has come only recently, and at a high price. For most of the 20th century, the role of fire in the nation's forests was dominated by the anthropocentric dogma of fire suppression. Indeed, all fire, whether wildfire or controlled, was to be avoided for the betterment of people as well as forests. Implemented dutifully by land managers in the years following World War I, this taming of nature appeared successful for many decades.

As a result, the average annual acreage affected by wildfire in the 11 Western States declined from the 1920s into a remarkably constant period in the 1950s and 1960s when approximately 500,000 acres were consumed annually. In the 1970s the trend turned upward, until the late 1980s and early 1990s, when the annual acreage consumed by wildfire began to increase dramatically. The 5 worst fire years since 1940 have occurred since 1985, but the variation is tremendous, with some years having almost no fire and others zooming to 6 times the earlier average.

To devotees of chaos theory, these fluctuations had the appearance of a system in dynamic transition between equilibrium states. The landscape was beginning to show the consequences of increased fuel loadings that result when fire is removed from a fire-dependent ecosystem. Such trends portend more severe wildfire years in the future.

Serving the Environmental Protection Agency as Chief of its Air Policy Branch in Washington, DC for most of the 1990s, I was responsible for looking over the horizon and identifying tomorrow's issues. Where the risk to public and/or ecosystem health warranted, the Air program office would mobilize to face the challenge. The challenge that brought wildfire to our attention was the health risk posed by fine particles such as those found in smoke. Research by Dr. Joel Schwartz estimated that as many as 40,000 people were dying each year due to exposure to airborne particulate matter.

[Haworth co-indexing entry note]: "Preface." Atkinson, R. Dwight. Co-published simultaneously in *Journal of Sustainable Forestry* (Food Products Press, an imprint of The Haworth Press, Inc.) Vol. 11, No. 1/2, 2000, pp. xxiii-xxv; and: *Mapping Wildfire Hazards and Risks* (ed: R. Neil Sampson, R. Dwight Atkinson, and Joe W. Lewis) Food Products Press, an imprint of The Haworth Press, Inc., 2000, pp. xvii-xix. Single or multiple copies of this article are available for a fee from The Haworth Document Delivery Service [1-800-342-9678, 9:00 a.m. - 5:00 p.m. (EST). E-mail address: getinfo@haworthpressinc.com].

Suspicions focused on small particles, those having a diameter of 2.5 microns or less, because they could penetrate deeper into the most sensitive regions of the lungs than the larger, so-called coarse particles. (By comparison, pollen is roughly 15 microns in diameter.) Although it was known that fine particles were associated with combustion, no one had inventoried these sources in order to determine their relative contributions. We commissioned that study and set in motion a chain of events that have culminated in this book, as well as many other ongoing activities.

The results of that fine particle emissions inventory indicated that wildfires in 1990, a relatively moderate fire year, released more fine particles into the air than the combined total from all diesel engines and coal-fired power plants in the US. In 1990, wildfires emitted over 500,000 tons vs. slightly under 400,000 for diesels (on-road plus off-road) and 100,000 for coal-fired power plants. In 1994, a severe fire year, wildfires were responsible for approximately 1,300,000 tons of fine particles.

John Core, the Executive Director of the Western States Air Resources Council (WESTAR), helped bring those concerns into focus with fire experts from the federal agencies in the west. Experts like Pete Lahm and Janice Petersen, US Forest Service, Brian Mitchell, National Park Service, and Scott Archer, Bureau of Land Management helped us understand topics like historical vs. contemporary fire return intervals, tree densities and hydrophobic soils, ladder fuels and stand replacing fires, conifers with serotinous cones and much more. The federal agencies were developing plans to reintroduce fire to the landscape via prescribed fire, but those plans raised the following questions:

"How many acres need this type of treatment?"

"Thirty million, probably more. We don't know exactly."

"How long will it take to reintroduce fire to thirty million acres?"

"It took many decades to make this problem, it is going to take many to fix it."

"If the problem is that large and will take decades to solve, how do you set priorities? How do you know that a certain place in Montana is more in need of reintroduced fire than somewhere in Idaho?"

Finally, a question that had no answer. There was, in fact, no grand scheme to identify and rank areas in need of priority consideration. Pondering this challenge, we began to see the possibility of developing a multi-layered geographic information system (GIS) to convey this complex data in a way that would be useful to decisionmakers and the public. Because the need was so urgent, such a system would be based on the principles of adaptive management: "go now with existing data, but make the system flexible enough to facilitate new science and data as they become available."

Putting together such a wide array of information would require close cooperation with numerous Federal and State agencies. We needed to pilot the concept on a smaller scale, demonstrate its feasibility, learn from our inevitable mistakes, and then move on to other areas. The idea to focus on Colorado came from discussions between Jim Hubbard, State Forester for Colorado, Neil Sampson of American Forests, and Joe Lewis of the Forest Service.

A main subject in planning meetings was how to make the data useful to public decisionmakers faced with setting priorities. We could generate maps that were so sophisticated and technical that, for most people, they would be useless. We needed to translate and index that information so it would show areas having high, medium, or low risks. Such a ranking scheme could be displayed in colored maps in a way that could be readily communicated. Indeed, each layer of the GIS needed this indexing. How, then, to develop such indices? It was here that the workshop strategy emerged–a place where, over the course of a week or so, specialists in each field (i.e., each GIS layer) would assemble and develop an "expert opinion" methodology for indexing wildfire hazards and risks.

What follows is the culmination of that workshop, the ideas and energy of an extremely dedicated group of scientists, specialists, and public servants. As you read through their discussions and study their GIS outputs, note the diversity of the agencies and institutions represented by the authors: 6 Federal agencies, 4 State agencies, 4 universities, and 3 non-governmental organizations. In addition, note that financial support for the workshop came from an equally diverse partnership: USEPA, USFS, BLM, NPS, the State of Colorado, ESRI, and American Forests. Beyond the pioneering, technical insights of the workshop participants, an equally important yet subtle message of the workshop is its testimony to the principles of reinvented government, where many diverse parts of the Federal government work with each other and with State and local governments and institutions to respond to a crisis faced by the people we all serve. In closing the workshop, I predicted to the participants: "Twenty years from now, on the day of your retirement, you will reflect back on your career and ask, 'What have I accomplished?' Your memory will return to Pingree Park and what you have accomplished this week. You will smile to yourself and say, 'Job well done.' " I hope you, the reader, will agree.

Dr. R. Dwight Atkinson

Acknowledgments

The workshop was sponsored by American Forests, the U.S. Environmental Protection Agency, USDA Forest Service, USDI Bureau of Land Management, USDI National Park Service, Colorado State Forest Service, and Environmental Systems Research Institute (ESRI). We are deeply indebted to all these organizations.

In addition, we received significant support and assistance from Colorado State University's Department of Forest Resources. Their expertise in Geographic Information Systems, and computer support, was invaluable. We also benefitted greatly from the outstanding facilities at Pingree Park, a mountain campus operated by the University.

Because of the workshop's total reliance on computer systems, GIS programs, and rapid compilation of often-incompatible data bases, the staff that installed the machinery, kept it running, and worked virtually around the clock for a week to meet the participant's needs, deserves special recognition. They include:

John Steffenson, ESRI, who made ARCINFO software available to the workshop.

Dr. Denis Dean, Colorado State University, who organized the GIS team and gave invaluable advice about methods, sources, and direction.

Russell (Rusty) Scott and Kevin Swab, Colorado State University, who assembled a combination of rented, purchased, and borrowed equipment into a functioning computer center and kept it working for the workshop's duration.

Jock Blackard, Andy Jan, and Alicia Lizarraga of Colorado State University; Jim Menakis of the USDA Missoula Fire Lab; Brian Banks and Eric Butler, of the Forest Service's Denver Regional Office; and Roger Stocker of WESTAR, who worked long hours to make the GIS programs serve the needs of the workshop.

In the weeks following the workshop, we continued to rely upon all of their skills as the working groups refined their maps and requested new products to reflect the group's progress. As the products neared publication quality, we turned to Eva Strand at the University of Idaho's College of Forest, Wildlife & Range Sciences Remote Sensing GIS Laboratory, who has produced the camera-ready maps found on the following pages. We are deeply indebted to her for her high-quality work, and to Liza Fox, Lab

Administrator, who provided support and technical advice whenever it was required.

We also want to reserve a special thanks to Dr. Graeme P. Berlyn, Professor of Forestry at the Yale University School of Forestry and Environmental Sciences, and Editor of the *Journal of Sustainable Forestry*. Dr. Berlyn supported this effort from the outset and waited patiently throughout the two years it took us to get the manuscripts worked into final condition. We are deeply grateful for his patience, and for the support of The Haworth Press in the final stages of the project.

EDITORS

R. Neil Sampson, Senior Fellow, American Forests, Washington, DC and President, The Sampson Group, Alexandria, VA.
R. Dwight Atkinson, US Environmental Protection Agency, Washington, DC
Joe W. Lewis, USDA Forest Service, Washington, DC.

ASSISTANT EDITOR

Maia J. Enzer, Forest Policy Center, American Forests, Washington, DC

SPONSORING COMMITTEE

R. Dwight Atkinson, US Environmental Protection Agency, Washington, DC
James E. Hubbard, State Forester, Colorado State Forest Service, Fort Collins, CO
Joe W. Lewis, USDA Forest Service, Washington, DC
R. Neil Sampson, American Forests, Washington, DC

INVITED SCIENTISTS

Bob Averill, USDA Forest Service, Golden, CO
Paul Beier, Northern Arizona University, Flagstaff, AZ
Don Brady, U.S. Environmental Protection Agency, Washington, DC
Coleen Campbell, Colorado Air Pollution Control Division, Denver, CO
Pamela Case, USDA Forest Service, Lakewood, CO
Kermit Cromack, Jr., Oregon State University, Corvallis, OR
Denis Dean, Department of Forest Sciences, CSU, Fort Collins, CO
Don Despain, National Biological Service, Yellowstone National Park, WY
Bruce Durtsche, Bureau of Land Management, Susanville, CA
Richard Everett, Wenatchee Forest Sciences Laboratory, USFS, Wenatchee, WA
Christian Giardino, Department of Forest Sciences, CSU, Fort Collins, CO
Ron Gosnell, Colorado State Forest Service, Lyons, CO
Colin Hardy, USDA Forest Service Fire Laboratory, Missoula, MT

Leah Juarros, Boise National Forest, USDA Forest Service, Boise, ID
Pete Lahm, USDA Forest Service, Phoenix, AZ
Lee MacDonald, Dept of Earth Resources, Colorado State University, Fort Collins, CO
Deborah Martin, U.S. Geological Survey, Boulder, CO
Roy Mask, USDA Forest Service, Gunnison, CO
Jim Menakis, USDA Forest Service Fire Laboratory, Missoula, MT
Melanie Miller, Bureau of Land Management, Boise, ID
Leon Neuenschwander, College of Forest, Wildlife & Range, University of Idaho, Moscow, ID
Bruce Polkowsky, US Environmental Protection Agency, Research Triangle Park, NC
Helen Rigg, Idaho Division of Environmental Quality, Boise, ID
Rob Sampson, Natural Resources Conservation Service, Roseburg, OR
Rose Smyrski, Forest Products Laboratory, Madison, WI
Tom Stephens, Colorado Natural Heritage Program, Fort Collins, CO
Roger Stocker, WESTAR, Portland, OR
Cathy Tate, U.S. Geological Survey, Denver, CO
Tracey Woodruff, U.S. Environmental Protection Agency, Washington, DC
Ron Zeleny, CO State Forest Service, Fort Collins, CO

SECTION I

Chapter 1

Indexing Resource Data
for Forest Health Decisionmaking

R. Neil Sampson
R. Dwight Atkinson
Joe W. Lewis

SUMMARY. A workshop involving 35 scientists was held at Pingree Park, Colorado in October, 1996, to develop procedures for evaluating the hazards and risks associated with extreme wildfire events in central and western Colorado. The basic question was: "Can the existing resource data be used to estimate and portray wildfire hazards and risks in sufficient spatial detail to be useful as a strategic planning and commu-

R. Neil Sampson is affiliated with American Forests, Washington, DC, and The Sampson Group, Alexandria, VA 22310. R. Dwight Atkinson is affiliated with U.S. Environmental Protection Agency, Washington, DC 20410. Joe W. Lewis is affiliated with USDA Forest Service, Washington, DC 20250.

[Haworth co-indexing entry note]: "Chapter 1. Indexing Resource Data for Forest Health Decisionmaking." Sampson, R. Neil, R. Dwight Atkinson, and Joe W. Lewis. Co-published simultaneously in *Journal of Sustainable Forestry* (Food Products Press, an imprint of The Haworth Press, Inc.) Vol. 11, No. 1/2, 2000, pp. 1-14; and: *Mapping Wildfire Hazards and Risks* (ed: R. Neil Sampson, R. Dwight Atkinson, and Joe W. Lewis) Food Products Press, an imprint of The Haworth Press, Inc., 2000, pp. 1-14. Single or multiple copies of this article are available for a fee from The Haworth Document Delivery Service [1-800-342-9678, 9:00 a.m. - 5:00 p.m. (EST). E-mail address: getinfo@haworthpressinc.com].

1

nications tool for decisionmakers? The answer, we believe, is "Yes, with some qualifications." Illustrations of relative wildfire hazards and the related risks to a variety of resources can be developed, and these can guide discussions about strategic planning for wildfire mitigation efforts. The products can be developed within a period of a few months at reasonable cost, and the lessons learned in this exercise can help make the effort more efficient. *[Article copies available for a fee from The Haworth Document Delivery Service: 1-800-342-9678. E-mail address: <getinfo@haworthpressinc.com> Website: <http://www.haworthpressinc.com>]*

KEYWORDS. Risk assessment, wildfire, hazard, western United States

INTRODUCTION

The increasing frequency, size and intensity of wildland fires in the western United States is a matter of significant national policy concern (USDI/USDA 1996). In addition to inflicting major suppression costs on public agencies and losses of life and property on the local residents caught in a large event, these wildfires are causing significant environmental impact. Ecosystems that developed under a history of frequent, low-intensity fires are now being impacted by high-intensity events that alter soil conditions, affect vegetative successional patterns, and create large landscape patterns that may reduce habitat diversity (Covington et al. 1994).

Across the 11 western states, the average annual acres burned has tripled since the 1950s and 1960s, when those averages are calculated for each decade, as shown in Table 1. Large wildfire years such as 1988, with the Yellowstone fires, and 1994, with the tragic incident at Storm King Mountain in Colorado, garner major public attention. National policymakers are increasingly aware that federal budget outlays for fire fighting can be in the range of $1 billion or more in a bad wildfire season.

For the people who live in the region, the dangers in the situation are significant. Not only are their homes and communities at risk when wildfires occur nearby, their health may be at risk when smoke affects air quality. In 1994, it was estimated that over 1.35 million tons of $PM_{2.5}$ (particulate matter less than 2.5 microns in diameter) were emitted from 65,700 fires that burned 3.8 million acres of federal lands in the west (Babbitt et al. 1994). Communities such as Wenatchee, Washington, and Boise, Idaho, experienced many days when air pollution exceeded health-based standards as a result of those events (Core 1995). As the Environmental Protection Agency and the States consider revisions of air quality standards and regulations, both in terms of their impact on human health and visibility, the role of wildland fire is a major issue (Rigg et al. 2000).

TABLE 1. Wildfire acres in 11 western United States, expressed as annual average by decade, 1950-1994 (USDA Forest Service 1992 and subsequent annual wildfire reports).

State	Forest Area	Annual Wildfire Averages				
		1950-59	1960-69	1970-79	1980-89	1990-94
		(Acres)				
Arizona	19,595,000	26,739	24,705	75,881	77,630	120,604
Colorado	21,338,000	14,901	11,550	13,691	40,759	28,956
Idaho	21,621,000	50,249	103,515	141,330	256,339	318,251
Montana	22,512,000	5,587	25,004	29,134	129,834	145,159
Nevada	8,938,000	15,616	21,290	41,477	158,163	93,807
New Mexico	15,296,000	29,391	18,638	42,630	85,637	222,984
Utah	16,234,000	12,502	17,226	32,603	120,867	78,651
Wyoming	9,966,000	3,481	9,019	26,520	179,921	53,465
INTERMOUNTAIN	135,500,000	158,737	230,948	403,266	1,049,149	1,061,877
California	37,263,000	259,948	170,723	276,204	298,286	301,218
Oregon	27,997,000	39,256	39,361	49,324	176,477	110,958
Washington	21,432,000	18,980	22,839	37,154	29,231	75,411
PACIFIC NORTHWEST	86,692,000	318,184	232,923	362,682	503,993	487,587
TOTAL WEST	222,192,000	476,920	463,871	765,948	1,553,142	1,549,464

THE CHALLENGE FACING PUBLIC OFFICIALS AND LAND MANAGERS

Public officials and land managers in the region are faced with a serious dilemma. Fire suppression efforts over many decades have contributed to millions of acres of forests, brushlands, and other wildland ecosystems that are now so heavily laden with flammable vegetation that large, high-intensity wildfires are virtually inevitable. In the western states, the vast majority of these fire-prone areas are on federal lands. Many of these areas are remote, but some are adjacent to major urban areas or rural communities or increasingly, intermixed with homes and other developments.

While there are many ways to treat these lands to reduce potential wildfire size and/or intensity, making decisions on where and how to focus land treatment efforts is not an easy process. Among the obstacles likely to be encountered by land managers and public officials are:

- Land treatment can be an expensive process, particularly when it requires removal of surplus biomass that is not economically marketable;
- There is a huge backlog of land needing treatment. Budget cutbacks, agency personnel reductions, and the enormity of the task make for difficult choices;

- One of the main methods of returning historical fire regimes to the landscape is through the use of prescribed fire. Prescribed fire poses the risk of escape, so is often avoided by land management agencies due to public opposition;
- Prescribed fire creates smoke that must be managed to protect public health. State regulatory agencies are caught between the need to protect air quality and the very real likelihood that limiting prescribed fires may increase the extent of wildfires, thus producing higher health hazards rather than lower;
- Land treatment is invasive, and public opposition to thinning or timber harvest may limit projects on federal lands, particularly where they involve building roads;
- Some of the lands where wildfire hazards are high are within reserved areas such as National Parks, where long-standing policies of non-intervention preclude most mechanical fuel treatments. In situations where existing fuel conditions have grown so serious that prescribed fire is too risky, the responsible land management agency faces a dilemma.

To overcome such obstacles, public officials and federal land managers need to develop more scientific procedures for identifying treatment projects, focusing on high-priority areas where the investment will be cost-effective. They must also be able to adequately explain their proposals to the public to prevent opposition from halting efforts. This calls for strategic decision tools that can identify the resources and values most likely to be placed at risk unless treatment is undertaken. Modern scientific data, ranging from traditional soil, water and vegetation surveys to satellite images of vegetative conditions, can be harnessed to the task. Displaying these data in a geographic map not only helps people understand the risks involved, but also is essential in illustrating how different areas and people may be affected, both by the current hazards and the impacts of proposed treatments.

In late 1996, 35 scientists, land managers, and resource experts gathered at the Pingree Park Campus of Colorado State University to test the extent to which existing and available resource data could be adapted to the creation of large-scale strategic decision tools. The test area was the State of Colorado—primarily the western portion of the state that contains most of the state's forest and shrub ecosystems. The challenge was reasonably straight-forward: could we identify areas where the future likelihood of large, intense wildfire events were highest, and could we then explain some of the economic and ecological values that would be put at greatest risk if such wildfires were experienced? Would such an effort result in a product that could help people decide how to set priorities for land treatment or other action that might reduce risks and lower future damages?

The results of that effort are described in the following papers, which explain in detail the workshop's efforts and conclusions. In general, however, the answers to the basic questions are positive. Areas of highest wildfire hazards and risks can be identified, at least in a relative way. The risks inherent in large, intense fire events can be characterized, and they differ significantly from place to place. The use of geographic information systems (GIS) technology to analyze and illustrate available information is essential to some of the interpretations, and provides an excellent way to display situations in an understandable way.

Using available information over large geographic areas provides a rapid and relatively low-cost method of informing strategic decisions, but this approach also creates significant limits to the utility of the product. While it may be possible to illuminate general areas where wildfire hazards are high and point out the types of risks that may be most important in different places, it is not always possible to pinpoint locations or establish treatment specifics. Those decisions will require a closer look at specific situations and landscapes.

The utility of the materials and models developed will be, of course, determined as they are tested, refined, and used. We hope that the results presented here will help scientists, decisionmakers, and the general public address some of the significant challenges posed by a growing human population in a setting dominated physically by the mountains, forests, grasslands, deserts and streams whose continued sustainability and ecological integrity underlies much of the future of the region.

THE COMMUNICATIONS DILEMMA

Fire is a natural process in wildland ecosystems, but one that can generate an enormous range of economic and ecological impacts. From a benign process that recycles carbon and nutrients with little or no lasting harm, to a super-hot inferno that kills all vegetation, damages soil with excessive heat, and alters ecological succession significantly, the impact of fire varies widely, according to the amount of fuel consumed, the heat and flame length of the fire as it goes through the vegetation, and the amount and duration of heat that is forced into the soil. Fire ecologists use the term *intensity* to indicate the flame length burning in above-ground vegetation, and *severity* to indicate the degree and impact of soil heating (Neuenschwander et al. 2000; Cromack et al. 2000).

Wildland fires can vary greatly in both intensity and severity within the fire perimeter, with some areas virtually untouched while others are severely impacted. Data on intensity and severity is often collected on modern wildfires, but historical fire data consists mainly of ignition date, cause, location of ignition, and an estimate of the final size. From such information, the ecological impacts of historical fires are difficult to determine. While scientists can make some judgments on the basis of prior vegetative cover, weather conditions, and fire reports, the diverse impacts of wildland fire make it very difficult to communicate fire situations accurately, and the ability to compare the impact of modern fires with those of the past is limited.

A fire event that could easily be described as catastrophic by people whose lands are damaged or homes destroyed might, to an ecologist, be a fairly normal event in terms of its impact on the wildland ecosystems it affects. A surface fire under a mature ponderosa pine forest may be what that forest needs to survive. A more intense fire may kill trees that have survived dozens of less-intense fires, so ecologists may judge it to be outside the historical range of conditions, and therefore, an ecologically damaging event. A fire fed by large amounts of ground fuels may smolder for days with no flames, but drive heat deeply into the soil, causing long-term site deterioration (Cromack et al. 2000). Whether or not any of these events is a catastrophe lies mainly in the value judgments applied by people describing the event.

One thing is certain: the events described above are different in terms of their economic and ecological effects. Yet, they will often be called by the same name – wildfire. Using a single term to describe dramatically different events makes effective communications impossible.

Thus, policymakers and land managers face a difficult challenge in explaining the need to increase one type of wildland fire while reducing a more damaging and costly type. The idea that some fires are essential to the sustainability of many forest types–and the survival of fire-dependent species–seems contradictory to the ongoing efforts that people make to prevent wildfires and protect life and property from fire damage. The many faces of wildland fire exceed our limited vocabulary.

CHARACTERIZING RANDOM EVENTS

Evaluating the hazards and risks involved with wildfire forces us to confront the randomness of natural events and develop methods of characterizing fire in ways that are more useful to decisionmakers (Botkin 1990). Wildland systems go through years (or centuries) of vegetative succession and development prior to a fire event, and the effects of seemingly disconnected soil patterns, climate trends, weather events, insect population dynamics, lightning strikes, diseases, and human actions shape the arrangement of live and dead vegetation in complex patterns across the landscape. Conditions for a major event are established when large amounts of flammable fuels allow a fire to develop high heat releases and expand readily over a large area.

Weather conditions immediately prior to and during a fire event may be the most critical factor in determining whether an ignition will result in a major wildfire or not. The vast majority of wildland fire ignitions result in fires of less than one acre (Neuenschwander et al. 2000). They either run out of fuel, are extinguished by rain or snow, or are suppressed by firefighting agencies. Wildfires that reach 100 acres in size, however, have a high probability of growing so large that suppression efforts are ineffective (Neuenschwander et al. 2000). When dry weather has made available fuels extremely flammable, the probability of an ignition growing into a major event go up.

Weather conditions during the fire event drive the fire itself. Hot, dry winds turn insignificant fires into major events, while cool, moist conditions have the opposite effect. Weather records can show the most common weather conditions under which the majority of larger fires occur (Rigg et al. 2000). In the fire's aftermath, weather events determine how rapidly the site begins to recover, and the amount of soil and watershed damage that may occur (MacDonald et al. 2000; Cromack et al. 2000).

Whether or not a large wildfire becomes a serious problem depends on the time and place where it occurs. While it is impossible to predict future ignition points, past history shows that some areas, largely those impacted by humans, experience significantly more ignitions than average (Neuenschwander et al. 2000). The proximity of large numbers of people, or large numbers of susceptible people or homes, to those ignition-prone areas can be used as a basis for establishing relative future risks that people face (Case et al., 2000). The location of those areas in relation to prevailing wind conditions establishes relative risks of health or visibility impacts from the smoke produced by an event (Rigg et al. 2000). Likewise, the presence and fragility of sensitive wildlife habitats, or threatened or endangered species, creates different relative concerns between areas (Despain et al. 2000).

The potential effects of a fire event, therefore, are shaped by the complex interactions of hundreds of events–some random and unpredictable; others planned and carried out by people–over long time periods. Understanding and characterizing such a chaotic process, let alone attempting to predict its future likelihood, is an exercise in assessing probabilities, uncertainty, and risk.

While this type of analysis creates a difficult exercise for scientists, and imposes limits on the uses of the resulting product, we are convinced that the information and insight that emerges can help people better understand the real nature of wildland management. Hopefully, it can also result in a better-informed debate on the many options, opportunities, and challenges posed by existing conditions.

INDEXING RESOURCE DATA

We used the term "indexing" to indicate how a scientific team would categorize the available data to produce a relative risk map. At what point should the map change colors, from green to yellow, or yellow to red? When is an impact an inconvenience, and when does it become a serious problem? These value judgments are based in scientific expertise, but are seldom clearly defined by research. An assertion that research has identified a distinct difference in impact between a slope of 49% and one of 51% is probably not supportable. On the other hand, it is much easier for five experts to agree on the steepness of slope where more serious impacts occur.

These are the kind of expert opinions represented in the following indices, and they are the most significant contribution that results from conducting this type of effort in a workshop setting. In a collaborative setting, experts representing different scientific disciplines, work experience, and agency outlook can debate, compare, and come to an agreement on the most important facts for a decision maker. Tens of thousands of pieces of data, and hundreds of scientific studies and observations, are condensed into a map of five simple colors. To the extent that the right pieces of data are skillfully integrated, the result can represent a significant improvement in communication between the scientific and the policy communities.

WILDFIRE HAZARD AND RISK

Extreme wildfires are the result of a combination of risk factors occurring in an area containing conditions that constitute hazards. Suter (1993) defined a hazard as "a state that may result in an undesired event, the cause of risk." That describes the fuel conditions in many western forests. An arrangement of vegetation and fuels that will support the development of fires that are larger, more intense, and more severe in their impacts on vegetation and soils than the majority of historical wildfires constitutes a hazard. If the fuels are re-arranged so that the same ignition, in the same weather, acts more like the historical norm, the hazard will have been reduced or eliminated.

A risk is something that "adds to the hazard and its magnitude the probability that the potential harm or undesirable consequence will be realized" (National Research Council 1989). It is generally expressed as a probability (Suter 1993). Ignition and weather conditions fit into this definition. Fires obviously need an ignition, and the weather conditions at the moment of ignition, and for hours and days thereafter, are a key to fire behavior. Both ignition and fire-supporting weather conditions are random events that can be characterized by the probability of their mutual occurrence, based on historical records.

So extreme wildfires are undesirable events whose impacts are shaped by a combination of a risk element (ignition in dry, windy weather) operating in a hazardous condition (abnormal fuel amounts and arrangements). This characterization offers two benefits: (1) it allows "extreme wildfires" to be discussed as a different kind of wildland fire than other fires that might be, from an ecological viewpoint, a neutral or positive event; and, (2) it allows these events to be categorized numerically, in terms of the probability of the risk occurring in an area where high hazards exist.

This characterization poses an interesting dilemma, however, between ecological definitions and human perceptions. On subalpine and boreal forests, fuels normally build up to the point where a high-intensity crown fire will result, on a return interval that may stretch to hundreds of years. Thus, for the most part, few if any of those forests would be, on an ecological basis, classified as having abnormally high-hazard conditions today. But if people have built homes in these areas, or established high-profile recreational facilities such as Yellowstone Park, a fire that represents a fairly normal ecological event can become a major disaster in human terms. Thus, it becomes important to characterize what other values are at stake in the event of a wildfire, since these will often be what drive the actions of land and fire management actions.

Ignition Risks

Ignition risks were categorized by analyzing the location and ultimate size of the fires that had been recorded in Colorado during the past decade (Neuenschwander et al. 2000). The ignitions were converted into a density function (number of ignitions over 10 years per 10,000 acres) to identify areas of higher relative probability for ignition. This usually identifies areas of suitable fuels subject to frequent lightning strikes or where human activities or fire-igniting facilities (power lines, railroads, etc.) are a common source. The assumption is that the conditions causing the past ignition density will likely continue. In that case, some areas will continue to have a higher ignition density than others.

Most ignitions result in little or no real damage. Some simply go out for lack of sufficient dry fuels. Many, particularly in mixed urban-wildland areas, are suppressed. But the ignition density is an important indicator of wildfire hazard-risk. Even if only 0.1% of the ignitions escape to cause a major event, the total number of ignitions will be the most reliable indicator as to where that one event is most likely to occur.

Fire behavior is driven primarily by the weather, supported by the type, amount, and arrangement of forest fuels. In some forest and range types, fuels dry to extreme flammability every year and thus merely await ignition. If ignition and extreme fire weather such as high winds coincide, the result is likely a large, extreme wildfire. In the moist forests, it is probably safe to assume that extreme events are linked to fairly specific and episodic weather conditions. The regional history indicates that large fires are grouped in certain years–most likely drought years with frequent thunderstorms and high winds. While those conditions can't be predicted, they can be identified. Thus, we can describe weather trends that raise the likelihood of extreme fire events, and calculate the probability of their occurrence on the basis of past history.

Fuel Hazards

Fuel conditions need to be identified and geographically located so that they can be overlain with ignition risks. Over large areas, this can only be done in a general way, from such tools as aerial image interpretation, where vegetative conditions that lead to large, extreme wildfires can be mapped. One of the current problems in developing large-area models is the inability of currently-available satellite imagery to characterize the amount and structural arrangement of the fuels located within and beneath the forest canopy. For this model, the wildfire working group used their knowledge of forest conditions in the study area to assign fuel characteristics on the basis of the general vegetation type maps that were available (Neuenschwander et al. 2000).

Fuels, in contrast to ignition and weather, can be modified by management. These modifications may significantly reduce the probability of a large, high-intensity stand-replacing wildfire in some habitat types, given similar ignition densities and weather conditions. The GIS maps, by illustrating the geographic areas where conditions are most likely to support an extreme wildfire, provide a strategic tool upon which to base plans for mitigation activities. The basic question is: "If there are only limited resources for doing mitigation work, where would those resources be most effectively utilized?" Treating high-hazard areas first does not guarantee that a major event will not occur, but it provides the best opportunity for reducing the risks associated with a major wildfire.

THE RISK ASSESSMENTS

These model portions define risk in a slightly different manner than what was described in connection with ignition and weather occurrences. Here, a resource is "at risk" from an extreme wildfire if it is located in the fire's path. The more densely the resource occurs within an area subject to an extreme wildfire, the higher the probability that it will suffer damage and thus the higher risk rating. In this usage, a stadium full of people creates more "at risk" people in the event of an earthquake than an empty stadium, even though the probability of the earthquake remains the same.

Table 2 does not tell us whether it is more important to try to reduce one

TABLE 2. Relative risk ratings to selected resource values, for 10 selected watersheds in western Colorado.

Hydrologic Unit Code[2]	Wildfire Ignition[3]	Human Health[4]	Visibility[5]	Relative Risk Ratings[1] Soil Erosion[6]	Stream Sediment[7]	T&E Species[8]	People and Homes[9]
10190005	5	4	5	2	3	0	3
10190002	4	5	2	3	2	4	5
14080107	5	2	5	3	2	5	1
10190006	4	4	3	2	2	4	2
14040109	3	2	3	1	2	5	1
14050006	4	3	1	1	1	0	1
14040106	4	2	3	1	1	5	1
14050007	3	2	3	1	2	4	1
14050002	3	3	3	1	2	4	1
14080202	3	1	1	2	2	3	2

[1]Risk ratings are: 0-none; 1-very low; 2-low; 3-medium; 4-high; 5-very high. [2]Figure 2.1 (Sampson and Neuenschwander 2000); [3]Figure 3.4 (Neuenschwander et al., 2000); [4]Figure 6.4 (Rigg et al., 2000); [5]Figure 6.6 (Rigg et al., 2000); [6]Figure 4.6 (MacDonald et al., 2000); [7]Figure 4.8 (MacDonald et al., 2000); [8]Figure 5.2 (Despain et al., 2000); [9]Figure 7.6 (Case et al., 2000).

risk or another, but it does provide useful information for a public debate or a manager's decision. In addition to helping people decide where to focus risk-reducing efforts, Table 2 illustrates the potential benefits from such efforts. The decision to spend scarce resources doing a fuel reduction project in a community watershed, as opposed to the same kind of effort aimed at protecting an endangered species site elsewhere, may be supported by a scientific assessment of the relative values at risk in each area. The decision on where to focus treatment, however, is based primarily on value judgments, not scientific determinations, and that difference must be recognized by all users.

The scientific teams can, however, make a very important contribution to those political discussions and judgements. By selecting the resource data to be utilized and compared, by indicating when the extent of an impact is likely to mean a significant degree of harm, and by showing how these relative impacts appear on the landscape, the scientific teams have provided a useful level of understanding. Some of the risk ratings (health risk from smoke transport, for example) are shaped not only by conditions in the immediate vicinity, but may be affected by conditions far removed. This type of "distributed risk" analysis is only feasible with fairly large scale analyses supported by sophisticated GIS models (Stocker 2000). For some people who are expert in the areas being studied, the information presented in Table 2 may not be new. But public decisionmakers are seldom experts in landscape ecology or wildfire behavior, and if the information developed by scientific expertise is clearly presented, it can be highly valuable to them.

CONCLUSIONS

As Table 2 and the maps in the following papers illustrate, a display of the relative risks in different geographic areas can be presented with this type of assessment. When displayed on maps, the results can help focus discussion on the highest-priority resource concerns and, consequently, the highest-priority areas for considering some sort of mitigation or treatment.

Some of the ways in which such an assessment may be useful include:

- Locating areas where more intensive efforts should be made to identify mitigation projects;
- Illustrating the "distributed risk" that may come from an event some distance away;
- Communicating the concepts of environmental hazards and risks to the public; and,
- Building public consensus over the location, type, and extent of fuel treatments, prescribed fires, or other practices designed to reduce the probability or extent of an extreme wildfire event.

There is one type of conclusion that we intentionally did not attempt to derive from this exercise. It has been suggested that, in Table 2, one could add up the index numbers horizontally in order to arrive at a "total risk score" for each watershed. We avoided this on the grounds that such a score would represent public value judgments rather than informed scientific judgment. If such an exercise is attempted, there are two ways to do it. One is simple addition, giving a score to each watershed. That is, in fact, assigning an equal weight to each risk factor. Is visibility as important as human health, or watershed integrity, in the public's decision making? If that is not the case, the other way to reach a score is to assign a value factor to each column in Table 2 that illustrates which risk elements are to be given more weight in the final score. Assigning those factors is, we concluded, more a political than a technical task; one better carried out by elected leaders in a democratic process. The maps and tables produced by this exercise can be helpful to guide that public debate, but they should not attempt to over-ride it with pre-formed value judgments.

REFERENCES

Babbitt, R.E., D.E. Ward, R.A. Susott, W.M. Hao and S.P. Baker. 1994. Smoke from western wildfires, 1994. *Proceedings, 1994 Annual Meeting, Interior West Fire Council*, Coeur d' Alene, Idaho, November, 1994.

Botkin, Daniel B. 1990. *Discordant Harmonies: A New Ecology for the Twenty-First Century*. New York, Oxford University Press. 241 p.

Case, Pamela, Brian Banks, Eric Butler, and Ronald Gosnell. 2000. Assessing Potential Wildfire Effects on People. In Sampson, R.N., R.D. Atkinson, J.W. Lewis (Eds.), *Mapping Wildfire Hazards and Risks.* Papers from the American Forests scientific workshop, September 29-October 5, 1996, Pingree Park, CO. The Haworth Press, Inc., New York.

Core, John E. 1995. Air quality regulations: treatment of emissions from wildfires vs. Prescribed fires. Paper presented at the Environmental Regulation & Prescribed Fire Conference, March 15-17, 1995, Tampa, Florida.

Covington, W.W., Everett, R.L., Steele, R., Irwin, L.L., Daer, T.A. and Auclair, A.N.D. 1994. Historical and anticipated changes in forest ecosystems of the Inland West of the United States, *Journal of Sustainable Forestry* 2(1&2): 13-64.

Cromack, Kermit Jr., Johanna D. Landsberg, Richard Everett, Ronald Zeleny, Christian Giardina, T. D. Anderson, Bob Averill, and Rose Smyrski. 2000. Assessing the Impacts of Severe Fire on Forest Ecosystem Recovery. In Sampson, R.N., R.D. Atkinson, J.W. Lewis (Eds.), *Mapping Wildfire Hazards and Risks.* Papers from the American Forests scientific workshop, September 29-October 5, 1996, Pingree Park, CO. The Haworth Press, Inc., New York.

Despain, Don G., Paul Beier, Cathy Tate, Bruce M. Durtsche, and Tom Stephens. 2000. Modeling Biotic Habitat High Risk Areas. In Sampson, R.N., R.D. Atkinson, J.W. Lewis (Eds.), *Mapping Wildfire Hazards and Risks.* Papers from the American Forests scientific workshop, September 29-October 5, 1996, Pingree Park, CO. The Haworth Press, Inc., New York.

MacDonald, Lee H., Robert Sampson, Don Brady, Leah Juarros, and Deborah Martin. 2000. Predicting Erosion and Sedimentation Risk from Wildfires: A Case Study from Western Colorado. In Sampson, R.N., R.D. Atkinson, J.W. Lewis (Eds.), *Mapping Wildfire Hazards and Risks.* Papers from the American Forests scientific workshop, September 29-October 5, 1996, Pingree Park, CO. The Haworth Press, Inc., New York.

National Research Council. 1989. *Improving Risk Communication.* Washington, DC:National Academy Press.

Neuenschwander, Leon F., James P. Menakis, Melanie Miller, R. Neil Sampson, Colin Hardy, Bob Averill and Roy Mask. 2000. Indexing Colorado Watersheds to Risk of Wildfire. In Sampson, R.N., R.D. Atkinson, J.W. Lewis (Eds.), *Mapping Wildfire Hazards and Risks.* Papers from the American Forests scientific workshop, September 29-October 5, 1996, Pingree Park, CO. The Haworth Press, Inc., New York.

Neuenschwander, Leon F. and R. Neil Sampson. 2000. A Wildfire and Emissions Policy Model for the Boise National Forest. In Sampson, R.N., R.D. Atkinson, J.W. Lewis (Eds.), *Mapping Wildfire Hazards and Risks.* Papers from the American Forests scientific workshop, September 29-October 5, 1996, Pingree Park, CO. The Haworth Press, Inc., New York.

Rigg, Helen Getz, Roger Stocker, Coleen Campbell, Bruce Polkowsky, Tracey Woodruff and Pete Lahm. 2000. A Screening Method for Identifying Potential Air Quality Risks from Extreme Wildfires. In Sampson, R.N., R.D. Atkinson, J.W. Lewis (Eds.), *Mapping Wildfire Hazards and Risks.* Papers from the American

Forests scientific workshop, September 29-October 5, 1996, Pingree Park, CO. The Haworth Press, Inc., New York.

Sampson, R. Neil and Leon Neuenschwander. 2000. Characteristics of the Study Area and Data Utilized. In Sampson, R.N., R.D. Atkinson, J.W. Lewis (Eds.), *Mapping Wildfire Hazards and Risks*. Papers from the American Forests scientific workshop, September 29-October 5, 1996, Pingree Park, CO. The Haworth Press, Inc., New York.

Stocker, Roger A. 2000. Methodology for determining wildfire land prescribed fire air quality impacts on areas in the western United States, In Sampson, R.N., R.D. Atkinson, J.W. Lewis (Eds.), *Mapping Wildfire Hazards and Risks*. Papers from the American Forests scientific workshop, September 29-October 5, 1996, Pingree Park, CO. The Haworth Press, Inc., New York.

Suter, Glenn W, II. 1993. *Ecological Risk Assessment*. Boca Raton, FL: Lewis Publishers.

U.S. Department of Agriculture Forest Service. 1992. *1984-1990 wildfire statistics*. Washington, DC: USDA Forest Service, State and Private Forestry, Fire and Aviation Management Staff.

USDI/USDA. 1996. *Federal Wildland Fire Management: Policy & Program Review*, Final Report, December 18, 1995. Washington, DC: U.S. Department of the Interior. 45 pp.

Chapter 2

Characteristics of the Study Area, Data Utilized, and Modeling Approach

R. Neil Sampson
Leon F. Neuenschwander

SUMMARY. A wildfire hazard and risk model was developed in a week-long workshop to test the feasibility of using existing resource data and combined expert assessment to help address the increasing concern over wildfires in the Western United States. The goal was to utilize readily-available resource data, translated into a GIS format, to develop and illustrate relative hazard and risk assessments in a format that would be useful to decision makers. This paper discusses the characteristics of the study area, the data that were assembled, and the methods utilized. The results of the hazard and risk assessments are discussed in the papers that follow. *[Article copies available for a fee from The Haworth Document Delivery Service: 1-800-342-9678. E-mail address: <getinfo@haworthpressinc.com> Website: <http://www.haworthpressinc.com>]*

KEYWORDS. Wildire hazard, risk model, geographic information systems, GIS

R. Neil Sampson is affiliated with American Forests, Washington, DC, and The Sampson Group, Alexandria, VA 22310. Leon F. Neuenschwander is affiliated with College of Forestry, Wildlife and Range Sciences, University of Idaho, Moscow, ID 83844.

[Haworth co-indexing entry note]: "Chapter 2. Characteristics of the Study Area, Data Utilized, and Modeling Approach." Sampson, R. Neil, and Leon F. Neuenschwander. Co-published simultaneously in *Journal of Sustainable Forestry* (Food Products Press, an imprint of The Haworth Press, Inc.) Vol. 11, No. 1/2, 2000, pp. 15-33; and: *Mapping Wildfire Hazards and Risks* (ed: R. Neil Sampson, R. Dwight Atkinson, and Joe W. Lewis) Food Products Press, an imprint of The Haworth Press, Inc., 2000, pp. 15-33. Single or multiple copies of this article are available for a fee from The Haworth Document Delivery Service [1-800-342-9678, 9:00 a.m. - 5:00 p.m. (EST). E-mail address: getinfo@haworthpressinc.com].

INTRODUCTION

We developed a GIS-based model that displays the probability of wildfire occurrence and severity, then addresses the relative risks to selected resources or values by asking "if an extreme wildfire occurs in this place, what amount of danger will be posed to these resources or values?" The goal is to identify geographic areas where extreme wildfires are the most likely to occur on the basis of past experience and current condition, and what potential for damage to high value resources exists within those areas if they burn in their current vegetative condition. The risks are compared on a relative scale, based on differences in occurrence and density of the selected values between geographic areas. The study area was the western two-thirds of the State of Colorado, and the geographic areas chosen for comparison within the study area were watersheds, as discussed below.

THE COLORADO STUDY AREA

We utilized the western two-thirds of the State of Colorado as the study location for the development of a wildfire hazard and risk model. The objective of the model was to display the relative hazards and risks facing the region, as a means of helping people debate the priorities and investments that might be considered to reduce hazards and mitigate risks (Sampson et al. 2000).

The State of Colorado is located in the west-central portion of the United States, including portions of the Rocky Mountains and the Great Plains, on the divide between the southern and northern regions of the Intermountain West (Figure 2.1). We selected a study area composed of the counties in the west and central portions of the state on the basis that each contained some forest cover, with wildfire as a significant management problem. We excluded the high plains of eastern Colorado, which are mostly in private agriculture and grassland (Figure 2.2). The result was a study area of some 46.3 million acres.

The eastern side of the study area is dominated by the Rocky Mountains, with over 250 peaks that reach elevations of 13,000 feet or higher. The foothills range in elevation from 5,500 to 8,000 feet, with ponderosa pine, pinyon-juniper, and western hardwoods dominating the vegetation. The montane life zone, with mixed conifer, lodgepole pine, and aspen forests, extends to about 9,500 feet in elevation. The subalpine zone, with spruce-fir forests, is found up to about 11,500 feet. Above that lies the alpine zone, with tundra vegetation similar to that found much further north (Figure 2.3).

To the west of the Rocky Mountains lies the Colorado Plateau, a dry region which is 6,000 to 7,000 feet in elevation, marked by desert shrub and

FIGURE 2.1. The conterminous United States, with the State of Colorado highlighted.

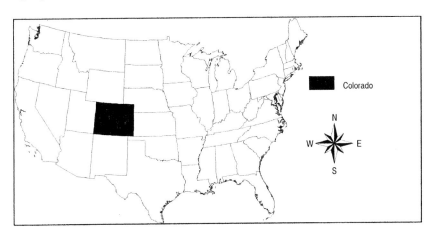

pinyon-juniper woodlands. In the far northwest corner of the state, the Green River Basin and the Uinta Mountain range are dominated by sagebrush steppe landscapes. The mountains of Colorado form the headwaters for four major rivers that contribute water to 18 states: the Colorado, the Rio Grande, the Arkansas, and the South Platte (Grolier 1998).

The climate is in the semiarid continental zone, but elevation significantly affects both temperature and precipitation. For the study area, average annual precipitation ranges from around 7 inches per year at Alamosa, near the border of New Mexico, to over 42 inches per year in the high mountains. Low elevation precipitation is highest in the spring, while the summer and fall seasons tend to be dry. Much of the mountain precipitation is received as snowfall between October and May.

The study area also has some of the fastest-growing areas in the West, leading to significant interactions between wildfires and a human population that is often characterized as loving rural areas to live and recreate in, but largely urban in its orientation (Case et al. 2000).

Using Colorado as the test area for developing a wildfire hazard and risk model creates some outcomes that may be driven by the geography and ecology of the state itself. Even though Colorado has experienced an increase in wildfire numbers and sizes since 1980, the state has experienced far less wildfire than neighboring states (Sampson et al. 2000, Table 1). Whether this is due to some inherent characteristic in Colorado's weather or vegetation patterns, or simply a measure of the randomness of wildfire activity, we did not try to determine. This was, in part, seen by the workshop organizers as a benefit, since the political controversy over salvage of fire-killed timber, road building and other public lands issues was not as intense as elsewhere, giving the workshop participants a chance to test the hazard and risk modeling approach in a more neutral political atmosphere.

THE WORKSHOP PROCESS

Participants for the workshop were selected from Universities, Federal and State agencies, and private organizations active in resource assessment and management (see Acknowledgments). An attempt was made to match the expertise of each participant with a working group assignment so that each working group could consider one aspect of the hazard and risk model. After two days of presentations that illustrated the capabilities and interests of each participant, the workshop spent the remainder of the week working in small groups to develop their individual interpretive layers for the model.

Two major challenges affected the outcome. First was the fact that the risks associated with a major fire event were difficult to address until the location, size, effect, and probability of such an event could be established. Thus, the Wildfire Risk working group needed to produce its products before the other groups could begin much of their work. This was not possible in such a short workshop, so the remainder of the working groups left the week-long effort with considerable work left to finish. If we had, in retrospect, given the Wildfire group a few days' head start, it would have been useful.

The other challenge was the integral role of existing databases and GIS technology in helping the participants see the effect of the risk indices they were developing. Suggesting that a certain data parameter (50% or greater slopes, for example) will be the best way to identify areas of greatest risk or concern for a given resource impact may be logical in terms of the research and data available, but when those areas are portrayed on a map, they may not be as effective as envisioned. Where that is the case, another data index may be tested, until the various data sets work best to provide a realistic and useful portrayal of the situation. That may not be apparent to the model's developers until they see their ideas tested in the GIS system.

That means that the GIS must be fully loaded with the available data sets and those data sets must be assembled in a compatible and comparable format. That is an enormous task; one that takes a great deal of skill and time. At Pingree Park, it was three days of almost round-the-clock work on the part of a skilled team of GIS experts before the data sets were ready for developing effective comparisons, analyses, and illustrations. That task, too, could have been most effectively addressed with a more ample head start before the workshop.

FIGURE 2.2. Watersheds of the Colorado study area.

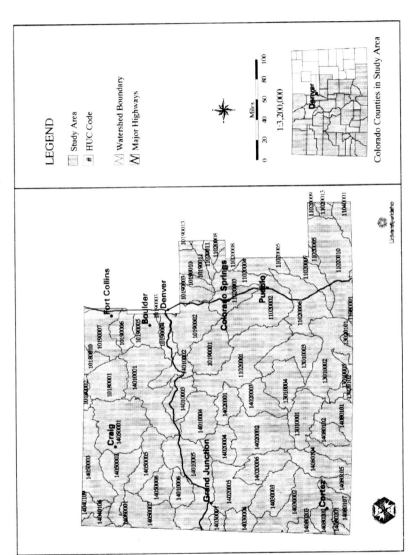

FIGURE 2.3. Vegetation types of the Colorado study area, as interpreted from AVHRR satellite imagery interpreted by the USDA Forest Service.

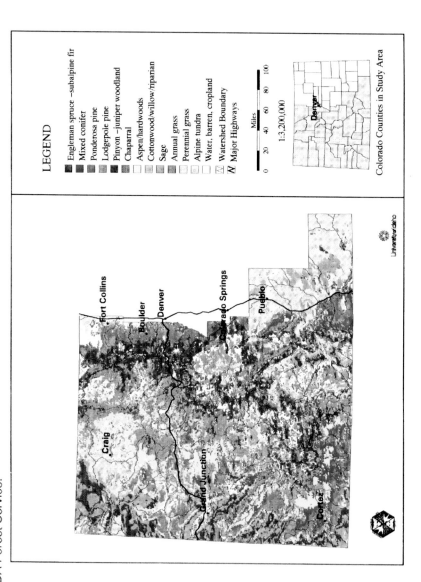

LEGEND

- Engleman spruce –subalpine fir
- Mixed conifer
- Ponderosa pine
- Lodgepole pine
- Pinyon –juniper woodland
- Chaparral
- Aspen/hardwoods
- Cottonwood/willow/riparian
- Sage
- Annual grass
- Perennial grass
- Alpine tundra
- Water, barren, cropland
- Watershed Boundary
- N Major Highways

Miles

0   20   40   60   80   100

1:3,200,000

Colorado Counties in Study Area

TABLE 1. Area of vegetation types within the Colorado study area, proportion of the study area by vegetation type, ignitions experienced, acres burned, and percent of each vegetation type burned in 1986 through 1995.

Dominant Vegetation Type	Total Area	Percent of Study Area	Wildfire History, 1986-1995		
			Ignitions Recorded	Acres Burned	Percent Burned
Spruce/Fir	6,258,733	13.51	1,180	19,428	0.31
Mixed conifer	2,135,809	4.61	793	7,951	0.37
Ponderosa pine	3,667,950	7.92	1,886	49,587	1.35
Lodgepole pine	7,610,317	16.42	1,142	11,961	0.16
Pinyon/Juniper	3,024,268	6.53	1,551	51,834	1.71
Aspen/Hardwood	2,872,857	6.20	521	9,522	0.33
Sage	3,399,708	7.34	1,429	207,500	6.10
Perennial grass	10,183,316	21.98	2,071	243,968	2.40
Cottonwood/Willow	60,762	0.13	36	18	0.03
Annual grass	2,977,338	6.43	1,540	47,507	1.60
Alpine tundra	104,728	0.23	3	1	0.00
Water	27,170	0.06	37	15	0.05
Barren	8,892	0.02	1	3	0.03
Agriculture	684,437	1.48	272	5,268	0.77
Grassland/Crop	3,323,138	7.17	796	22,839	0.69
TOTAL	46,339,423	100.00	13,258	677,400	1.46

While the Wildfire and GIS teams were getting their initial work completed, the remaining working groups (see below) focused on deciding what data they would utilize, and what methods of comparison were most effective, to portray the different risks inherent in the current forest health and wildfire hazard situation. Jointly-written working group papers to explain their chosen approach were outlined, and writing begun. By the end of the week, each working group had a technical paper underway, to be completed by correspondence following the workshop. Those papers appear as following chapters that reflect the workshop's efforts.

THE MODELING APPROACH

Geographic Extent and Map Scale

Basic decisions about the GIS mapping exercise involved considerations of the task at hand, the data sets available, and the very limited time available for the exercise. The first decision was to fix the geographic extent of the maps to be produced. With the State of Colorado chosen as the test area, we decided to limit the data sets to focus on those counties that had some forest within their boundaries. That produced a study area of roughly 46 million acres, occupying the western two-thirds of the state. The study area included all of the Front Range counties and most of the associated urbanized areas (Figure 2.2). For some interpretations such as air pollution impacts, however, that area was deemed too limited, and the maps included the entire state.

The scale chosen was 1:1,000,000–as this would produce data resolution appropriate for strategic decision-making. Where data were available in grid format, the most common scale was 1 km^2 grids, which fits reasonably well with the 1:1,000,000 scale.

It was decided at the workshop to present all data in English units, as opposed to metric, since the intended audience for the model is policymakers, not scientists. Either format presents some difficulties, as the use of English units means that the output from the GIS must be converted from metric.

The Landscape Unit

In order to compare the relative hazards and risks between areas, it was necessary to determine a landscape unit for the basic comparison. That unit could have been a county, township, or other geographic area that could be usefully compared. For this study, the landscape unit chosen was the eight-digit hydrologic unit code (HUC) watershed assigned by the U.S. Geological Survey (Seaber et al. 1987). This produced 72 units ranging in size from 1,000 to 2,000,000 acres (Figure 2.2). The average size was somewhat more uniform than the range indicates, since most of the very small units were created by clipping the watersheds to conform to the state and county boundaries of the study area.

The Ecological Unit

The ecological unit selected for this exercise was the vegetation type as identified by the 3rd revision of AVHRR satellite cover type classification (Loveland et al. 1991). The AVHRR classification was modified by the Columbia River Basin Science Team and the Fire Emissions Project of the Grand Canyon Visibility Transport Commission, and provided by the USDA Forest Services' Intermountain Fire Sciences Laboratory at Missoula, Montana. It has a pixel size of 1 km^2, and delineates 15 cover types (Figure 2.3). Table 1 indicates some of the characteristics of the study area, including the total area of the different vegetation types as indicated by the satellite imagery used in the workshop process. It also illustrates how fire had affected the different vegetation types in the decade 1986-1995.

The Data Sets

Geo-referenced data layers were obtained from a variety of sources (Table 2). Data preparation was by the GIS teams from the Department of Forest

TABLE 2. Geographic data sets utilized in the GIS model construction.

Data Set	Type	Source	Prepared by
Political boundaries	Raster	ESRI	CSU
Watersheds	Raster	USGS	CSU
Digital Elevation Model	Grid	USGS	CSU
Soils (STATSGO)	Grid	NRCS	CSU
Vegetation (AVHRR)	Grid	NASA	USFS Missoula Fire Center
Fire ignitions	Point	Fire agencies	BLM Service Center, Denver
Threatened & Endangered Species	Point	Colorado Natural Heritage Program	CNHP
Population, social measures	Census block	U.S. Census Bureau	USFS Denver Regional Office

Sciences at Colorado State University (CSU), and two USDA Forest Service teams; one at the Denver Regional Office and the other at the Missoula Fire Laboratory. Table 2 indicates the data sets used and their source.

The scale and resolution of the data limited the detail that could be produced within the modeling exercise, but was consistent with the objective of producing a strategic rather than operational model. It is also the case that these are the data sets most typically available to decision makers at the state scale. The goal was to provide decisionmakers with general ideas about relative hazard and risk levels. This is, at times, inconsistent with the desire of scientists to be precise, and the needs of land managers to identify specific project activities and locations. Those kinds of detail need, however, to be done with data sets at scales and resolutions appropriate to the questions being asked. This will normally mean focusing on a much smaller geographic area, to keep the analysis within the capability of the computing power available.

THE WILDFIRE HAZARD AND RISK ASSESSMENT

We were provided a historical fire data set for Colorado that had been assembled as part of an interagency effort, the Colorado Fire Project. The set contained some 14,500 fire records over the ten-year period 1986 through 1995. The Bureau of Land Management's Denver Service Center had spent considerable time working with the different federal, state and private data sets, so the data were in a consistent format. In the process, a significant number of the state and private records were left out of the data set because they lacked useable location information. Since many of these omitted records were associated with the large private land base in the counties on the eastern plains, we felt the omissions were not a serious hindrance with respect to the wildfire history in the forested areas. When the fire data records were "clipped" to the western portion of the state, and obvious duplicate records removed, there were a total of 13,258 fire ignitions for the study area (Figure 3.1, Neuenschwander et al. 2000) .

These data were converted into grid format by the GIS team, resulting in a data set of the number of ignitions per 1 km^2 grid over the 10-year period. On the assumption that areas experiencing the most ignitions in the past will continue to experience the highest relative ignition rates, the GIS team identified the watersheds most likely to experience future wildfire ignitions (Neuenschwander et al. 2000).

Comparing the geographically-located ignition data to the vegetative cover type map produced an estimate of the number of ignitions per vegetative cover type (Table 1). Ignition dates were grouped to array the ignition history by month of ignition, which is useful in determining weather conditions typical of the highest-occurrence period and predicting future fire behavior (Figure 3.2, Neuenschwander et al. 2000). Since the data set indicated the final size of each ignition occurrence, it was also possible to develop estimates of the relative probability of a fire ignition growing into a large (1,000 acres and more) wildfire, and compare those to vegetative types as well (Figure 3.3, Neuenschwander et al. 2000). The season for large wildfires in Colorado ranges from June through September, with July being the month of highest occurrence (Neuenschwander et al. 2000).

The initial analysis of the historical fire data identified nine watersheds which have a high probability of large wildfire events, on the basis of the number of ignitions per watershed in the historical record. After the workshop concluded we re-evaluated the data to construct an ignition density index which identified ten watersheds that had experienced greater than 5 ignitions per 10,000 acres over the decade (Figure 3.4, Neuenschwander et al. 2000). We also identified the most likely ignition point within each watershed by identifying the map grid where past ignitions were highest.

Comparisons with vegetative cover types and fuel models provided the basis for estimates of the total amount of vegetation likely to be burned in a large fire event, and the amount of air pollution released by the fire in tons of PM$_{10}$ released per acre burned and per 24-hour event (Figure 2.4). Since the most likely period for a major fire is July, meteorologic data for that month can be used to develop an array of likely weather conditions. This allows development of a stochastic model that can examine linked probabilities for

FIGURE 2.4. Schematic for development of wildfire hazard-risk model.

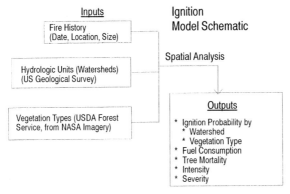

wildfire, pollution production, and dispersal in a geographic context. Thus, the historical fire data provided several useful pieces of information that would have been missing had the data not been analyzed within the context of a GIS.

Defining the "Extreme Wildfire"

The vast majority of fire ignitions (98.5%) in Colorado remain small. In the past decade, only 0.2% of the ignited fires achieved sizes larger than 5,000 acres, and only 0.03% (4 fires out of 13,258) achieved a size of 25,000 acres or more. This contrasts with some of the other western states, which have experienced several fires over 50,000 acres in recent years (Neuenschwander and Sampson 2000). In Colorado, the large fire events have been almost entirely wind-driven, so the characteristic fire shape is long, narrow, and oriented more with wind direction than with topographic or vegetative patterns.

In spite of their relative rarity, the workshop focused on fires of 25,000 acres or more to portray the big event that might characterize the most important hazards and risks in Colorado. This was done for several reasons, including:

- A consensus among the scientists that the ecosystems in the region are increasingly likely to support large fire events due to ongoing vegetative change and the lack of fuel management on large areas (An indication of this can be seen by studying Table 1. If only 1.35% of the ponderosa pine ecosystem has burned in the past 10 years, and the ecological fire return interval for ponderosa pine is 10 to 25 years, this system has clearly been fire-deprived in recent years. Other systems can be assessed in a similar manner. This exercise shows, for example, that the sage vegetation type has been burning most nearly to its historical range, while the other vegetation types are burning at much longer return intervals than history would suggest is normal.);

- A sense that it is the large events that have the most serious potential for adverse economic and ecological impacts, so that, even if they are rare, they are important enough so that if treatment can reduce their likelihood, it is worth considering; and,

- The broad-area resolution of the data utilized in the exercise could not accurately discern small areas, and therefore would not reliably identify small events.

Considering the risks to human and social institutions in Colorado may require, however, a look at the effect of multiple small events (Case et al. 2000). Communities and fire management agencies seldom see one large event in isolation. Normally, a weather period associated with a large fire event is characterized by a significant number of events that, together, may overwhelm institutional capacity. If the initial response capacity is limited to 10 fires and there are 20 simultaneous ignitions, the local fire agencies must choose which ones to attack. A wrong choice, or bad luck in the face of extreme fire conditions, can lead to a fire that quickly grows too big and intense to control.

The increasing concern across the western States in recent years is that, in the absence of an active wildland treatment program to reduce risks in high-hazard areas, these clusters of ignitions would increasingly overwhelm suppression capability and lead to a higher number of events that were large and unmanageable. The major policy recommendation of the National Commission on Wildfire Disasters was that, in the absence of effective work to treat wildland ecosystems and make them more fire-tolerant, a policy based solely on improving suppression capability was likely to be overwhelmed and ineffective (NCWD 1994). In 1995, a year-long federal policy study agreed with that conclusion, saying "agencies and the public must change their expectation that all wildfires can be controlled or suppressed" (USDI/USDA 1996).

The identification of the 10 most likely extreme wildfire ignition locations, provided the basis for a comparative assessment as to where such an event is most likely to occur in the future. On the premise that a major wildfire event occurs at a time when there are many fires going in the state, and suppression capacities are fully occupied, we were able to pose the question: How would a major fire in these 10 places compare, in terms of risks to natural resources, communities, and people?

It is clear from Figure 3.4 that the relative risks of a major wildfire differ significantly between the 10 watersheds. Some of the western watersheds are in remote, mountainous regions, while those on the Front Range lie adjacent to Colorado's major population centers. The task of the working groups was to use the available data to express the risks posed by a wildfire, in terms of relative differences between watersheds.

Establishing Relative Risk Ratings

The goal of the five scientific teams was to develop a process comparing the different landscape and/or ecological units in terms of the relative risks an extreme wildfire would pose to selected resources and values. Their approaches and findings are described in the working group papers that follow. A brief summary is described below.

Risks to People

Wildfires can destroy human life and property in spite of the best protection efforts. Most wildfires are suppressed or go out before they become a major threat. Whenever there are several simultaneous ignitions in extreme fire weather conditions, however, local suppression capability may becoming overwhelmed, resulting in an extreme event. U.S. Census data can be utilized to develop some indicators of relative risk to human populations. The elements chosen for this analysis were the total population and population density within each watershed, the existence of homes served by individual water supplies or wood heat (on the basis that they may be more rural and dispersed, and thus harder to protect), and populations that might be placed at additional risk due to their circumstances (children, elderly, low-income) (Figure 2.5).

These factors were chosen as examples of a method, not as an exhaustive

FIGURE 2.5. Schematic for development of human and community risk indices.

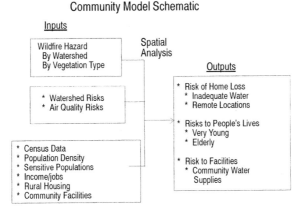

listing of relevant social information. The method used, however, indicates that taking a structured approach to considering information about the ways in which people experience wildfire, and how they may be affected by it, can provide decisionmakers with useful information in making judgments about prioritizing mitigation and protection programs (Case et al. 2000).

Air Quality Risks

The working group on air quality utilized the location, fuel consumption, and PM_{10} emission estimates developed by the wildfire working group to compare the relative impact of a 25,000 acre event in each of the high-likelihood watersheds. The two major impacts considered were the effects of wildfire-produced PM_{10} on human health and visibility. The outcomes are significantly different. Not surprisingly, the human health risk is greatest in areas of high population located downwind from the wildfire. Threats to visibility were more likely in the rural areas where Class I visibility areas (national parks, monuments, etc.) were located (Figure 2.6).

A meteorological simulation model was utilized to determine conditions that would result in transport of pollution from the simulated wildfire locations to receptor areas. Important receptor areas are population centers, particularly those that have been identified as non-attainment areas, and non-degradation areas such as National Parks and Wilderness Areas. An air quality simulation model (REMSAD) capable of handling secondary aerosol production, wet and dry deposition, and meteorological inputs was utilized to map 24-hour maximum PM_{10} concentrations from each of the modeled wildfire events (Rigg et al. 2000).

FIGURE 2.6. Schematic for development of air quality risk indices.

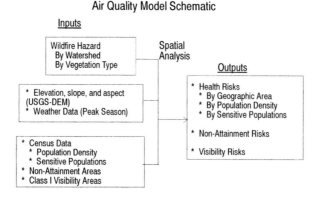

Watershed Risks

Watershed risks were evaluated in a model designed to predict post-fire erosion and sedimentation risks on a pixel, catchment, and landscape scale within the study area. The model for predicting post-fire erosion includes a hydrophobicity risk index (HRI) based on vegetation type, predicted fire severity, and soil texture. The risk of post-fire erosion was then assessed for each pixel by combining the HRI, slope, soil erodibility, and a factor representing the likely increase in soil wetness as a result of vegetation mortality. Stream sedimentation was estimated as a function of stream gradient, and the composite soil erosion and stream sedimentation risks were indexed and compared across all 72 watersheds (Figure 2.7).

Predicted surface erosion risks were low in most of the study area, with the highest risk in areas of steep slopes containing both coarse soils and vegetation types that could support high-intensity and high-severity fires. Risk of developing hydrophobic soils following a wildfire was predicted to be highest in the heavily-vegetated mid-elevation montane areas characterized by coarse-textured soils. Catchments with a higher proportion of lower-gradient reaches where deposition of eroded sediments is possible were rated as the highest sedimentation risks (MacDonald et al. 2000).

Critical Habitats

The goal of this working group was to identify means of indexing areas where large wildfires might threaten the habitat of species of plants and

FIGURE 2.7. Schematic for development of the watershed risk model.

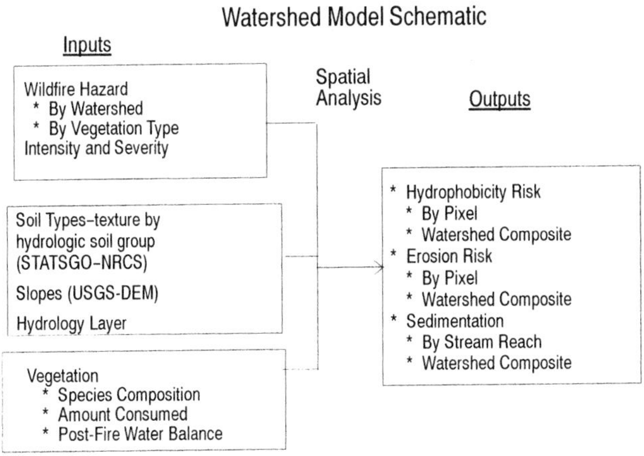

Watershed Model Schematic

animals of special concern to society. While recognizing that large, intense wildfires in some areas pose hazards to some species, it was also recognized that fire is necessary to improve habitat conditions for other species. The habitat index is thus an attempt to illustrate methods of determining where, and under what conditions, these risks and benefits can be compared (Figure 2.8).

The parameters selected for study included threatened and endangered species, Mexican spotted owls, cutthroat trout, and elk. The results were mixed. There are areas where a wildfire poses a relatively high risk to threatened and endangered species, including spotted owls and trout, on the basis that the area contains a high density of fragile sites upon which the species depends. The extent of that risk is related to the specificity of the habitat requirements of the species in question. For a highly site-dependent species like the cutthroat trout, sedimentation and the loss of riparian vegetation could mean a serious impact and, if there are few remaining populations because of other land use changes, protecting these sites becomes a high priority. On the other hand, mobile generalists such as elk are not likely to be greatly inconvenienced, let alone threatened, by wildfire. In fact, the most likely wildfire scenarios probably improve elk habitat because they will increase early successional vegetation types (Despain et al. 2000).

Ecosystem Recovery

This working group tackled one of the more undocumented aspects of wildfire risk. The question was: "Are there areas where, because of their environmental attributes and current condition, there is a relatively higher risk that an extreme wildfire may significantly alter ecosystem recovery

FIGURE 2.8. Schematic for development of wildlife habitat risk indices.

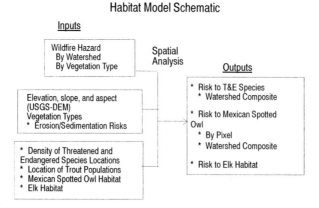

following a major disturbance?" The group identified those factors which place a system at risk, primarily soils that are likely to be damaged by erosion, nutrient loss, and heat damage to structure and texture (Figure 2.9). Some types of vegetation, such as ponderosa pine, may be at higher risk because of the flammability of its needles, but generally vegetation types are widely-enough distributed that fire does not pose a wide-area threat. On a specific site, a wildfire lethal to large trees, or two fires in a brief time span that precludes recovery of sexually-mature trees to provide new seed sources, may alter the successional pathways for a period of decades or longer. In general, however, the most significant and persistent damage will occur when a fire's effects consume soil organic matter, vaporize nutrients, and fuse soil clays, which can permanently alter the soil (Cromack et al. 2000). Further damage may be caused by soil erosion if precipitation is sufficient to cause surface runoff before post-fire vegetation has re-occupied the soil to provide protection.

CONCLUSIONS

The models developed in this study represent an initial effort to provide improved scientific information to the debate on how to cope with the current wildfire threat in the western United States. These models clearly need to be improved by additional research and development. They are, for example, only static, in that they displays the conclusions of one set of assumptions and condition assessments. Further work should be carried out to develop dynamic models that can answer the many "what-if" questions that decisionmakers must address.

FIGURE 2.9. Schematic for developing the ecosystem recovery indices.

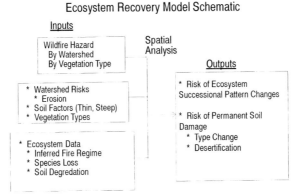

Ecosystem Recovery Model Schematic

A major research and development need is the ability to accurately catego-rize fuel loadings, arrangements, and conditions over large areas in an eco-nomic manner. Fire behavior models are available, but they depend on quan-titative data on fuel amount and arrangement which can only, at this time, be collected reliably with field methods. In this exercise, scientific judgment was used to estimate the fuel data needed. This approach will not be adequate for the large-scale modeling needed to address wildfire hazards and risks. As a result, new data sources, most likely from satellite imagery, need to be developed and refined.

A further goal is to take this broad-area strategic model down in scale to a single watershed or county, where higher-resolution data could be utilized to identify more specific areas of high hazard and risk, leading to more specific plans and recommendations for treatment. Looking at a smaller area may change the relative importance of the physical, biological, and ecological processes being studied. This may, in turn, require a careful look at the structure of the model, and revisions or adjustments made as needed. Such efforts can use many of the approaches suggested here, but more site-specific models will also require an economically-feasible method of acquiring the necessary GIS data layers at the proper scale, especially the fuel loading data needed for developing wildfire hazard ratings.

REFERENCES

Case, Pamela, Brian Banks, Eric Butler, and Ronald Gosnell. 2000. Assessing Poten-tial Wildfire Effects on People. In Sampson, R.N., R.D. Atkinson, J.W. Lewis (Eds.), *Mapping Wildfire Hazards and Risks.* Papers from the American Forests scientific workshop, September 29-October 5, 1996, Pingree Park, CO. The Ha-worth Press, Inc., New York.

Cromack, Kermit Jr., Johanna D. Landsberg, Richard Everett, Ronald Zeleny, Chris-tian Giardina, T. D. Anderson, Bob Averill, and Rose Smyrski. 2000. Assessing the Impacts of Severe Fire on Forest Ecosystem Recovery. In Sampson, R.N., R.D. Atkinson, J.W. Lewis (Eds.), *Mapping Wildfire Hazards and Risks.* Papers from the American Forests scientific workshop, September 29-October 5, 1996, Pingree Park, CO. The Haworth Press, Inc., New York.

Despain, Don G., Paul Beier, Cathy Tate, Bruce M. Durtsche, and Tom Stephens. 2000. Modeling Biotic Habitat High Risk Areas. In Sampson, R.N., R.D. Atkin-son, J.W. Lewis (Eds.), *Mapping Wildfire Hazards and Risks.* Papers from the American Forests scientific workshop, September 29-October 5, 1996, Pingree Park, CO. The Haworth Press, Inc., New York.

Grolier Interactive, Inc. 1998. *The 1998 Grolier Multimedia Encyclopedia* (CD-ROM).

Loveland, T.R. J.M. Merchant, D.O. Ohlen, and J.F. Brown. 1991. Development of a land-cover characteristics database for the conterminous U.S. *Photogrammetric Engineering and Remote Sensing* 57(11): 1453-1463.

MacDonald, Lee H., Robert Sampson, Don Brady, Leah Juarros, and Deborah Martin. 2000. Predicting Erosion and Sedimentation Risk from Wildfires: A Case Study from Western Colorado. In Sampson, R.N., R.D. Atkinson, J.W. Lewis (Eds.), *Mapping Wildfire Hazards and Risks*. Papers from the American Forests scientific workshop, September 29-October 5, 1996, Pingree Park, CO. The Haworth Press, Inc., New York.

Neuenschwander, Leon F., James P. Menakis, Melanie Miller, R. Neil Sampson, Colin Hardy, Bob Averill and Roy Mask. 2000. Indexing Colorado Watersheds to Risk of Wildfire. In Sampson, R.N., R.D. Atkinson, J.W. Lewis (Eds.), *Mapping Wildfire Hazards and Risks*. Papers from the American Forests scientific workshop, September 29-October 5, 1996, Pingree Park, CO. The Haworth Press, Inc., New York.

Neuenschwander, Leon F. and R. Neil Sampson. 2000. A Wildfire and Emissions Policy Model for the Boise National Forest. In Sampson, R.N., R.D. Atkinson, J.W. Lewis (Eds.), *Mapping Wildfire Hazards and Risks*. Papers from the American Forests scientific workshop, September 29-October 5, 1996, Pingree Park, CO. The Haworth Press, Inc., New York.

NCWD. 1994. *Report of the National Commission on Wildfire Disasters*, R. Neil Sampson, Chair. Washington, DC: American Forests, April 1994. 29 pp.

NRCS. 1994. State Soil Geographic (STATSGO) Data Base: Data Use Information. USDA–Natural Resources Conservation Service. Miscellaneous Publication 1492, Washington, DC 112 pp.

Rigg, Helen Getz, Roger Stocker, Coleen Campbell, Bruce Polkowsky, Tracey Woodruff and Pete Lahm. 2000. A Screening Method for Identifying Potential Air Quality Risks from Extreme Wildfires. In Sampson, R.N., R.D. Atkinson, J.W. Lewis (Eds.), *Mapping Wildfire Hazards and Risks*. Papers from the American Forests scientific workshop, September 29-October 5, 1996, Pingree Park, CO. The Haworth Press, Inc., New York.

Sampson, R. Neil, Dwight Atkinson, and Joe Lewis. 2000. Indexing Resource Data for Forest Health Decisionmaking. In Sampson, R.N., R.D. Atkinson, J.W. Lewis (Eds.), *Mapping Wildfire Hazards and Risks*. Papers from the American Forests scientific workshop, September 29-October 5, 1996, Pingree Park, CO. The Haworth Press, Inc., New York.

Seaber, Paul R., F. Paul Kapinos, and George L. Knapp. 1987. *Hydrologic Unit Maps*. U.S. Geological Survey Water-Supply Paper 2294. Denver, CO: USGS, Books and Open-File Reports Section.

USDI/USDA. 1996. *Federal Wildland Fire Management: Policy & Program Review*, Final Report, December 18, 1995. Washington, DC: U.S. Department of the Interior. 45 pp.

SECTION II

Chapter 3

Indexing Colorado Watersheds to Risk of Wildfire

Leon F. Neuenschwander
James P. Menakis
Melanie Miller
R. Neil Sampson
Colin Hardy
Bob Averill
Roy Mask

Leon F. Neuenschwander is affiliated with College of Forestry, Wildlife and Range Sciences, University of Idaho, Moscow, ID 83844. James P. Menakis is affiliated with Fire Sciences Laboratory, USDA Forest Service, Missoula, MT 59807. Melanie Miller is affiliated with Bureau of Land Management, Boise, ID 83709. R. Neil Sampson is affiliated with American Forests, Washington, DC 20006. Colin Hardy is affiliated with Fire Sciences Laboratory, USDA Forest Service, Missoula, MT 59807. Bob Averill is affiliiated with USDA Forest Service, Golden, CO 80401. Roy Mask is affiliated with USDA Forest Service, Gunnison, CO 81230.

[Haworth co-indexing entry note]: "Chapter 3. Indexing Colorado Watersheds to Risk of Wildfire." Neuenschwander, Leon F. et al. Co-published simultaneously in *Journal of Sustainable Forestry* (Food Products Press, an imprint of The Haworth Press, Inc.) Vol. 11, No. 1/2, 2000, pp. 35-55; and: *Mapping Wildfire Hazards and Risks* (ed: R. Neil Sampson, R. Dwight Atkinson, and Joe W. Lewis) Food Products Press, an imprint of The Haworth Press, Inc., 2000, pp. 35-55. Single or multiple copies of this article are available for a fee from The Haworth Document Delivery Service [1-800-342-9678, 9:00 a.m. - 5:00 p.m. (EST). E-mail address: getinfo@haworthpressinc.com].

SUMMARY. We utilized 10 years of fire data from the Colorado Fire Project, in connection with several GIS databases, to illustrate a method of assigning large-wildfire risk indices to the watersheds of the mountainous western side of Colorado. This was done to identify high-risk areas so that other working groups could utilize wildfire locations, sizes, probabilities, and probable effects as a basis for indexing the risks posed to environmental and cultural resources in the State. The basic questions were: (1) where are large wildfires most likely to be experienced in the future, and (2) what kinds of effects might such fires cause?

With the data and time available, we are able to answer those questions in a static manner, identifying three regions of the state where clusters of watersheds share higher wildfire risks than elsewhere. We can give general levels of impact on the basis of vegetation types and fuel models, but further detail in the geographic data, vegetative conditions, and the fire weather during the major fire season would move the model from static to dynamic, making it more useful as a decision making tool. *[Article copies available for a fee from The Haworth Document Delivery Service: 1-800-342-9678. E-mail address: <getinfo@haworthpressinc. com> Website: <http://www.haworthpressinc.com>]*

KEYWORDS. Wildfire, hazard-risk assessment, risk assessment, geographic information system

INTRODUCTION

This paper describes the development of a methodology for indexing the relative risk of major wildfires on the mountainous western side of Colorado. It is done by combining recent (1986-1995) fire ignition frequency data with observed fire sizes within the vegetative types in the study area. We indexed the level of risk according to the number of wildfires that are likely to be ignited within a watershed, on the basis that the more ignitions that occur, the greater the risk that one will expand into a major fire event.

The outputs sought from the GIS model at the workshop include:

- A relative probability or indexing of fire ignitions by watersheds, to identify those areas with the highest wildfire hazards in the study area; and,
- A test to illustrate several "prototype" large wildfire events within those high-hazard watersheds, indicating where they might ignite, what area they might cover, and what fuel they would consume, as a means of providing estimates for assessing their likely impacts on soils, watersheds, vegetation, wildlife habitat, air quality, and neighboring communities.

In order to produce these outputs, it was necessary to develop input data and factors for estimating fire characteristics and effects by vegetation type, including:

- The most likely point of origin for a wildfire within each watershed;
- A probabilistic but deterministic model for fire size at the point of origin;
- The development of burn severity, tree mortality, and fire patchiness ratings for vegetative cover types; and,
- The production of PM_{10} produced by wildfire within each vegetation cover type.

METHODS AND DEFINITIONS

The primary task in preparing the model comes in the calculation of the input data and factor tables that the model utilizes in the spatial analysis. Some of the information for the inputs was developed with database management methods, while some required spatial analysis within a GIS tool. This section describes how these data were developed at the workshop. In several places, while we used the best available data, it is noted that other data and methods are preferable where they are available.

Fire History Data

The basic data for studying Colorado fire history was a data set containing the date, location, and final size of 14,463 fires that had been identified by Colorado land management and fire agencies over the period 1986 to 1995. The data was brought together into a single format, and the events recorded in Township, Range, Section and quarter section were converted to latitude/longitude for use in the ARCINFO geographic information system (GIS).[1]

The fire data set was reviewed to remove duplicate records, and then clipped within ARCINFO to the geographic extent of the study area, which was the counties in the western portion of the State that contain forest cover. Figure 3.1 illustrates the study area, with fire locations between 1986 and 1995 plotted as dots. This resulted in a final count of 13,258 fires for analysis within the study area. This data was combined with a spatial map of Colorado using 1 km^2 (247 acre) cells, and the number of ignitions per km^2 over the period of record was calculated.

1. The fire occurrence data were compiled by Susan Goodman, USDI Bureau of Land Management, National Applied Resources Science Center, Denver, CO.

FIGURE 3.1. Fire ignitions in Colorado, 1986-1995.

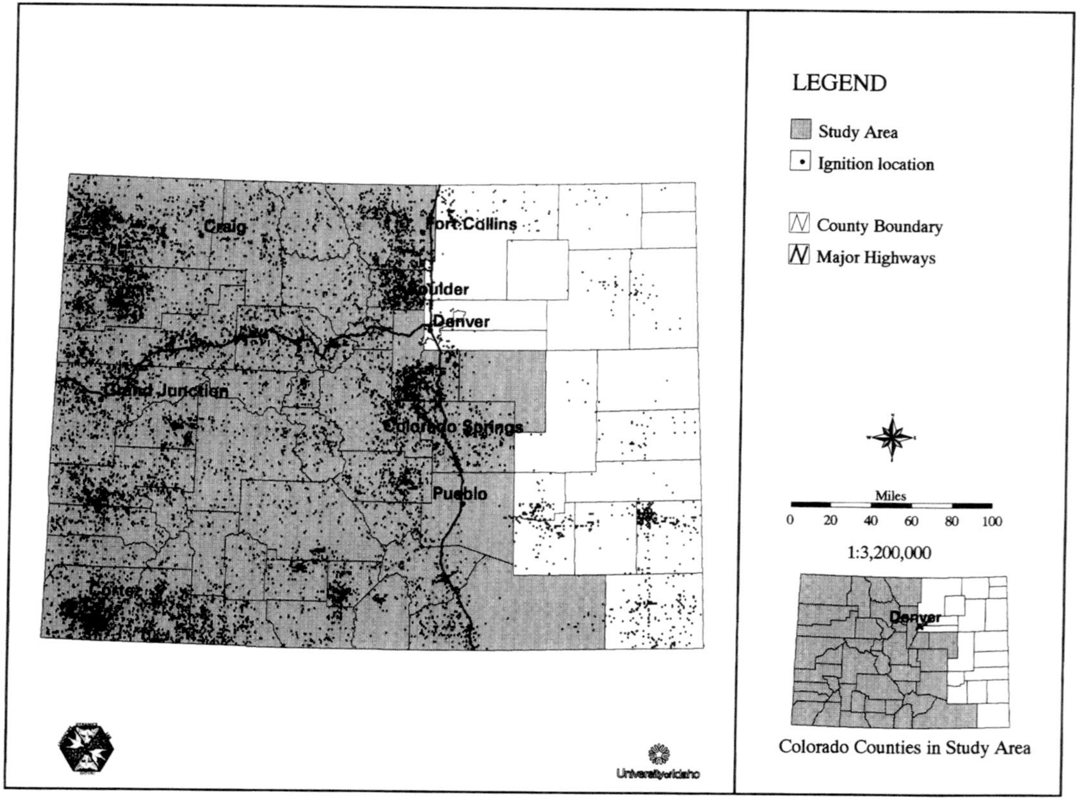

Probability and Spatial Location of Ignitions

Using the assumption that the probability of future ignitions can be represented by the frequency distribution of the past decade, we used the ignition count in each cell divided by the total ignitions over the study area as the probability of future ignitions in each grid cell. This was calculated for each 1 km^2 cell within a watershed and within a vegetative cover type, then adjusted to a common base area for relative comparison. In this model, the base comparison is the number of ignitions per 10,000 acres over the 10-year period of record.

The number and probability of ignitions was compared to both a geographical unit (e.g., watershed) and a biological unit (e.g., ponderosa pine) to be used in the spatial analysis. For this study, the geographical unit utilized is the eight-digit hydrologic unit code (HUC) watershed assigned by the U.S. Geological Survey (Seaber et al. 1987). The biological unit is the 3rd revision of AVHRR satellite cover type classification (Loveland et al. 1991). The AVHRR classification was modified by the Columbia River Basin Science Team and the Fire Emissions Project of the Grand Canyon Visibility Transport Commission, and provided by the Intermountain Fire Sciences Laboratory. It also has a pixel size of 1 km^2, and delineates 15 cover types.

The factors generated by the GIS analysis are:

- Ignitions per 10,000 ac. for each watershed over the decade of record; and,
- Ignitions per 10,000 ac. for each vegetation cover type over the same period.

The watersheds with the most ignitions over the past decade will, we assume, be at a higher risk to future wildfires. The vegetative cover types in those locations will affect fire behavior and support estimates of fire characteristics and effects.

Likelihood of Fire Origin

The likelihood of fire origin is estimated by selecting the pixel within each watershed that had experienced the highest number of ignitions over the past decade. Because of the short time available, we used a random number generator to select one pixel where two or more within a watershed had experienced the same number of ignitions. This was done to provide a geographic reference point to other working groups (e.g., air quality) from which to evaluate potential smoke transport and impact from prototype fires, and not to suggest any predictive confidence.

Likelihood of Fire Date

In order to estimate the most likely weather conditions for estimating the effects of future fires, we grouped the fire history data by 10-day periods. The results are illustrated in Figure 3.2, indicating that the month of July has historically been the time when most large-acreage fires have occurred. Without access to actual fire weather data for the area in those dates, we assumed the prototype fires in the model would burn under fuel moisture conditions typical of that period (Rothermal 1983). With access to local fire weather data, we could have generated random weather sets to provide an improved set of conditions for the model (Neuenschwander and Sampson 2000).

Probability of Fire Size

This is determined by analyzing the fire history database and developing a cumulative distribution of fires by ultimate size. For any given size class, the probability of a fire growing into that size is the number of fires that reached that size or greater, divided by the total number of fires. This can be expressed as a percent and summarized for a given vegetation type and/or the overall study area to be used in a stochastic hazard risk model. It can also be tallied for a given watershed or other geographic unit within the spatial analysis. For example, this probability could be calculated as the number of large (1,000 acres and larger) fires in the past decade within a watershed

FIGURE 3.2. Number of ignitions and acres burned in 1,000-acre and larger fires, Colorado, 1986-1995 (March through October).

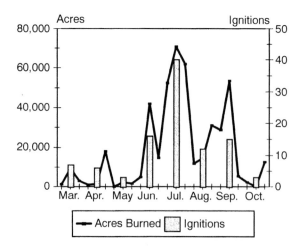

divided by the total number of large fires in the study area. This is not spatial data but it is good attribute data to supplement the ignition frequency attribute.

The 1986-95 data (Table 1) indicate that Colorado wildfires have a 97.75% chance of burning less than 1 km^2 (250 acres), a 0.9% chance of burning 4 km^2 (1,000 acres), and only a 0.2% chance of burning 20 km^2 (5,000 acres) or more. However, once a fire has burned 250 acres, it has a 39% chance of growing to the 1000-acre size, and once that size is achieved, it has 17% chance to burning to 5,000 acres or larger (Figure 3.3).

This is an important aspect for fire managers, who must consider the probability of an existing fire growing into a major event. Since the probability of a fire spreading from 250 acres to a much larger size is high, any fire of that size, once burning, must be considered very dangerous, and likely to grow exponentially if conditions are right. Fires spreading to 5,000 or even 25,000 acres will be influenced by changes in weather, fuels and terrain more than factors associated with the initial ignition probability.

Fire Characteristics and Effects

We developed input tables to estimate fire characteristics and effects by vegetation cover types, using available information on interactions among the vegetation community type, structural condition, fuel model, topography, fuel moisture and wind (Table 2). The calculation process was the same as used to estimate emissions for the Grand Canyon Visibility Transport Commission (WGA, in press). Vegetation classification was based on an AVHRR classifi-

TABLE 1. Total ignitions by vegetation type (1986-95) and the percent that resulted in fires larger than 50, 520, and 1,000 acres, Colorado.

Vegetation Type	Total Ignitions	Percent > 50 Ac.	Percent > 250 Ac.	Percent > 1,000 Ac.
Spruce/Fir	1,180	2.46	1.10	0.51
Mixed conifer	793	2.90	1.13	0.38
Ponderosa pine	1,886	3.29	0.90	0.48
Lodgepole pine	1,142	3.06	0.96	0.35
Pinyon/Juniper	1,551	4.77	2.26	0.77
Aspen/Hardwood	521	3.07	0.77	0.58
Cotton/Will/Riparian	36	0.00	0.00	0.00
Sage	1,429	7.84	4.06	1.75
Annual grass	1,540	3.51	1.88	0.52
Perennial grass	2,071	10.24	5.17	2.03
Alpine tundra	3	0.00	0.00	0.00
Water	37	0.00	0.00	0.00
Barren	1	0.00	0.00	0.00
Agriculture	272	1.47	0.37	0.37
Grassland/Crop	796	3.64	1.76	0.50
TOTAL	13,258	4.90	2.25	0.88

FIGURE 3.3. Probability of an increase in wildfire size.

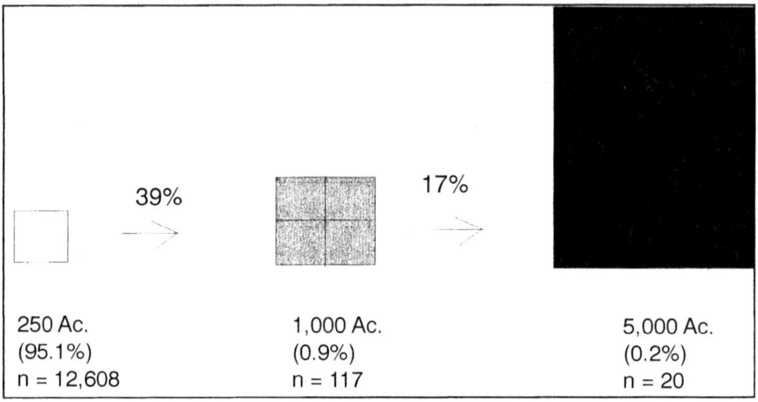

39%	17%	
250 Ac.	1,000 Ac.	5,000 Ac.
(95.1%)	(0.9%)	(0.2%)
n = 12,608	n = 117	n = 20

TABLE 2. Vegetative cover type, ownership, fuel loading, and fuel moisture content.

Cover Type	Ownership	3''+ Fuel Loading (tons/ac)	Duff Loading (tons/ac)	Moisture Content 3''+ fuels (percent)
Spruce-fir	Federal	34.6	20.4	17
Spruce-fir	St/Priv	27.6	15.9	17
Mixed Conifer	Federal	20.5	20.4	16
Mixed Conifer	St/Priv	14.6	15.9	19
Ponderosa pine	All	15.8	6.0	13
Lodgepole pine	All	16.0	9.1	19
Pinyon-Juniper	All	0.0	1.5	9
Aspen/Hardwoods	All	0.0	2.3	18
Cot/Wil/Riparian	All	0.0	1.5	10
Sage/Oak/Grass	All	0.0	1.5	7
Annual grass	All	0.0	0.8	11
Perennial grass	All	0.0	0.8	11

cation at a 1 kilometer scale (Lahm and Peterson, in press). Representative National Fire Danger Rating System fuel models were adjusted by adding or decreasing the fuel by one third (Hardy et al. 2000), so each fuel model was represented by low, medium, or high (L, M, H) loadings. We used the fuel loadings and calculations made by Hardy et al. (2000) to estimate fire characteristics for each vegetative cover type, based on the fuel moistures likely to occur at 90th percentile fire season weather.

We used these fuel loading and moisture values to calculate fuel and forest floor (duff) consumption, proportion of fuel consumed in flaming and smoldering combustion, and production of particulate emission (PM_{10}). Table 3 lists the estimated values for consumption of fuel greater than 3 inches in diameter, duff, and total fuel consumption, as calculated by Hardy et al. (2000). These values are provided because they relate most closely to burn severity, tree mortality, and particulate production.

The attributes of burn severity, tree mortality, and fire patchiness were also assigned to each vegetation cover type. Burn severity was estimated by comparing pre-fire loading of woody debris greater than $3''$ in diameter ($3''+$) and duff to the amount of consumption in these fuel classes predicted by the fire model as a means of estimating the residual material likely to remain after a fire event. Burn severity is an indicator of the heat transmitted into litter, duff, and soil layers, and the plant structures within them. It relates to the formation of hydrophobic soil layers, which are most likely to develop where accumulations of woody debris or deep duff layers are consumed and soil temperatures are raised to $220°C$ or higher (Giovannini 1994). It also relates to subsequent soil erosion in areas where hydrophobicity may not be present, as removing the protective duff and debris layer exposes topsoil to

TABLE 3. Estimated fuel consumption and fire effects by vegetative cover type and ownership.

Cover Type	Ownership	Fuel Consumption			Burn Severity	Tree Mortality	Patchiness	PM_{10}
		$3''$ + dia.	Duff	Total				
		(Tons/acre)						(lb/acre)
Spruce-fir	Federal	17.5	10.6	36.0	M	M	Y	755.3
Spruce-fir	St/Priv	13.3	8.2	26.5	M	M	Y	572.8
Mixed Conifer	Federal	10.0	11.4	28.7	M	M	Y	636.0
Mixed Conifer	St/Priv	6.8	8.2	21.1	M	M	Y	464.1
Ponderosa pine	All	8.1	3.1	20.8	M	M	Y	378.7
Lodgepole pine	All	7.4	4.7	15.0	M	H	N	325.5
Pinyon-Juniper	All	0.0	1.5	5.7	L	M	Y	98.2
Aspen/Hardwoods	All	0.0	1.2	6.0	L	L	Y	101.7
Cot/Willow/Riparian	All	0.0	0.8	4.1	L	L	Y	74.5
Sage/Grass	All	0.0	1.5	4.8	L	H	N	98.2
Annual grass	All	0.0	0.8	2.8	L	L	N	47.7
Perennial grass	All	0.0	0.8	2.5	L	L	N	43.3

additional rainfall impact which loosens soil particles and increases erosion (MacDonald et al. 2000).

Burn severity under the conditions in the study area was estimated to be moderate for forest types dominated by coniferous species, and low for other types. Associated fire model results suggest that about one-half of the coarse woody debris and duff would remain after fires in forested areas. Localized areas of high burn severity would occur, but do not represent "typical" effects. Deciduous forest types, pinyon-juniper, and shrub and/or grass types were assigned a low severity because the total amount of fuel consumed was estimated at less than 6 tons per acre. The subsurface heating that would result from the burning of this amount of fuel would not be significant, except on a small percent of the area where a localized concentration of fuel burned, such as the litter/duff layer beneath a juniper tree.

Most of the large (1,000 acres and larger) fires occur in July, but Figure 3.2 also shows a significant amount of fire in late August and September. These late summer fires are likely to be of higher severity due to drier fuel moisture conditions and greater total fuel consumption. Watersheds impacted by high-severity events are susceptible to damaging soil erosion in the event of a high-intensity storm before soil and vegetation recovery, so high-severity fires that occur earlier in the year may place an area at greater risk to damage due to the increased probability of experiencing an intense summer storm after the fire (MacDonald et al. 2000).

Mortality is an estimate of the proportion of the tree canopy that is killed. It can be caused by lethal heating of the canopy, bole, or roots. (It is recognized that death of the deciduous overstory can be accompanied by significant amounts of post-fire sprouting from roots.) Mortality is estimated from fuel consumption, pre-fire fuel distribution, and the degree of fire resistance of tree species dominant in each vegetation type. Mortality classes are: Low–0 to 20%; Moderate–21 to 80%; and, High–81 to 100%.

Mortality was estimated to be moderate (21 to 80 percent), when averaging the entire fire area in coniferous forest and woodland types (with the exception of lodgepole pine), because only about one half of large diameter fuels and duff are expected to be consumed, and due to the patchy nature in which fires occur in these fuel types. In our opinion, fires in the lodgepole pine type in Colorado are more likely to be wind-driven events that would kill a high proportion of this moderately fire resistant species. Tree mortality in deciduous forests is expected to be low under the modeled conditions because levels of fuel consumption are quite low, fires are likely to be quite patchy, and it is not expected that enough heat would be present to girdle many tree stems.

Fire patchiness is an estimate of the unburned inclusions likely to be left by fire, and is related to tree mortality. Significant patches of unburned area that vary in size are expected to remain after fires in most forest vegetation types, except for ponderosa pine. In Colorado's ponderosa pine forests, the fire regime appears to have changed, as it has throughout ponderosa's western range, from one of frequent, non-lethal fires to one of infrequent, lethal fires (Covington et al. 1994). Although fuel consumption is fairly low in sage/grass and grass communities, the high rate of heat release and rate of spread of these fires tend to result in few unburned areas.

PM_{10} is the amount of particulate matter less than 10 microns in diameter that is released in a fire, in pounds per acre of burned area. It is derived from estimated fuel consumption and an emission factor (pounds of particulate/ton of fuel consumed) for each fuel type, and can be grouped into two levels. Spruce-fir, mixed conifer, ponderosa pine, and lodgepole pine produce three to six times as much particulate as other vegetation types. Pinyon-juniper, deciduous forests, and sage/grass produce about one and one half to two times as much particulate as annual and perennial grass types.

The values produced by these calculations and estimates are contained in Table 3, and these values were then assigned to each cell in the 1 km^2 grid according to the vegetation type and ownership attributes for that cell.

There are precautions required in using these estimates, however. They are made from a coarse modeling approach, conducted at a very broad scale, using 90th percentile weather. Site specific effects resulting from fires burning under these conditions would have a considerable range relative to the average figures depicted. Impacts ranging from none to severe could be expected in coniferous forest types because the amount of fuel consumed could result in considerable localized heating in some areas. We believe these types of estimates are appropriate, however, to compare the relative fire characteristics and effects that might be anticipated under the different vegetative conditions within the watersheds of the study area. While this produces information too generalized for predicting the behavior of any one event, it produces useful comparisons for strategic planning purposes.

RESULTS AND DISCUSSION

Probability of Ignitions

The indexing of Colorado watersheds based on ignition frequencies was aided by the wide range in the ignition history between the watersheds (from 0 to 827 fires over 10 years) that was revealed in the spatial analysis. High risk watersheds reported more than 400 fires in the decade, while 23 watersheds reported less than 10 fires during the same period. The range for ignitions per 10,000 acres over the decade was from 0 to 18 (Table 4). For a

TABLE 4. Watersheds in the Colorado study area, by size (clipped to state and county boundaries), total wildfire ignitions (1986-1995), ignitions per 10,000 km^2 per decade, and ignitions and average final fire size in three size classes.

HUC	Area (km^2)	Ignitions, 1986-95	Ignitions/ 10,000 Ac.	Under 50 acres		50-1,000 acres		Over 1,000 acres	
				No.	Acres	No.	Acres	No.	Acres
10180001	3704	87	1	79	3.3	6	187	2	4175
10180002	493	14	1.1	14	0.8	0	0	0	0
10180010	1015	22	0.9	20	2	0	0	2	2650
10190001	4165	189	1.8	188	1	1	72	0	0
10190002	4539	827	7.4	815	1.1	12	134	0	0
10190003	1428	85	2.6	83	0.8	2	425	0	0
10190004	1431	60	1.7	60	0.4	0	0	0	0
10190005	2173	608	11.3	603	0.9	2	189	3	2276
10190006	1930	367	7.7	365	1.6	2	163	0	0
10190007	3819	359	3.8	343	0.9	15	202	1	1967
10190008	50	3	2.4	3	13.9	0	0	0	0
10190010	820	6	0.3	6	1.5	0	0	0	0
10190011	1847	3	0.1	3	5	0	0	0	0
10190013	298	0	0	0	0	0	0	0	0
10250001	22	0	0	0	0	0	0	0	0
11020001	7919	375	1.9	370	1.2	5	182	0	0
11020002	5956	425	2.9	408	1.5	14	237	3	1789
11020003	2409	203	3.4	198	1.2	5	313	0	0
11020004	1919	143	3	116	4.7	24	237	3	1023
11020005	2385	19	0.4	13	4.6	5	232	1	2000
11020006	4822	243	2	241	1.5	2	300	0	0
11020007	2549	39	0.6	38	2.8	1	80	0	0
11020008	908	38	1.7	28	14.6	10	133	0	0
11020009	380	1	0.1	1	1	0	0	0	0
11020010	7541	170	0.9	154	1.2	8	156	8	1673
11020011	864	4	0.2	1	25	2	350	1	1500
11020012	828	4	0.2	2	17	2	235	0	0
11020013	602	5	0.3	3	20.7	2	133	0	0
11040001	1408	7	0.2	6	8.4	1	100	0	0
11040004	38	0	0	0	0	0	0	0	0
11080001	146	4	1.2	3	0.2	0	0	1	2500
13010001	3461	79	0.9	78	2.6	1	288	0	0
13010002	6513	380	2.4	377	1.2	3	152	0	0
13010003	4135	178	1.7	176	1.4	0	0	2	7044
13010004	3490	99	1.1	98	1.6	1	57	0	0
13010005	1390	40	1.2	38	1	2	201	0	0
13020101	321	1	0.1	1	0.1	0	0	0	0
13020102	214	4	0.8	4	1.3	0	0	0	0
14010001	7518	289	1.6	274	1.4	14	216	1	1104
14010002	1768	120	2.7	120	0.2	0	0	0	0

HUC	Area (km²)	Ignitions, 1986-95	Ignitions/ 10,000 Ac.	No. Ign. 10,000 ac.	Average SC1	No. Ign. SC2	Average size, SC2	No. Ign. SC3	Average size, SC3
14010003	2523	273	4.4	266	1.9	7	317	0	0
14010004	3758	122	1.3	118	1.8	4	187	0	0
14010005	8030	798	4	753	1.9	38	242	7	2766
14010006	1821	80	1.8	77	1.4	3	153	0	0
14020001	1990	39	0.8	37	1.1	2	94	0	0
14020002	6231	149	1	138	0.8	11	103	0	0
14020003	2858	54	0.8	49	2.5	5	119	0	0
14020004	2503	207	3.3	199	0.9	7	135	1	3848
14020005	4275	222	2.1	216	1.8	6	245	0	0
14020006	2945	172	2.4	170	1.6	2	121	0	0
14030001	730	48	2.7	46	1.9	1	120	1	1332
14030002	4930	244	2	230	2.5	9	280	5	2204
14030003	3991	452	4.6	438	1	13	154	1	4068
14030004	1632	83	2.1	77	2.8	5	198	1	1112
14030005	5	0	0	0	0	0	0	0	0
14040106	889	183	8.3	167	1.7	10	343	6	11758
14040109	1146	163	5.8	152	1.2	9	288	2	3250
14050001	6795	228	1.4	201	2.6	23	256	4	2225
14050002	4036	584	5.9	503	2.5	63	303	18	6344
14050003	4441	244	2.2	213	1.1	26	332	5	2888
14050005	3510	224	2.6	206	2.2	13	218	5	4052
14050006	2366	502	8.6	482	1.8	18	296	2	1184
14050007	3811	553	5.9	500	2.6	45	267	8	2482
14060001	195	28	5.8	26	2.4	2	640	0	0
14080101	4205	378	3.6	370	2.3	8	128	0	0
14080102	1748	184	4.3	179	1.7	3	77	2	3580
14080104	2969	311	4.2	306	1.3	4	126	1	6706
14080105	1111	124	4.5	113	2.2	11	308	0	0
14080107	1927	827	17.4	781	2.2	31	295	15	6061
14080201	397	19	1.9	11	8.2	3	222	5	19400
14080202	1665	237	5.8	228	1.9	9	271	0	0
14080203	954	26	1.1	26	1.4	0	0	0	0

map indicating the location of the watersheds by HUC, see MacDonald et al. (2000, Figure 4.1).

On the basis of that data, we determined that Colorado has 11 watersheds at highest relative risk to large wildfires (Figure 3.4). These watersheds have the highest incidence of past fire activity and are most likely to experience more risk in the future. When extreme fire weather conditions coincide with ignitions, large wildfires are likely to occur.

The correlation between the number of fire ignitions and the total acres burned in Colorado's forested watersheds is low (Table 5), as one would

TABLE 5. Ignitions per 10,000 acres, by vegetation type, and ignitions and acres burned by fire size class, 1986-95, Colorado study area.

Vegetation Type	Ignitions per 10,000 Ac.	< 50 Acres		50 to 999 Acres		> 1,000 Acres	
		Ignitions	Acres	Ignitions	Acres	Ignitions	Acres
Spruce/Fir	1.89	1,151	1,381	23	4,485	6	14,862
Mixed conifer	3.71	770	1,078	20	3,800	3	4,611
Ponderosa pine	5.14	1,824	2,736	53	8,586	9	42,939
Lodgepole pine	1.50	1,107	1,771	31	5,177	4	7,992
Pinyon/Juniper	5.13	1,477	2,511	62	14,818	12	39,180
Aspen/Hardwood	1.81	505	657	13	1,625	3	8,415
Cotton/Willow/Rip	5.92	36	18	0	0	0	0
Sage	4.20	1,317	2,634	87	24,621	25	186,650
Annual grass	5.17	1,486	2,378	46	13,432	8	34,440
Perennial grass	2.03	1,859	4,648	170	47,770	42	203,700
Alpine tundra	0.29	3	1	0	0	0	0
Water	13.62	37	15	0	0	0	0
Barren	1.12	1	3	0	0	0	0
Agriculture	3.97	268	268	3	402	1	5,000
Grassland/Crop	2.40	767	1,381	25	5,625	4	17,268
TOTAL		12,608	21,478	533	130,341	117	565,057
PERCENT		95.10	3.00	4.02	18.18	0.88	78.82

expect. With less than 1 percent of the ignitions responsible for almost 80 percent of the total acres burned, it suggests that the number of ignitions is a better indicator of fire risk than total burned acres due to the conditional probability of an ignition growing into a large event. On the basis of the past decade's experience, Colorado has about one major wildfire (1,000 acres and larger) for each 100 ignitions. The study area has experienced just over 13,000 ignitions in the past decade (Table 5). If that rate continues, one should anticipate an average of 100-120 large fires per decade.

Fire spread depends on many factors, including the vegetation type and structure, fuel matrix, continuity, and moisture, weather, and topography, as well as management factors such as access, response time, equipment, and fire fighter availability that influence whether or not an ignition is suppressed while it is still small and manageable.

Suppression efforts have clearly been responsible for the size achieved by past fires. Once fires surpass 50 acres in size (as 4% of those in the past decade have done), they have a greater chance of growing much larger, and are harder to suppress and more dangerous to both firefighters and residents. Where fuel loads are heavy, their probability of high-severity ecological impact goes up, as well. This places a heavy burden, and a paradoxical one, on fire suppression agencies. If they fail to suppress virtually every ignition when it is small, the risks of major damage are significant. Where they succeed in suppressing the fire, surrounding fuel conditions can become increasingly dangerous as these systems grow and store both live and dead biomass. Successful suppression must be linked to vegetative and fuel reduction treatments if it is to provide real protection to people and ecosystems.

A large number of ignitions, particularly if they occur at the same time, may overwhelm suppression capability. Thus, the larger the number of ignitions, the higher the probability that one of them will become large. However, it should be noted that a large fire event can occur anywhere on the landscape where ignition, fuel conditions, weather, and ineffective suppression response happen to coincide.

The number of ignitions per 10,000 acres of each vegetation type shown in Table 5 also gives a sense of the importance of human activities. The highest ignition frequency is around water-related areas and in cottonwood/willow riparian areas–areas likely to be associated with homes or recreation sites. While these areas showed a high frequency, the fires were few in total number (less than 1% of all ignitions) and small in size (less than 0.5 acres on average), indicating areas where fire spread is limited, probably by aggressive suppression efforts.

For the forested areas, ignition frequencies are highest for ponderosa pine, pinyon-juniper and sage brush vegetation types. The sage category includes oak brush and mountain shrubs, which are a particular problem in Colorado fire management (Zeleny pers. comm.). These three vegetation types, along with perennial grass, exhibited the greatest number of fires which grew to over 1,000 acres in size (Table 5). About 2 percent of all ignitions in perennial grass and sage grow to 1,000 acres or larger, with over 10 percent of the ignitions in perennial grass vegetation reaching 50 acres or more. These tables clearly indicate which vegetation types are most flammable, and most likely to fuel a large fire event when ignitions occur.

Likelihood of Fire Origin

This was generated as part of the GIS output, as shown in Figure 3.5. To help other groups develop risk estimates based on fire effects, we generated prototype events within the watersheds with the highest probability of fire ignition. The area covered by each event was modeled on the basis of July weather conditions. A random number generator chose how fire might spread from grid to grid on the map. In Colorado, most large wildfires are wind-driven, thus most likely to spread in the northeasterly direction common under July wind conditions. The resulting shapes were not modified by any model of vegetation or topography. The actual spread and shape of a real fire event is, of course, determined by weather, fuels, and topography, but the data bases being used in this strategic model were not sufficiently detailed to support these refinements.

As Figure 3.5 indicates, under extreme fire weather conditions, the communities along Colorado's Front Range, as well as those in the urban-wildland intermix in the southwest and northwest corners of the state, are at the greatest risk. This will doubtless come as no surprise to the residents of those areas, since the identification was based on their most recent 10 years of experience. The relative differences between those watersheds and other similar areas was, however, larger than we expected to find.

Fire Characteristics and Effects

Figure 3.5 illustrates the location of four high-risk vegetation types–ponderosa pine, pinyon-juniper, sage, and perennial grass–in relation to the watersheds with high ignition frequencies. The high-risk watersheds on the northwest and southwest corners of Colorado are dominated by grass, brush, and woodland vegetation types, while the high-risk watersheds along the Front Range contain ponderosa pine at the upper levels, grading to brush and grass on the lower slopes. Clearly, the high-ignition areas on the map, and the vegetation types associated with high probabilities for large fires, fall on the same landscape in many places. This lends credence to their identification as areas at higher risk of large wildfire events.

We generated estimates of the amount of smoke that would be produced if a large wildfire occurred in the watersheds that had experienced the highest incidence of ignitions within the past decade. We selected a very large event (25,000 acres) to match the needs of the Air Quality model (Riggs et al. 2000).

Since the exact location and spread of the fire was not available, the amount of vegetation burned was based on the proportion of acres of each vegetation type in the watershed. Table 6 illustrates how the method would return estimates for three sample watersheds. Smoke (PM_{10}) was calculated by taking pounds per acre of PM_{10} for each vegetation type (Table 3) times the amount of acres burned in that vegetation type (Lahm 1994). To simulate a wind event for air quality modeling, a 24-hour burn rate was calculated by multiplying the total PM_{10} produced by 80 percent.

The wide variation in per-acre average and total pollution emitted from similarly-sized wildfires in different watersheds, as illustrated in Table 6, supports the hypothesis that different relative risks can be effectively portrayed by this method.

CONCLUSIONS

The methods described provide a basis for indexing Colorado watersheds on the basis of ignition frequencies and vegetation types. Such an index

FIGURE 3.4. Probability of wildfire ignitions, by watershed, Colorado study area, in number of ignitions per 10,000 acres over 10 years.

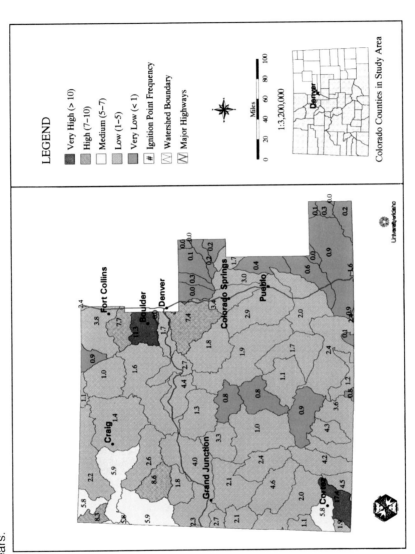

FIGURE 3.5. Location of high-probability large wildfires in Colorado study area, with most likely affected vegetation types.

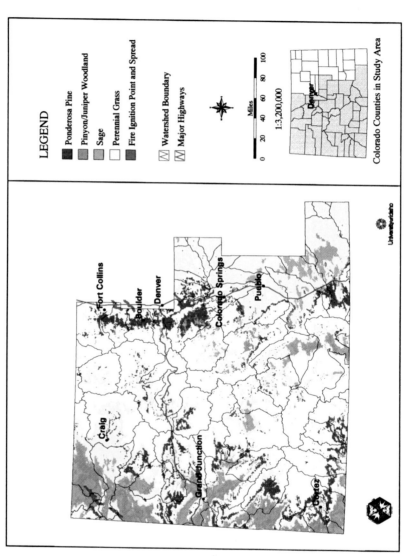

TABLE 6. Estimated PM_{10} emissions resulting from a 25,000-acre wildfire in three samples watersheds, based on average vegetation distribution within each watershed and the estimated emission from each vegetation type.

Vegetation Type	PM_{10} lbs/ac.	HUC 10190002			HUC 14050002			HUC 14080107		
		Percent of HUC	Acres Burned	PM_{10} (tons)	Percent of HUC	Acres Burned	PM_{10} (tons)	Percent of HUC	Acres Burned	PM_{10} (tons)
Spruce-fir (Feds)	755	21	5,275	1,991	0	0	0	0	0	0
Spruce-fir (Non Fed)	573	6	1,550	444	0	0	0	1	150	43
Mixed Conifer (Feds)	636	2	575	183	0	0	0	3	700	223
Mixed Conifer (Non Fed)	464	3	800	186	0	0	0	5	1,300	302
Ponderosa pine	379	39	9,775	1,852	2	450	85	13	3,125	592
Lodgepole pine	326	7	1,625	265	1	175	29	3	700	114
Pinyon/Juniper	98	7	1,625	80	10	2,450	120	21	5,350	262
Aspen/Hard	102	2	475	24	2	450	23	8	2,075	106
Cotton/Willow/Riparian	74	0	0	0	0	0	0	0	0	0
Sage	98	0	0	0	19	4,750	233	23	5,750	282
Annual grass	48	7	1,725	41	4	1,025	25	16	3,950	95
Perennial grass	43	1	275	6	61	15,300	329	5	1,350	29
Other	0	5	1,250	0	2	500	0	2	500	0
Total for HUC		100	25,000	5,072	100	25,000	843	100	25,000	2,047
Average PM_{10} per acre of wildfire (lbs/acre)				406			67			164

provides a basis upon which other risks can be assessed, and priorities and strategies for mitigation efforts, allocation of suppression resources, and other measures can be discussed by policymakers and the public.

Taken to a finer level of resolution for project planning purposes, it can assist in identifying large areas of dangerous fuel conditions where initial treatment efforts should be focused. Such an effort should provide excellent guidance in identifying cost-effective treatment opportunities to reduce the risk and impact of large wildfires. The wildfire risk model strongly suggests that such studies should be initiated in three areas of Colorado.

The data provided by the Colorado fire agencies and counties is indispensable in this type of analysis, but its utility could be improved in the future by more accurate reporting of fire location in latitude and longitude. This would provide the basis for easy combination with other spatial data sets for additional study and interpretation.

This prototype model, developed in the span of a one-week workshop, could be vastly improved by several types of information which, if available, were not on hand at the time. These include better cover type mapping, including vegetative structural condition attributes, fuel model mapping, higher resolution DEM's (slope, elevation and aspect models), and daily fire weather records from which stochastic weather conditions could be generated. Fire perimeter maps for the large fires, along with maps indicating the

patterns of intensity and severity experienced within these fires, would greatly aid in developing data-based predictive models. These additional tools
allow the modeler to generate spatial probabilities of fire size and intensity,
along with improved estimations of fuel consumption and smoke emission,
fire severity, and other fire effects that can help analysts evaluate the risks and
hazards associated with various events.

Additional, and more detailed, data layers could also assist in bringing the
model down to a watershed or county scale that would be useful for more
specific hazard-risk assessment, as well as provide a better decision tool to
focus project activities.

With the information available, we were, however, able to provide an
index as to the watersheds most likely to experience wildfire. Using this tool
as a basis for additional interpretation, and improving it as new data becomes
available, provides Colorado planners with useful insight into the states's
wildfire situation. Building on this methodology could provide an equally
useful decision tool in the other Western states, all of whom face similar, if
not worse, hazards and risks.

REFERENCES

Covington, W.W., R.L. Everett, R. Steele, L.L. Irwin, T.A. Daer, and A.N.D. Auclair.
    1994. Historical and anticipated changes in forest ecosystems of the Inland West
    of the United States. *Journal of Sustainable Forestry* 2(1/2/3/4):13-63.
Giovannini, G. 1994. The effect of fire on soil quality. In Sala, M. & J.L. Rubio
    (Eds.), *Soil Erosion and Degradation as a Consequence of Forest Fires.* Papers
    from the International Conference on Soil Erosion and Degradation as a Conse-
    quence of Forest Fires, Barcelona, 1991. Logroño: Geoforma Ediciones. pp. 15-27.
Hardy, C.H., Burgan, R.E., and R.D. Ottmar. 2000. A database for spatial assess-
    ments of fire characteristics, fuel profiles, and PM_{10} emissions. In Sampson, R.N.,
    R.D. Atkinson, J.W. Lewis (Eds.), *Mapping Wildfire Hazards and Risks:* Papers
    from the American Forests scientific workshop, September 29-October 5, 1996,
    Pingree Park, CO. The Haworth Press, Inc., New York.
Lahm, Peter. 1994. *1989 prescribed fire emissions inventory: Methodology and as-
    sumptions.* Prepared for the Grand Canyon Visibility Transport Commission.
    Technical Committee. Emissions Subcommittee. 19 pages plus appendices.
Lahm, P.W. and J.L. Peterson. In press. Final Report of the Fire Emissions Project.
    Tucson, AZ: USDA Forest Service.
Lahm, P.W. and J.L. Peterson. Unpublished Fire Emissions Project Data. Tucson,
    AZ: USDA Forest Service
Loveland, T.R. J.M. Merchant, D.O. Ohlen, and J.F. Brown. 1991. Development of a
    land-cover characteristics database for the conterminous U.S. *Photogrammetric
    Engineering and Remote Sensing* 57(11): 1453-1463.
MacDonald, Lee H., Robert Sampson, Don Brady, Leah Juarros, and Deborah Mar-
    tin. 2000. Predicting erosion and sedimentation risk from wildfires: A case study

from Western Colorado. In Sampson, R.N., R.D. Atkinson, J.W. Lewis (Eds.), *Mapping Wildfire Hazards and Risks:* Papers from the American Forests scientific workshop, September 29-October 5, 1996, Pingree Park, CO. The Haworth Press, Inc., New York.

Neuenschwander, Leon F. and R. Neil Sampson. 2000. A wildfire and emissions policy model for the Boise National Forest. In Sampson, R.N., R.D. Atkinson, J.W. Lewis (Eds.), *Mapping Wildfire Hazards and Risks:* Papers from the American Forests scientific workshop, September 29-October 5, 1996, Pingree Park, CO. The Haworth Press, Inc., New York.

Rigg, Helen Getz, Roger Stocker, Coleen Campbell, Bruce Polkowsky, Tracey Woodruff and Pete Lahm. 2000. A Screening Method for Identifying Potential Air Quality Risks from Extreme Wildfires. In Sampson, R.N., R.D. Atkinson, J.W. Lewis (Eds.), *Mapping Wildfire Hazards and Risks:* Papers from the American Forests scientific workshop, September 29- October 5, 1996, Pingree Park, CO. The Haworth Press, Inc., New York.

Rothermal, R.C. 1983. How to predict the spread and intensity of forest and range fire. Gen. Tech. Rep. INT-143. USDA Forest Service, Intermountain Research Station. 161 pp.

Seaber, Paul R., F. Paul Kapinos and George L. Knapp. 1987. *Hydrologic Unit Maps.* U.S. Geological Survey Water-Supply Paper 2294. Denver, CO: USGS, Books and Open-File Reports Section.

WGA. In press. First Revision to the Grand Canyon Visibility Transport Commission's Assessment. Denver, CO: Western Governor's Association.

PERSONAL COMMUNICATIONS

Zeleny, R. 1996. Oral presentation on Colorado wildfire conditions and challenges, Workshop on Indexing Resource Data for Forest Health Decisionmaking, Pingree Park, CO, Oct. 1, 1996.

Chapter 4

Predicting Post-Fire Erosion and Sedimentation Risk on a Landscape Scale: A Case Study from Colorado

Lee H. MacDonald
Robert Sampson
Don Brady
Leah Juarros
Deborah Martin

Lee H. MacDonald is affiliated with the Department of Earth Resources, Colorado State University, Fort Collins, CO 80523. Robert Sampson is affiliated with Natural Resources Conservation Service, Boise, ID 83710. Don Brady is affiliated with U.S. Environmental Protection Agency, 401 M Street, S.W. 4503 F, Washington, DC 20460. Leah Juarros is affiliated with U.S. Forest Service, Boise National Forest, 1750 Front Street, Boise, ID 83702. Deborah Martin is affiliated with U.S. Geological Survey, 3215 Marine Street, Boulder, CO 80303.

This work was conducted with funding from the Environmental Protection Agency and the logistical support of *American Forests*. Dr. Denis Dean and his staff, particularly Jihn-Fa Jan, provided technical support for analyses, and the authors appreciate their assistance. Dwight Atkinson, Deirdre Dether, Allen Gellis, Robert Jarrett, John Moody, Philip Omi, Wayne Owen, and Peter Wohlgemuth provided reviews of an earlier draft of this paper, and the authors are grateful for their many helpful comments and suggestions.

[Haworth co-indexing entry note]: "Chapter 4. Predicting Post-Fire Erosion and Sedimentation Risk on a Landscape Scale: A Case Study from Colorado." MacDonald, Lee H. et al. Co-published simultaneously in *Journal of Sustainable Forestry* (Food Products Press, an imprint of The Haworth Press, Inc.) Vol. 11, No. 1/2, 2000, pp. 57-87; and: *Mapping Wildfire Hazards and Risks* (ed: R. Neil Sampson, R. Dwight Atkinson, and Joe W. Lewis) Food Products Press, an imprint of The Haworth Press, Inc., 2000, pp. 57-87. Single or multiple copies of this article are available for a fee from The Haworth Document Delivery Service [1-800-342-9678, 9:00 a.m. - 5:00 p.m. (EST). E-mail address: getinfo@haworthpressinc.com].

SUMMARY. Historic fire suppression efforts have increased the likelihood of large wildfires in much of the western U.S. Post-fire soil erosion and sedimentation risks are important concerns to resource managers. In this paper we develop and apply procedures to predict post-fire erosion and sedimentation risks on a pixel-, catchment-, and landscape-scale in central and western Colorado.

Our model for predicting post-fire surface erosion risk is conceptually similar to the Revised Universal Soil Loss Equation (RUSLE). One key addition is the incorporation of a hydrophobicity risk index (HYRISK) based on vegetation type, predicted fire severity, and soil texture. Post-fire surface erosion risk was assessed for each 90-m pixel by combining HYRISK, slope, soil erodibility, and a factor representing the likely increase in soil wetness due to removal of the vegetation. Sedimentation risk was a simple function of stream gradient. Composite surface erosion and sedimentation risk indices were calculated and compared across the 72 catchments in the study area.

When evaluated on a catchment scale, two-thirds of the catchments had relatively little post-fire erosion risk. Steeper catchments with higher fuel loadings typically had the highest post-fire surface erosion risk. These were generally located along the major north-south mountain chains and, to a lesser extent, in west-central Colorado. Sedimentation risks were usually highest in the eastern part of the study area where a higher proportion of streams had lower gradients. While data to validate the predicted erosion and sedimentation risks are lacking, the results appear reasonable and are consistent with our limited field observations. The models and analytic procedures can be readily adapted to other locations and should provide useful tools for planning and management at both the catchment and landscape scale. *[Article copies available for a fee from The Haworth Document Delivery Service: 1-800-342-9678. E-mail address: <getinfo@haworthpressinc.com> Website: <http://www.haworthpressinc.com>]*

KEYWORDS. Wildfire, soil erosion, sedimentation, geographic information system, risk assessment

INTRODUCTION

Numerous plot and watershed-scale studies have documented the increase in runoff and erosion following wildfires (Tiedemann et al. 1979). The fire-flood-erosion cycle has been most thoroughly documented in chaparral environments (e.g., Rice 1974; Laird and Harvey 1986; Wells 1987). The risk to life and property are particularly apparent in places such as Southern California (e.g., McPhee 1989; Forrest and Harding 1994) and the San Francisco

Bay Area (Booker et al. 1993), where rapid development has encroached on ecosystems with short fire return intervals.

Large increases in erosion rates have also been observed after wildfire in forested environments (e.g., Helvey 1980; Morris and Moses 1987; Amaranthus 1989; Scott and Van Wyk 1990). The magnitude of the post-fire increase in erosion appears to be highly correlated with fire intensity (Robichaud and Waldrop 1994; Scott 1993), where fire intensity is the relative amount of heat flux (Covington and Moore 1994; Whelan 1995). Post-fire increases in soil erosion rates have also been documented in grasslands (e.g., Cheruiyot et al. 1986; Emmerich and Cox 1994).

The observed increases in runoff and erosion following wildfires or high-intensity prescribed burns have been attributed to several processes. Removal of the vegetative canopy reduces interception and evapotranspiration losses, which then leads to increases in net precipitation, higher antecedent moisture conditions, and increased annual water yields (e.g., Tiedemann et al. 1979; O'Loughlin et al. 1982; Megahan 1983). In high-intensity burns the removal of the protective vegetation and litter increases rainsplash and surface sealing (e.g., DeBano et al. 1979; Beschta 1990; Onda et al. 1996). These changes may be exacerbated by a concomitant reduction in soil organic matter and resulting destruction of soil aggregates (Tiedemann et al. 1979; DeBano 1989; Prosser 1990). In the southwestern US it has been shown that the downslope transport of sediment by dry ravel greatly increases following wildfire (Krammes 1965; Florsheim et al. 1991; Wohlgemuth et al. 1996), and this has also been observed in the Oregon Coast Range (Bennett 1982, cited in McNabb and Swanson 1990).

In some vegetation types fire volatilizes the secondary compounds in the litter and soil organic matter, and the condensation of these organic substances on the underlying cooler soil creates a hydrophobic layer 1-10 cm below the surface (e.g., DeBano et al. 1970; Savage 1974; Wells et al. 1987; Scott and Van Wyk 1990). The strength of this hydrophobic layer increases with fire severity (DeBano and Krammes 1966) and is generally strongest in coarse-textured soils because of their lower surface area (DeBano et al. 1970; Campbell et al. 1977).

The strength and likelihood of a hydrophobic layer also varies with vegetation type. Much of the basic work on hydrophobicity has focused on chaparral communities, but the well-documented association between chaparral and post-fire hydrophobicity is probably partly due to the high intensity of chaparral fires and the propensity of chaparral to grow on coarse-textured soils.

Much less literature is available on the development of hydrophobic layers in other vegetation types, but the literature suggests that more xeric vegetation types have higher concentrations of the secondary compounds that contribute to the development of hydrophobic layers. Post-fire hydrophobic layers have been documented in some pine forests (particularly *Pinus ponderosa* and *Pinus radiata*) (Scott and Van Wyk 1990; Onda et al. 1996) and mixed conifer forests (Dyrness 1976). Hydrophobic layers might also be expected in other vegetation types with substantial surface fuel loadings and high concentrations of secondary compounds, such as oak woodlands, lodgepole pine, pinyon-juniper, and spruce-fir.

Fire-induced hydrophobic layers are of primary concern when they cause infiltration rates to drop below precipitation intensity. The resulting shift in runoff generation from subsurface flow to infiltration-excess overland flow will substantially increase runoff volumes and the size of peak flows from a given rainstorm (Nassari 1989; Scott 1993). More importantly, the combination of rainsplash and surface overland flow can increase erosion rates by one or more orders of magnitude (e.g., Wohlgemuth et al. 1996) and greatly increase the proportion of eroded material delivered from hillslopes to the stream channels (Sampson 1944; Prosser 1990). The resultant increase in sediment supply increases sediment yields and can cause severe aggradation in lower-gradient channels that are not simultaneously subjected to large increases in the size or duration of peak flows (Meyer et al. 1995).

High intensity rainfall events on burned slopes can also generate debris flows. In most cases these are rapidly-moving slurries of water, ash, and sediment triggered by a post-fire reduction in infiltration rates. Such flows can severely scour existing channels and deliver large amounts of material to debris fans or downstream reaches, often with severe consequences to life, property, and aquatic ecosystems (e.g., Klock and Helvey 1976; McPhee 1989; Wohl and Pearthree 1991; Rinne 1996).

While a number of studies have investigated the effects of wildfires on runoff and erosion, we know of no efforts to predict post-fire erosion and sedimentation hazards across a landscape or large catchment. Many ecologists have argued that fire suppression has greatly increased fuel loadings throughout the Western U.S., and consequently the risk of large-scale, high-intensity wildfires (Water Resources Center 1989; Agee 1994; Covington and Moore 1994). The increasing availability of geographic information systems (GIS) and geo-referenced databases make it possible to predict fuel loadings and fire intensity. By combining this information with geo-referenced data on soil texture, slope, precipitation intensity, and stream gradients, we believe that we can predict, at least on a relative basis, the post-fire surface erosion hazard and resultant sedimentation risk.

The approach developed in this paper draws upon the conceptual model that underlies the Revised Universal Soil Loss Equation (RUSLE) (Renard et al. 1997). RUSLE, which is simply an updated version of the Universal Soil Loss Equation, predicts the combined soil loss from rainsplash, sheetwash, and rill erosion as a product of rainfall erosivity, soil erodibility, slope length, slope steepness, vegetative cover, and management practices. We adapted this widely-used empirical model to post-fire situations by incorporating the additional hazards due to removing the canopy and surface vegetation cover, increasing soil moisture, and developing a hydrophobic layer. The susceptibility of stream segments to sedimentation was assessed from the estimated stream gradients. These broad-scale predictions allow land managers and policy makers to compare fire-induced erosion and sedimentation risks across watersheds, or evaluate relative risk within smaller catchments for planning purposes.

This paper reports on our efforts to: (1) develop a conceptual model for predicting surface erosion risk prior to and following wildfire; (2) apply this model across catchments in central and western Colorado; (3) assess the relative risk of post-fire surface erosion for each catchment; (4) rate the susceptibility of each stream segment to sedimentation; and (5) rank the catchments in western Colorado according to the proportion of stream segments at risk per unit catchment area. Although the results are specific to the study area, we believe that the methodology developed here could easily be adapted to other areas.

STUDY AREA AND DATA SOURCES

As described in Sampson and Neuenschwander (1999), the study area consists of approximately 180,000 km² in central and western Colorado. The high plains of eastern Colorado were excluded because these areas are mostly in private agriculture and generally not subject to wildfires. The study area was divided into 72 basins according to the eight-digit hydrologic unit codes (HUC) assigned by the U.S. Geological Survey (Figure 4.1; Table 1). Although many basins were truncated by the state line or the study boundary, the typical size for an entire catchment was 1500-8000 km². The average number of ignitions and acres burned for each of the 72 basins was determined from detailed data on wildfires in Colorado from 1986 to 1995 (Neuenschwander et al. 2000).

Geo-referenced data layers were obtained from a variety of sources (Sampson and Neuenschwander 2000). Digital elevation data were on a 90-m grid scale, and these data were used to calculate the average slope between grid squares. The state soils database (STATSGO) provided detailed information on each soil type, but the resolution of the STATSGO database is limited by the 6 km² minimum map unit (NRCS 1994). Fourteen vegetation and land cover types were derived from Advanced Very High Resolution Radiometer (AVHRR) coverage, and this was mapped on a one square kilometer grid.

FIGURE 4.1. Hydrologic unit codes (HUC) and catchments in the study area.

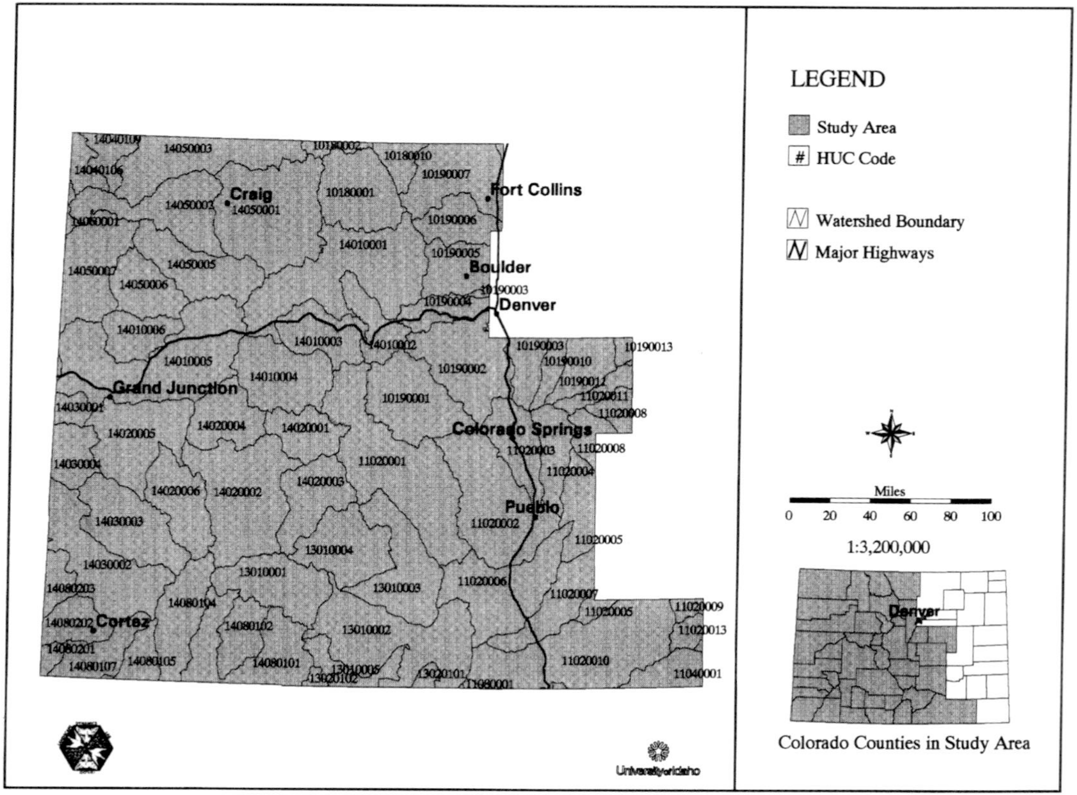

Fuel loading and predicted fuel consumption from wildfire were estimated by fuel type for each vegetation and cover type (Neuenschwander et al. 2000). The litter/duff fuel loadings and predicted consumption were used to estimate a qualitative fire severity for each vegetation and land cover type. Similar values were grouped, and this led to four fire severity classes (FI-SEV). We also assigned a propensity to form hydrophobic layers (HYPRO) to each vegetation type according to our best estimate of the amount of secondary compounds in the leaves and litter. Values for the latter index ranged from one to three, and both of these relative rankings were confirmed by foresters and fire scientists familiar with wildfires in Colorado.

For each vegetation type we also assigned a factor to represent the relative increase in soil wetness that would occur if the predominant vegetation was killed by a wildfire. This factor was included because hydrophobic layers are ineffective once they are wetted, so most or all of the winter and spring precipitation will infiltrate into the soil. The reduced evapotranspiration losses during the growing season result in higher soil moisture contents as compared to unburned areas. High amounts of soil moisture will increase the relative likelihood of overland flow due to a saturated soil profile, thus affecting the movement of sediment to the stream channel and the size of peak flows (e.g., O'Loughlin et al. 1982; Beschta 1990). The change in wetness was scaled according to the likely increase in annual water yield. Values for the major vegetation types were derived from paired-watershed experiments (e.g., Bosch and Hewlett 1982; Troendle et al. 1987), predictive models developed by the U.S. Forest Service and EPA (EPA 1980), and observed soil moisture changes in similar vegetation types (Sampson 1944; Klock and Helvey 1976; Helvey 1980). Little or no change in soil wetness was predicted for vegetation types in more arid areas or that would exhibit very rapid hydrologic recovery after a wildfire.

Perennial streams were mapped as vectors from 1:100,000 topographic maps. Pixels with streams were identified, and we assumed the stream length within a pixel to be 90 meters. The total length of streams in a watershed was estimated by multiplying the total number of pixels with streams by 90 m. Stream gradients were calculated as the difference in elevation between adjacent pixels that had been designated as having a perennial stream divided by the stream length. Drainage density was calculated by dividing the estimated kilometers of streams by the area of the catchment in square kilometers. Six catchments were excluded from this portion of the analysis because only the headwaters were located within the study area and the total channel length was too small to be considered representative (i.e., less than 10 km of stream channels).

TABLE 1. List of hydrologic unit codes (HUC), catchment characteristics, composite erosion risk, composite sedimentation risk, percent burned and number of ignitions.

HUC	Catchment Area (km²)	Drainage Density (km/km²)	Hydrophobicity Composite risk (CHYRISK)	Erosion Composite risk index (CSURFERO)	Percent Streams <2% gradient	Percent Streams >6% gradient	Sedimentation Risk index (CSEDRISK)	Percent Burned 1986-1995	Ignitions per 40 km² 1986-1995
10180001	3704	0.19	134.5	0.6	39	21	3.36	1.06	0.9
10180002	493	0.11	150.6	0.4	20	42	2.56	0.01	1.1
10180010	1015	0.16	115.5	0.2	26	38	2.76	2.13	0.9
10190001	4165	0.10	106.9	0.6	35	26	3.18	0.03	1.8
10190002	4539	0.11	137.7	2.0	30	31	2.99	0.22	7.3
10190003	1428	0.05	90.1	0.0	47	11	3.72	0.26	2.4
10190004	1431	0.09	120.9	2.5	30	35	2.91	0.01	1.7
10190005	2173	0.13	130.2	0.9	37	26	3.22	1.44	11.2
10190006	1930	0.14	103.6	0.8	35	31	3.09	0.19	7.6
10190007	3819	0.18	59.9	0.1	36	27	3.18	0.56	3.8
10190008*	50	0.19	0.0	0.0	39	9	3.60	0.34	2.4
10190010	820	0.03	55.8	0.0	45	10	3.69	0.00	0.3
10190011	1847	0.07	4.1	0.0	45	13	3.64	0.00	0.1
10190013	298	0.20	0.0	0.0	37	22	3.31	0.00	0.0
10250001*	22	0.03	0.0	0.0	67	0	4.33	0.00	0.0
11020001	7919	0.13	83.9	3.3	26	37	2.79	0.07	1.9
11020002	5956	0.14	45.4	1.7	35	25	3.20	0.63	2.9
11020003	2409	0.14	44.9	0.1	44	18	3.52	0.30	3.4
11020004	1919	0.06	10.8	0.0	57	3	4.08	1.96	3.0
11020005	2385	0.04	13.3	0.0	56	6	3.99	0.55	0.3
11020006	4822	0.11	76.7	1.9	35	23	3.24	0.08	2.0
11020007	2549	0.09	52.3	0.2	42	18	3.48	0.03	0.6
11020008	908	0.06	25.9	0.0	53	4	3.98	0.77	1.7
11020009	380	0.09	0.0	0.0	37	19	3.36	0.00	0.1
11020010	7541	0.09	51.2	0.8	34	24	3.20	0.80	0.9
11020011	864	0.10	2.9	0.0	45	13	3.65	1.04	0.2
11020012	828	0.08	0.3	0.0	49	7	3.84	0.25	0.2
11020013	602	0.06	0.0	0.0	37	15	3.43	0.22	0.3
11040001	1408	0.11	1.6	0.0	29	26	3.06	0.04	0.2
11040004*	38	0.03	0.0	0.0	40	0	3.80	0.00	0.0
11080001*	146	0.08	186.0	7.3	23	39	2.68	0.00	1.1
13010001	3461	0.09	125.1	15.0	28	35	2.88	0.06	0.9
13010002	6513	0.08	93.7	3.4	38	26	3.23	0.06	2.3
13010003	4135	0.05	80.8	2.1	38	25	3.27	1.40	1.7
13010004	3490	0.08	65.6	2.9	34	28	3.12	0.02	1.1
13010005	1390	0.10	125.4	12.4	34	32	3.04	0.13	1.2
13020101	332	0.03	55.0	0.0	70	10	4.21	0.00	0.1

ID									
13020102*	214		206.6	11.2	21	45	2.51	0.01	0.7
14010001	7518	0.08	126.7	0.9	20	45	2.51	0.24	1.5
14010002	1768	0.13	129.4	6.9	22	47	2.49	0.01	2.7
14010003	2523	0.12	139.8	7.0	20	47	2.45	0.44	4.3
14010004	3758	0.10	149.7	19.1	22	44	2.58	0.10	1.3
14010005	8030	0.07	86.3	2.8	27	34	2.86	1.51	4.0
14010006	1821	0.14	113.2	12.6	24	40	2.67	0.13	1.8
14020001	1990	0.13	103.6	1.0	24	42	2.65	0.05	0.8
14020002	6231	0.09	87.4	4.8	24	39	2.69	0.08	1.0
14020003	2858	0.10	94.4	4.2	28	34	2.87	0.10	0.8
14020004	2503	0.09	85.4	0.3	21	43	2.57	0.80	3.3
14020005	4275	0.11	64.7	0.8	25	37	2.77	0.18	2.1
14020006	2945	0.15	81.4	4.2	28	33	2.90	0.07	2.3
14030001	730	0.11	23.4	0.0	25	34	2.83	0.86	2.6
14030002	4930	0.20	62.2	1.4	27	34	2.86	1.16	2.0
14030003	3991	0.12	68.8	1.7	25	37	2.77	0.66	4.5
14030004	1632	0.13	37.1	0.0	20	45	2.51	0.57	2.0
14030005*	5	0.02	40.5	0.0	0	0	3.00	0.00	0.0
14040106	889	0.12	29.3	0.1	23	36	2.74	33.81	8.2
14040109	1146	0.17	23.3	0.0	26	34	2.84	3.28	5.7
14050001	6795	0.14	90.2	1.1	28	35	2.86	0.91	1.3
14050002	4036	0.14	23.7	0.0	31	31	3.01	13.49	5.8
14050003	4441	0.13	64.0	0.4	33	24	3.19	2.12	2.2
14050005	3510	0.15	75.7	1.1	24	40	2.69	2.71	2.6
14050006	2366	0.18	16.3	0.1	25	36	2.77	1.46	8.5
14050007	3811	0.17	26.2	0.2	26	35	2.83	3.52	5.8
14060001	195	0.21	8.9	0.0	19	50	2.38	2.79	5.7
14080101	4205	0.10	102.6	7.1	30	35	2.90	0.18	3.6
14080102	1748	0.07	81.7	5.2	29	34	2.90	1.78	4.2
14080104	2969	0.09	103.1	8.5	28	36	2.83	1.04	4.2
14080105	1111	0.11	74.0	3.0	32	27	3.10	1.32	4.5
14080107	1927	0.15	98.2	3.7	27	32	2.90	21.67	17.2
14080201	397	0.11	66.6	0.4	38	20	3.37	99.65	1.9
14080202	1665	0.14	103.6	0.7	33	25	3.16	0.70	5.7
14080203	954	0.13	53.5	0.0	33	24	3.18	0.02	1.1
Mean	2606	0.110	70.7	2.4	33%	28%	3.09	0.03	2.7
Std. Deviation	2111	0.044	49.0	3.9	11%	13%	0.45	0.13	2.9
Coeff. Variation	81%	40%	69%	165%	35%	46%	14%	426%	108%
Maximum	8030	0.21	206.6	19.1	70%	50%	4.33	99.65	17.2
Minimum	5	0.02	0.0	0.0	0%	0%	2.38	0.00	0.0
Skewness	0.92	0.15	0.31	2.38	0.89	-0.57	0.81	6.91	2.43

* Excluded from sedimentation analysis

65

MODEL DESCRIPTION

The first index we developed assessed the relative risk of generating a fire-induced hydrophobic layer by the function:

$$\text{HYRISK} = \text{FISEV} \times \text{HYPRO} \times \text{TEXTURE} \qquad (1)$$

where HYRISK was the hydrophobicity risk index, FISEV was the fire severity by vegetation type, HYPRO was the propensity to form a hydrophobic layer by vegetation type, and TEXTURE was soil texture. Assigned values for FISEV ranged from 0 to 6, while HYPRO ranged from 0 to 3 (Table 2). Given the large number of soils listed in STATSGO, we took the pragmatic approach of rating soil texture as coarse or fine according to the hydrologic soil group (Musgrave and Holtan 1964). High permeability soils (i.e., soils in hydrologic group A or B) were assumed to be coarse-textured and assigned a value of 3. Low permeability soils (i.e., soils in hydrologic groups C or D) were assumed to be fine-textured and assigned a value of 1, as the latter soils would be much less likely to develop a hydrophobic layer following a wildfire. We generally did not assign zero values for FISEV or HYPRO, as this would automatically result in an erosion hazard of zero regardless of slope, soil type, or fire severity. The calculated HYRISK values ranged from 0 to 36 as compared to a theoretical range of 0 to 54. These values were grouped into five classes for calculating risk assessments and three classes for mapping purposes.

TABLE 2. List of vegetation types and their associated fire severity (FISEV), propensity to form hydrophobic layers (HYPRO), and post-fire wetness factor (W).

Vegetation Type	Fire Severity	Propensity to Form a Hydrophobic Layer	Post-Fire Wetness
Spruce/fir	6	1	1.5
Mixed conifer	6	2	1.5
Ponderosa pine	3	3	1.3
Lodgepole pine	3	3	1.5
Pinyon/Juniper	1	3	1.0
Aspen/Hardwood	1	1	1.3
Cottonwood/Willow	1	1	1.3
Sage	1	3	1.1
Annual grass	1	1	1.0
Perennial grass	1	1	1.0
Alpine tundra	1	1	1.0
Water	0	0	1.0
Agriculture	1	1	1.0
Grassland/Cropland	1	1	1.0

The inherent erodibility of a site was defined as the soil erodibility index (SOILEROD), and this was calculated for each pixel by:

$$\text{SOILEROD} = K \times S \tag{2}$$

where K and S are the soil erodibility and slope factors, respectively, from RUSLE (Renard et al. 1997). K values were obtained directly from the STATSGO database and these generally ranged from 0.10 to 0.45. S was calculated according to the slope angle in degrees (θ) by the following equations (Renard et al. 1997):

$$S = 10.8 \sin\theta + 0.03 \text{ (for slopes less than 5.14 degrees)} \tag{3}$$

$$S = 16.8 \sin\theta + 0.50 \text{ (for slopes greater than 5.14 degrees)} \tag{4}$$

Relative surface erosion risk was predicted by:

$$\text{SURFERO} = \text{HYRISK} \times \text{SOILEROD} \times \text{W} \tag{5}$$

where SURFERO is the surface erosion risk index, HYRISK is the hydrophobicity risk index (equation 1), SOILEROD is the soil erodibility index (equation 2), and W represents the increase in soil wetness by vegetation type (Table 2). The resulting SURFERO values ranged from 0 to 295, and these were grouped into five classes to produce the pixel-scale map of surface erosion risk over the study area (Table 3).

A composite surface erosion risk index (CSURFERO) was determined for each basin by summing the fraction of area in each surface erosion risk class times its respective class rank (equation 6):

$$\text{CSURFERO} = \sum_{i=1}^{5} \left(\frac{\text{number of pixels in class } i}{\text{total number of pixels in catchment}} \times \text{SURFERO class} \right) \tag{6}$$

Similarly, a composite hydrophobicity risk index (CHYRISK) was determined for each watershed by summing the fraction of the watershed in each hydrophobicity class times its hydrophobicity risk class (Table 4):

TABLE 3. Classification of surface erosion risk (SURFERO) values.

Soil Erosion Risk Index	Class and Weighting
0-60	1
60-120	2
120-180	3
180-240	4
240-300	5

TABLE 4. Classification of hydrophobicity risk (HYRISK) values.

Hydrophobicity Risk Index	Class and Weighting
0-7	1
7-14	2
14-21	3
21-28	4
28-36	5

$$\text{CHYRISK} = \sum_{i=1}^{5} \left(\frac{\text{number of pixels in class } i}{\text{total number of pixels in catchment}} \times \text{HYRISK class} \right) \qquad (7)$$

The use of a RUSLE-type methodology ignores the possibility of mass movements and implies that most of the eroded material will be sand-sized or smaller. Since these particles are easily entrained, stream gradient can be used as an index of sedimentation risk (e.g., Benda and Dunne 1987; Montgomery and Buffington 1993). Stream segments with a gradient greater than six percent were assumed to have sufficient energy to transport the eroded sediment to the next downstream segment, and these high-gradient reaches were rated as having little sedimentation risk. Stream segments with a gradient of 2-6% were assumed to have a moderate risk of sedimentation, while segments with a gradient less than 2% were presumed to have a high risk of sedimentation. Sedimentation risk (SEDRISK) values of 5, 3, and 1 were assigned to stream segments with a high, moderate, or low risk of sedimentation, respectively.

These gradient breaks are consistent with studies of downstream sediment transport from debris flows (Benda and Dunne 1987). A two-percent gradient is also a critical break in two of the most commonly-used stream classification systems (Rosgen 1994; Montgomery and Buffington 1997). Montgomery and Buffington explicitly designate streams with a gradient of less than two percent as transport-limited or "response" reaches because of their sensitivity to sediment deposition.

A composite sedimentation risk (CSEDRISK) in each catchment was determined by summing the fraction of stream pixels in a gradient class times the sedimentation risk factor and dividing by the total number of stream pixels within a catchment (equation 8):

$$\text{CSEDRISK} = \sum^{3} \left(\frac{\text{number of pixels in class } i}{} \times \text{SEDRISK} \right) \qquad (8)$$

RESULTS

Landscape-Scale Assessments

The hydrophobicity risk map of the study area (Figure 4.2) indicates that the greatest risk is in the more densely vegetated mid-elevation montane areas. The higher values generally run parallel to the Continental Divide, although the highest elevations have less vegetation and therefore a lower risk.

There are generally more high-risk areas in the western part of the state and on the western side of the Continental Divide than on the eastern slopes. This pattern suggests that the highest risk of post-fire hydrophobicity is generally in the more mesic zones where high fuel loadings can generate high severity fires. Another zone with high risk areas extends to the western border in the central part of the state, and this includes areas similar to the infamous Storm King fire that killed 14 firefighters in July 1994. Patches of high risk areas can also be found in the mid-elevation zones along the Front Range. Lower elevation areas generally have a lower risk of hydrophobicity because of their lower fuel loadings.

The higher spatial resolution of the data used to develop the surface erosion risk map yielded a more detailed delineation of areas with a high erosion risk (Figure 4.3). Areas with the highest risk were in the steeper montane areas that also had a high hydrophobicity index. Areas with particularly high SURFERO values were in the Sawatch Range in the central part of the study area, the San Juan Mountains in the southern part of the state, and the area around the Roan Plateau (just north of Grand Junction in the west-central part of the study area).

Catchment-Scale Analysis

A statewide plot of sedimentation risk did not present any clear patterns, and a more explicit evaluation of the patterns of HYRISK, SURFERO, and SEDRISK was only possible from smaller scale maps. Thus, a complete set of maps was prepared for three catchments (HUC codes 11020002, 14030003, and 14050006) located in different parts of the state with moderate-to-high frequency of ignitions. For illustrative purposes this section will focus only on the middle Arkansas basin in the southeastern part of the study area (HUC code 11020002), as this basin had a moderate CSURFERO (25th of the 72 catchments analyzed).

The mapped hydrophobicity index clearly reflects the coarse scale of the vegetation and soil layers (Figure 4.4). The highest HYRISK values were in the southwestern part of the basin, and comparisons with other maps indicate these areas are generally between 2000 and 3000 meters elevation on lands belonging to the Bureau of Land Management or the San Isabel National Forest.

The areas with the highest surface erosion risk were also in the southwestern part of the basin (Figure 4.5). Although these areas were generally associated with a high hydrophobicity risk index, not all areas with a high hydrophobicity risk have a high surface erosion risk. These results suggest that slope has a relatively strong effect on surface erosion risk, and the comparatively high resolution of the digital elevation model allows a much finer-scale resolution of the relative surface soil erosion risk as compared to the hydrophobicity risk.

Sedimentation risk is also plotted on Figure 4.5. As expected, the smaller, headwater segments generally have the steepest gradients and hence the lowest sedimentation risk. The mainstem of the Arkansas River extends from west to east across this basin, and it consistently has a high sedimentation risk. The lower reaches of the other major streams within this catchment also tend to have a high sedimentation risk, but there often are intervening segments with intermediate sedimentation risks.

Interbasin Comparisons

A third set of analyses examined the interrelationships between different variables and the ranking of each of the 72 basins according to their composite hydrophobicity risk (CHYRISK), composite surface erosion risk (CSURFERO), and composite sedimentation risk (CSEDRISK). As indicated in Table 1, CHYRISK ranges from 0 to 207. The watersheds with the highest CHYRISK are generally in the south-central (San Juan) and north-central mountains. Watersheds in the eastern and northwestern parts of the study area generally had lower CHYRISK values, even though there were often some areas with high HYRISK values within these watersheds.

The distribution of composite surface erosion values was highly skewed, as most basins tended to have relatively low CSURFERO values and only a few basins had relatively high CSURFERO values (Table 1). The skewed distribution of CSURFERO values led us to place each catchment into one of five classes according to the logarithm of CSURFERO (Table 5). Only 23 basins were designated as having a moderate or higher composite post-fire surface erosion risk on a basin-wide scale, while the other 49 basins were classified as having no more than a slight risk for post-fire surface erosion.

Basins with the highest CSURFERO values tended to be clustered in the southern and central mountains, and, to a lesser extent, in the west-central part of the state (Figure 4.6). This pattern suggests that one reason for the relatively low CSURFERO values in many basins is the large amounts of range and agricultural lands. Low post-fire erosion rates on the large ex-

FIGURE 4.2. Predicted hydrophobicity risk (HYRISK) by pixel in the study area.

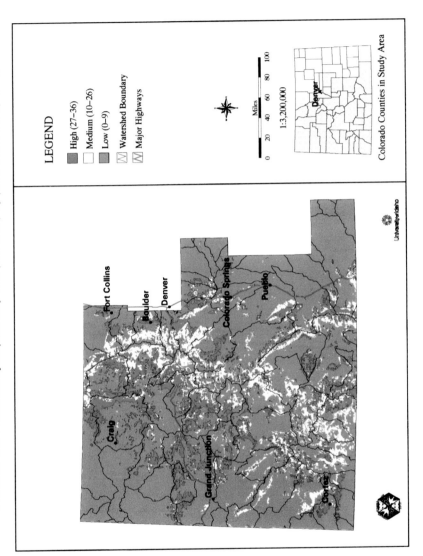

FIGURE 4.3. Predicted surface erosion risk (SURFERO) by pixel in the study area.

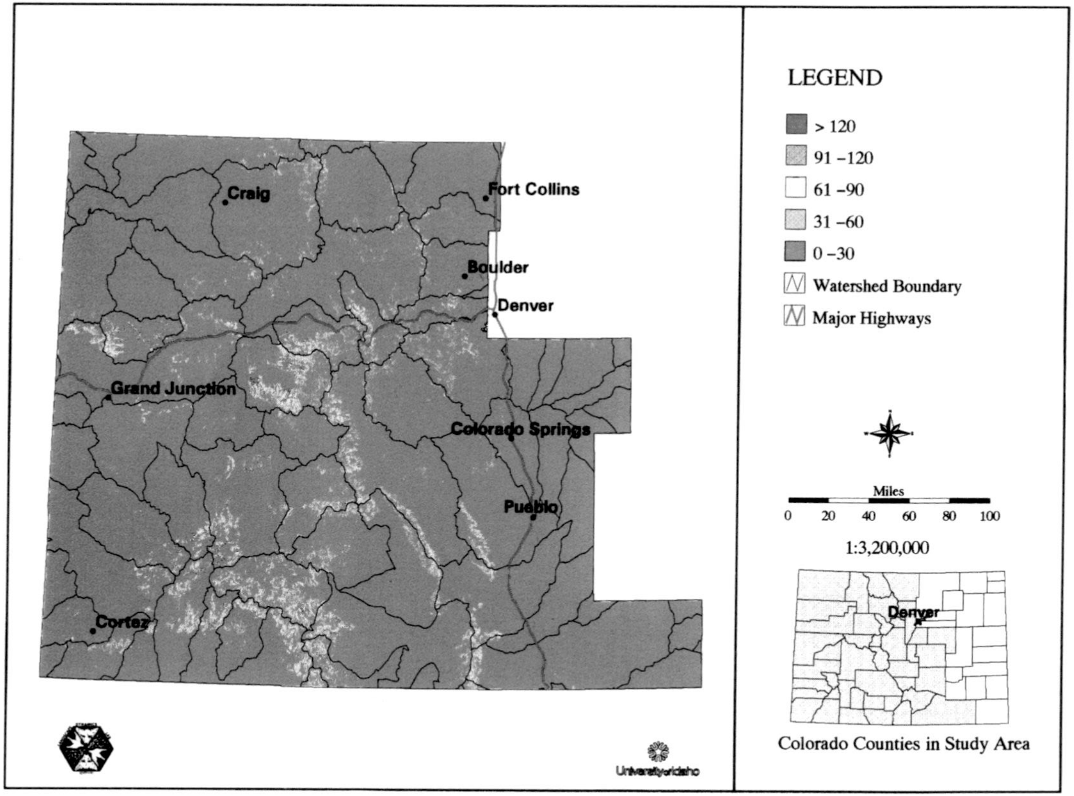

FIGURE 4.4. Predicted post-wildfire hydrophobicity in the middle Arkansas basin.

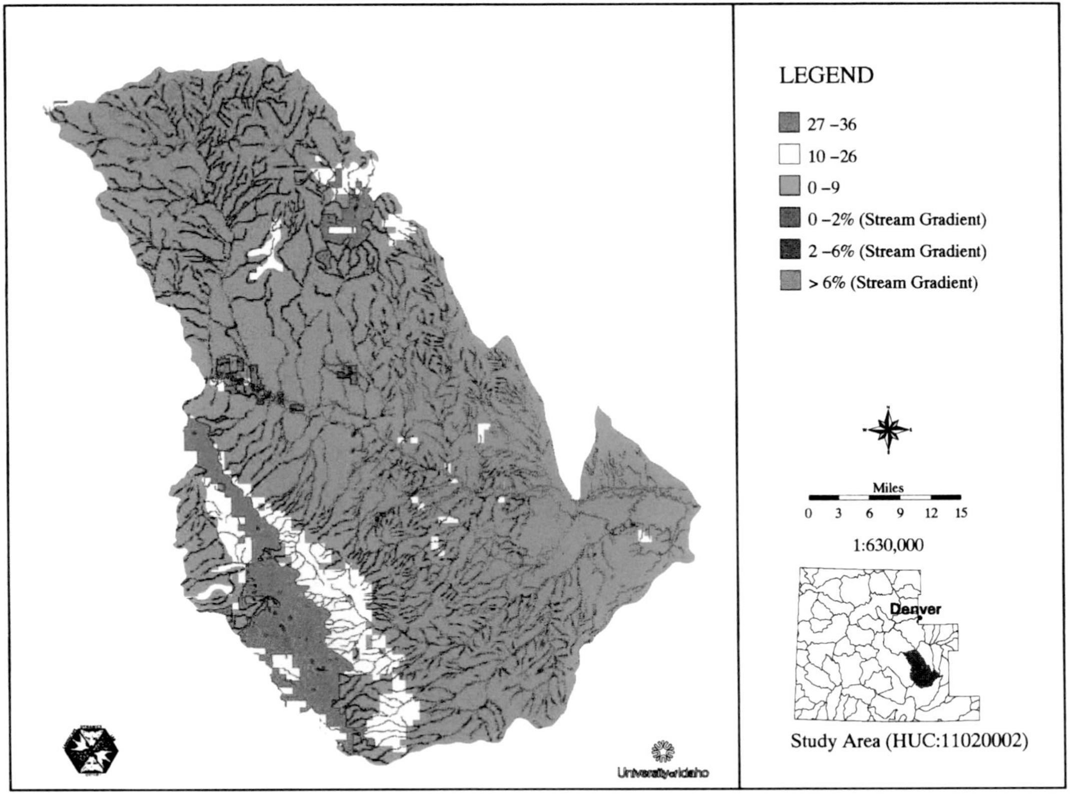

LEGEND

- 27 – 36
- 10 – 26
- 0 – 9
- 0 – 2% (Stream Gradient)
- 2 – 6% (Stream Gradient)
- > 6% (Stream Gradient)

Miles
0   3   6   9   12   15

1:630,000

Denver

Study Area (HUC:11020002)

University of Idaho

FIGURE 4.5. Predicted post-wildfire soil erosion and sedimentation risk in the middle Arkansas basin.

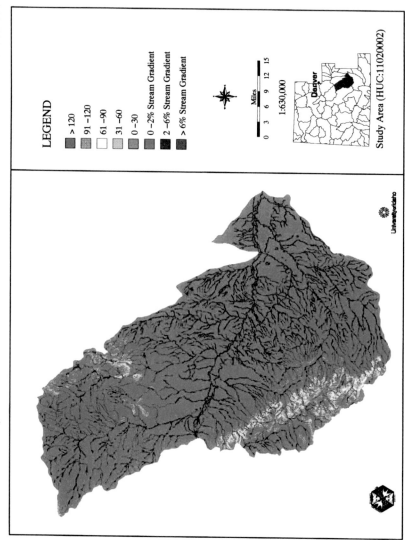

FIGURE 4.6. Map of composite soil erosion risk (CSURFERO) by catchment.

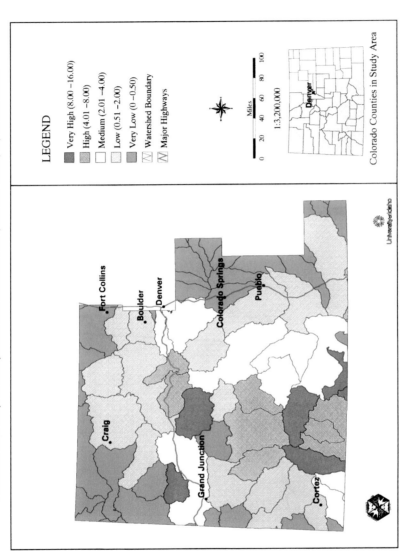

LEGEND

Very High (8.00 – 16.00)
High (4.01 – 8.00)
Medium (2.01 – 4.00)
Low (0.51 – 2.00)
Very Low (0 – 0.50)
Watershed Boundary
Major Highways

Miles
0   20   40   60   80   100

1:3,200,000

Colorado Counties in Study Area

Denver

University of Idaho

panses of these vegetation types would compensate for smaller areas with higher SURFERO values (e.g., Figure 4.5), thereby yielding a lower composite value.

This "dilution" of high SURFERO values raises the issue of whether CHYRISK and CSURFERO depend in part on basin size. Higher CSURFERO values, for example, might be more likely in smaller basins because there would be less area in lower elevation zones with lower SURFERO values. Many basins in the southern mountains were truncated by the state line (Figure 4.1), and this also might lead to disproportionately high CSURFERO values. However, the correlation coefficient between watershed area and CSURFERO was only 0.10, and only 0.23 between watershed area and CHYRISK. Scatter plots showed no consistent trend between CSURFERO or CHYRISK and catchment area.

The effect of other variables besides the predicted hydrophobicity on surface erosion risk was assessed on a watershed scale by plotting CSURFERO against CHYRISK (Figure 4.7). Overall, the composite hydrophobicity index explained only 40% of the variation in the composite erosion index. This correlation and a comparison of Figures 4.2-4.5 confirm that a moderate

TABLE 5. Classification of composite surface erosion (CSURFERO) values and the number of catchments in each class.

Composite Surface Erosion Risk	Class	Number of Catchments
0-0.5	Very little risk	34
0.5-2.0	Slight risk	15
2.0-4.0	Moderate risk	9
4.0-8.0	High risk	8
8.0-16.0	Very high risk	6

FIGURE 4.7. Plot of composite hydrophobicity risk (CHYRISK) versus composite soil erosion risk (CSURFERO) for each catchment in the study area.

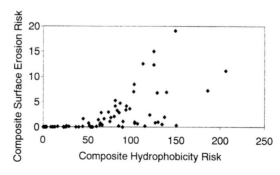

CHYRISK is necessary for a high CSURFERO, but a high CHYRISK does not necessarily result in a high CSURFERO.

Drainage Density and Composite Sedimentation Index

Calculated drainage densities ranged from 0.03 to 0.21 km/km^2 (Table 1). These relatively low values are due in part to the use of 1:100,000 maps to delineate channels, as well as the implicit assumption that each stream pixel has only 90 meters of channel. There was considerable variation between basins in the number of stream segments in the different gradient categories. For example, the proportion of streams with gradients less than two percent ranged from 19 to 57%, and the proportion of streams with gradients greater than six percent ranged from 4 to 47% (Table 1).

The calculated composite sedimentation risk (CSEDRISK) ranged from 2.4 to 4.2 (Table 1). Catchments with the highest sedimentation risk were generally in the eastern part of the study area where drainage densities were low and most streams had gradients of less than two percent (Table 1; Figure 4.8). Steeper catchments in the central and west-central parts of the study area had lower composite sedimentation risks.

DISCUSSION

The procedures developed in this paper for predicting erosion and sedimentation risks are consistent with the limited literature on post-fire erosion and existing conceptual models. Both the patterns of risk and the correlation analyses suggest that the results are credible as a first estimate. However, before the results can be used to assist in setting policy or guiding future management, several other issues must be recognized. These include the temporal and spatial scale of the analyses, model limitations and validation, and the use of relative versus absolute predictions.

Temporal and Spatial Scale of the Analyses

An important limitation of this work is that the estimated erosion and sedimentation risks do not consider the frequency or varying intensity of wildfires within a vegetation type, or the rate of recovery. To properly quantify risk, the predicted erosion and sedimentation rates need to be distributed over long-term average cycles of burning and recovery. Differences in the relative frequency of fire might cause some vegetation types, such as sagebrush or ponderosa pine, to have much higher long-term average erosion rates than vegetation types with higher post-fire erosion rates, but less frequent fires.

FIGURE 4.8. Map of composite sedimentation risk (CSEDRISK) by catchment.

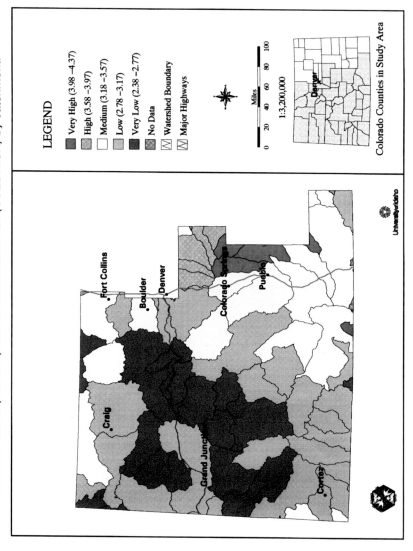

Different vegetation types will also have different rates of regrowth and hence different patterns of erosion rates over time. Resprouting species, such as aspen, should recover more quickly than spruce-fir. Recovery rates will also vary by erosion process. Surface erosion, for example, generally declines rapidly a fire, while mass movements triggered by a decrease in root strength will tend to be most frequent 5 to 10 years after a fire (USFS 1981). Ideally site- and process-specific recovery rates could be predicted from combining spatial data on key factors, such as aspect, vegetation type, elevation, and annual precipitation. The incorporation of fire frequency and recovery rates could alter the present catchment-scale rankings of erosion and sedimentation risks.

For this reason, we compared the mean number of ignitions and mean area burned by catchment to our catchment-scale estimates of composite soil erosion and sedimentation risks. While none of these plots showed any significant correlation, the highest CSURFERO values were associated with a very low probability of being burned and a relatively low probability of an ignition event. Although data on ignitions and area burned are only available for a 10-year period (1986-1995), these results indicate that our predictions of erosion and sedimentation risk need to consider the likelihood of burning.

In general, the catchments with the largest burned area were rangeland catchments. Relatively frequent, low-intensity fires in these basins may not result in much surface erosion or sedimentation risk for several reasons. First, these areas have limited amounts of precipitation and runoff. Second, these basins lack the fuel loading to generate an intense fire that could burn off much of the surface organic matter and possibly create a hydrophobic layer. Thus, basins with smaller, more severely burned areas could ultimately have much higher erosion and sedimentation rates than these large rangeland catchments that burn more frequently.

These arguments for lengthening the temporal scale are reversed with regard to the spatial scale. Basin-scale comparisons may be useful to identify higher-risk zones, but these will not have the necessary spatial resolution to guide where one should initiate a more intensive fuels treatment program.

Working at a spatial scale appropriate to vegetative manipulation or other types of intervention will also facilitate a more explicit spatial linkage between predicted erosion and sedimentation risk. At present sedimentation risk is assessed independently of the likely sediment production upstream of that location. In reality the upstream sediment production needs to be routed through the stream network to the segments of concern (Bunte and MacDonald in press). Not all streams will have the same resource value, and the incorporation of site-specific values (e.g., habitat for an endangered fish or a domestic water supply reservoir) can only be done on a smaller-scale with more specific data.

At the hillslope scale the location of a fire relative to the stream channel is an important control on the amount of sediment that is actually delivered to the stream channel (EPA 1980; Scott 1993). Higher resolution vegetation, soil, and topographic data are necessary to evaluate the delivery of sediment down a hillslope to a channel. If the analysis is limited to smaller areas, these spatial issues could be explicitly considered, and this would lead to more realistic assessments of potential environmental effects and better management guidelines.

Model Validation, Refinements, and Limitations

The predicted erosion risks are based on a relatively simple conceptual model of surface erosion from rainsplash, sheetwash, and rilling. In other areas wildfires have been shown to increase the number of landslides (Tiedemann et al. 1979) and the rate of dry ravel (Krammes 1965; Florsheim et al. 1991; Wohlgemuth et al. 1996). Consultations with fire and watershed scientists suggest that these other erosion processes are of lesser importance after wildfires in Colorado. Field observations of recently burned areas in the Colorado Front Range (Buffalo Creek and Pingree Park) indicate that dry ravel can be an important process, but surface runoff is usually the primary mechanism for delivering eroded sediment to the stream channels (Morris and Moses 1987).

Debris flows and channel erosion may also be important sediment sources, but these are much more difficult to predict. After the July 1994 South Canyon fire in western Colorado large amounts of sediment were delivered to the channel network by wind erosion, dry ravel, and rill and gully erosion. Heavy rains in early September then eroded more material from the burned hillslopes and scoured an estimated 70,000 m^3 of material from the hillsides and the channels as hyperconcentrated flows and debris flows (Cannon et al., 1998). Meyer et al. (1995) found a similar topographic sequence of rill erosion, debris torrents, and main channel scour after the 1988 Yellowstone fires.

In other areas it may be necessary to predict the likelihood of mass movements following wildfires. The development of such a model would need to incorporate other factors, such as root cohesion and pore water pressures, but the formulation could largely follow the methodology used by Dietrich et al. (1994) to predict susceptibility to landslides. Any effort to predict mass movements will require higher resolution data, as the occurrence of mass movements is highly dependent on topographic convergence and slope. Accurate mapping of these characteristics requires as small a pixel size as is feasible (Quinn et al. 1995). Higher resolution data are also needed because most slides will be smaller than the 90 × 90-meter pixels used here.

Several studies have noted that the post-fire surface erosion risk is highly dependent on the occurrence of a high intensity storm (e.g., Krammes 1965; Renard et al. 1996; Prosser and Williams 1998). This means that our erosion prediction model could be improved by adding either a deterministic or a stochastic precipitation component. Jarrett (1990) has already noted that the occurrence of large runoff events in Colorado varies with elevation, and it would be relatively easy to incorporate a precipitation factor, such as the 2-year 30-minute precipitation event, into the surface erosion risk index (equation 5).

Further refinements might include stochastic components to account for the seasonal timing of both fires and future precipitation events. Wildfires in the early summer, for example, are more likely to be followed by high intensity rainfall events as was observed at Storm King Mountain in 1994 and Buffalo Creek near Denver in 1996. Negative impacts from hydrophobic soil layers and soil erosion are much less likely if a late fall fire is first subject to snowfall rather than rainfall events. The likely timing of a wildfire will vary with vegetation type and location, just as the likelihood of a given precipita- tion event will vary over the course of a year and with location. Hence a more accurate assessment of post-fire erosion and sedimentation risks will require a combination of deterministic functions, based on location and vegetation type, with stochastic components to represent the relative likelihood of differ- ent fire and rainfall events.

A major limitation in the development and application of the model used in this study is the uncertainty over the strength and persistence of a fire-in- duced hydrophobic layer. Observations at Buffalo Creek and Pingree Park suggest that most of the erosion was due to the complete elimination of the surface vegetative cover and the pulverization of the soil by burning off the organic matter and breaking down the soil aggregates. The resulting fine-tex- tured, cohesionless surface layer was highly susceptible to rainsplash, soil sealing, and, on the steeper slopes, dry ravel. The observed high density of rills may have been due to a combination of surface sealing and the develop- ment of a hydrophobic layer. Although surface sealing and a hydrophobic layer reduce infiltration by different processes, the effect of each process on runoff is similar. A high fire intensity is critical to both processes and the two processes may be synergistic.

The persistence of fire-induced hydrophobic layers has not been rigorous- ly evaluated. A number of studies have documented a rapid decline in post- fire erosion rates, and this is taken as *de facto* evidence for a breakdown of the hydrophobic layer (e.g., Morris and Moses 1987; Prosser and Williams 1998). Root growth, animal burrowing, and a variety of other physical, bio- logical, and chemical processes all will act to break up a hydrophobic layer, and we are not aware of any study that has shown accelerated erosion for more than four years after a fire. In the absence of detailed studies, the more persistent increases in erosion could be ascribed to a reduction in cover rather than a persistent, fire-induced hydrophobic layer.

A serious limitation to the use of our surface erosion prediction models is the relative absence of plot-scale data from the study area on post-fire hydrophobicity, runoff, and surface erosion rates. The limited data from Morris and Moses (1987) are not sufficient to calibrate, much less validate, the models developed in this paper. Observations from Buffalo Creek and Pingree Park suggest that the models presented here may overemphasize the development of a hydrophobic layer, but the relative results may well be accurate because the factors controlling post-fire erosion and runoff-generation (i.e., loss of surface cover, surface sealing and hydrophobicity) are similar regardless of which process is reducing infiltration rates. Buffalo Creek also may not be typical, as a high-intensity fire in early summer was followed by a 1-hour rainfall event that may have a recurrence interval of 100 years or more (R. Jarrett, U.S. Geological Survey, pers. comm., 1996). The high intensity of this rainfall event may have masked the effect of a hydrophobic layer relative to less extreme rainfall events.

The other sources of post-fire erosion that are not considered in our model are the effects of fire suppression and rehabilitation. Efforts to control a wildfire typically involve the construction of fire lines in rugged terrain, and these are usually constructed with little regard to streamside management zones or post-fire erosion rates. The relative importance of erosion from suppression efforts will depend on the particular fire, landscape, and type of suppression activities, but the implications of fire suppression efforts must also be considered if one is predicting post-fire erosion and sedimentation risks. Similarly, one should also include the reduction in erosion associated with post-fire rehabilitation efforts (MacDonald 1989). Limited research suggests that fire rehabilitation efforts have had mixed success in reducing post-fire sediment production and delivery (Taskey et al. 1989; Booker et al. 1998; Wohlgemuth et al. 1996).

Relative vs. Absolute Predictions

A final issue is the context of our predictions. Should the predicted erosion and sedimentation risks be evaluated relative to pre-disturbance erosion and sedimentation rates in the area of interest, or on a more absolute scale, as in this paper? Data and time limitations forced us to utilize an absolute scale in this paper, but land managers may wish to focus their efforts on areas and stream segments predicted to have the greatest change in erosion and sedimentation relative to pre-fire conditions. On the other hand, a large percentage increase in erosion may be relatively meaningless in areas where the pre-fire erosion rate is low. Thus an approach that evaluates both the absolute increase and the increase relative to background is preferable to accurately assess risk and evaluate management options.

CONCLUSIONS

The increasing availability of geo-referenced databases makes it possible to develop and apply conceptual soil erosion and sedimentation models across a range of spatial scales. In central and western Colorado the predicted post-fire surface erosion rates were highest in steep areas with vegetation types that could support high-intensity fires. Predicted surface erosion risks were low in most of the study area.

Sedimentation risks were based on stream gradient. Areas with a higher proportion of low-gradient streams had the greatest sedimentation risk.

The methods and models developed here can be adopted for use elsewhere, but there is an urgent need to assess these predictions against field data. Several possible improvements in the models and the approach were identified, and these include the addition of both deterministic and stochastic components. The predictions also could be improved by using higher-resolution topographic, soils, and vegetation data. In the absence of specific field data, we believe that the models developed here can provide useful comparisons of surface erosion and sedimentation risks across a range of spatial scales. Such information is a necessary first step to guide future analysis and management activities.

REFERENCES

Agee, J.K. 1994. Fire and weather disturbances in the terrestrial ecosystems of the eastern Cascades. USDA Forest Service General Technical Report PNW-GTR-320. 52 pp.

Amaranthus, M.P. 1989. Effect of grass seeding and fertilization on surface erosion in two intensely burned sites in southwest Oregon. In: *Proceedings of the Symposium on Fire and Watershed Management*. USDA Forest Service, Pacific Southwest Forest and Range Experiment Station, General Technical Report PSW-109. pp. 148-149.

Benda, L. and T. Dunne. 1987. Sediment routing by debris flows. In: *Erosion and Sedimentation in the Pacific Rim*, R.L. Beschta, T. Blinn, G.E. Grant, G.G. Ice and F.J. Swanson, eds., Proc. Corvallis Symp., IAHS Publ. no. 165. pp. 213-223.

Bennett, K.A. 1982. *Effects of slash burning on surface soil erosion rates in the Oregon Coast Range*. M.S. thesis, Oregon State University, Corvallis, Oregon. 70 pp.

Beschta, R.L. 1990. Effects of fire on water quantity and water quality. In *Natural and Prescribed Fire in the Pacific Northwest*. J.D. Walstad, S.R. Radsevich and D.V. Sandberg, eds. Oregon State University Press, Corvallis, Oregon. pp. 219-232.

Booker, F.A. 1998. *Landscape and management response to wildfires in Colorado.* M.S. thesis, Dept. of Geology and Geophysics, University of California, Berkeley, CA. 436 pp.

Booker, F.A., W.E. Dietrich, and L.M. Collins. 1993. Runoff and erosion after the Oakland firestorm. *California Geology.* November/December:159-179.

Bosch, J.M. and J.D. Hewlett. 1982. A review of catchment experiments to determine the effect of vegetation changes on water yield and evapotranspiration. *Journal of Hydrology* 55:3-23.

Bunte K. and L.H. MacDonald. In press. Scale consideration and the detectability of sedimentary cumulative effects. Technical Bulletin, National Council for Air and Stream Improvement, New York, NY. 300 pp.

Campbell, R.E., M.B. Baker, P.F. Folliott, F.R. Larson, and C.C. Avery. 1977. Wildfire effects on a ponderosa pine ecosystem: an Arizona case study. USDA Forest Service Research Paper RM-191, Rocky Mountain Forest and Range Experiment Station, Fort Collins, CO. 16 pp.

Cannon, S.H., P.S. Powers, and W.Z. Savage. 1998. Fire-related hyperconcentrated and debris flows on Storm King Mountain, Glenwood Springs, Colorado, USA. *Environmental Geology* 35(2-3): 210-218.

Cheruiyot, S.K., W.H. Blackburn, and R.D. Child. 1986. Infiltration rates and sediment production of a Kenya bushed grassland as influenced by prescribed burning. *Trop. Agric.* (Trinidad). 63(2):177-180.

Covington, W.W. and M.M. Moore. 1994. Postsettlement changes in natural fire regimes and forest structure: ecological restoration of old-growth ponderosa pine forests. In: *Assessing Forest Ecosystem Health in the Inland West,* R.N. Sampson and D.L. Adams, eds. The Haworth Press, Inc., New York. 461 pp.

DeBano, L.F. 1989. Effects of fire on chaparral soils in Arizona and California and post-fire management implications. In: *Proceedings of the Symposium on Fire and Watershed Management.* USDA Forest Service, Pacific Southwest Forest and Range Experiment Station, General Technical Report PSW-109. pp. 55-62.

DeBano, L.F. and J.S. Krammes. 1966. Water repellent soils and their relation to wildfire temperatures. *Bulletin of the International Association of Scientific Hydrology* 11(2):14-19.

DeBano, L.F., L.D. Mann, and D.A. Hamilton. 1970. Translocation of hydrophobic substances into soil by burning organic litter. *Soil Science Society of America Proceedings* 34:130-133.

DeBano, L.F., R. M. Rice, and C.E. Conrad. 1979. Soil heating in chaparral fires: effects on soil properties, plant nutrients, erosion and runoff. USDA Forest Service Research Paper PSW-145, Pacific Southwest Forest and Range Exp. Station, Berkeley, CA. 21 pp.

Dietrich, W.E., R. Reiss, and D.R. Montgomery. 1994. A process based model for colluvial soil depth and shallow landsliding using digital elevation data. *Hydrologic Processes* 9(314):383-390.

Dyrness, C.T. 1976. Effect of soil wettability in the high Cascades of Oregon. USDA Forest Service Research Paper PNW-202, Pacific Northwest Forest and Range Experiment Station, Corvallis, OR. 18 pp.

Emmerich, W.E., and J.R. Cox. 1994. Changes in surface runoff and sediment production after repeated rangeland burns. *Soil Science Soc. Am. J.* 58:199-203.

EPA. 1980. An approach to water resources evaluation of non-point silvacultural sources. Environmental Research Laboratory, U.S. Environmental Protection Agency, Athens, Georgia. EPA-600/8-80-012.

Florsheim, J.L., E.A. Keller, and D.W. Best. 1991. Fluvial sediment transport in response to moderate storm flows following chaparral wildfire, Ventura County, southern California. *Geological Society of America Bulletin* 103:504-511.

Forrest, C.L. and M.V. Harding. 1994. Erosion and sediment control: Preventing additional disasters after the Southern California fires. *Journal of Soil and Water Conservation.* 49(6):535-541.

Helvey, J.D. 1980. Effects of a north central Washington wildfire on runoff and sediment production. *Water Resources Bulletin* 16(4): 627-634.

Jarrett, R.D. 1990. Hydrologic and hydraulic research in mountain rivers. *Water Resources Bulletin* 26(3):419-429.

Klock, G.O. and J.D. Helvey. 1976. Debris flows following wildfire in north central Washington. In: *Proc. Third Federal Inter-Agency Sedimentation Conference*, Water Resources Council. pp. 91-98.

Krammes, J.S. 1965. Seasonal debris movement from steep mountainside slopes in southern California. In: *Proceedings, Federal Interagency Sedimentation Conference, 1963.* USDA Misc. Publication 970. pp. 85-89.

Laird, J.R. and M.D. Harvey. 1986. Complex-response of a chaparral drainage basin to fire. In *Drainage Basin Sediment Delivery*, R.F. Hadley, ed., Int. Assoc. of Hydrologic Sciences Publication no. 159. pp. 165-183.

MacDonald, L.H. 1989. Rehabilitation and recovery following wildfires: A synthesis. In: *Proceedings of the Symposium on Fire and Watershed Management.* USDA Forest Service, Pacific Southwest Forest and Range Experiment Station, General Technical Report PSW-109. pp. 141-144.

McNabb, D.H. and F.J. Swanson 1990. Effects of fire on soil erosion. In: *Natural and Prescribed Fire in the Pacific Northwest.* J.D. Walstad, S.R. Radsevich and D.V. Sandberg, eds. Oregon State University Press, Corvallis, Oregon. pp. 55-62.

McPhee, J.M. 1989. *The Control of Nature.* The Noonsday Press, New York, NY. 272 pp.

Megahan, W.F. 1983. Hydrologic effects of clearcutting and wildfire on steep granitic slopes in Idaho. *Water Resources Research* 19(1):811-819.

Meyer, G.A., S.G. Wells, and A.J.T. Jull. 1995. Fire and alluvial chronology in Yellowstone National Park: Climatic and intrinsic controls on Holocene geomorphic processes. *Geological Society of America Bulletin* 107:1211-1230.

Montgomery, D.R. and J.M. Buffington. 1993. Channel classification, prediction of channel response and assessment of channel condition. Report TFW-SH10-93-02, Department of Natural Resources, Olympia Washington. 84 pp.

Montgomery, D.R., and J.M. Buffington. 1997. Channel-reach morphology in mountain drainage basins. *Geological Society of America Bulletin* 109(5):596-611.

Morris, S.E., and T.A. Moses. 1987. Forest fire and the natural soil erosion regime in the Colorado Front Range. *Annals of the Assoc. of American Geographers* 77(2):245-254.

Musgrave, G.W. and H.N. Holtan. 1964. Infiltration; In: *Handbook of Applied Hydrology*. V.T Chow, ed. McGraw Hill, New York.

Nassari, I. 1989. Frequency of floods from a burned chaparral watershed. In: *Proceedings of the Symposium on Fire and Watershed Management*. USDA Forest Service, Pacific Southwest Forest and Range Experiment Station, General Technical Report PSW-109. pp. 68-71.

Neuenschwander, Leon F., James P. Menakis, Melanie Miller, R. Neil Sampson, Colin Hardy, Bob Averill and Roy Mask. 2000. Indexing Colorado Watersheds to Risk of Wildfire. In Sampson, R.N., R.D. Atkinson, J.W. Lewis (Eds.), *Mapping Wildfire Hazards and Risks*: Papers from the American Forests scientific workshop, September 29-October 5, 1996, Pingree Park, CO. The Haworth Press, Inc., New York.

NRCS. 1994. State Soil Geographic (STATSGO) Data Base: Data Use Information. USDA-Natural Resources Conservation Service. Miscellaneous Publication 1492, Washington, DC 112 pp.

O'Loughlin, E.M., N.P. Cheney, and J. Burns. 1982. The Bushrangers experiment: hydrological response of a eucalypt catchment to fire. In *Symposia on Forest Hydrology*, 11-13 May, 1982, Melbourne. pp. 132-138.

Onda, Y., W.E. Dietrich, and F. Booker. 1996. The overland flow mechanism in Mt. Vision fire area, Northern California. *Eos*. 77(46): F209.

Prosser, I. 1990. Fire, Humans, and Denudation at Wangrah Creek, Southern Tablelands, N.S.W. *Australian Geographical Studies* 28:77-95.

Prosser, I.P., and L. Williams. 1998. The effect of wildfire on runoff and erosion in native *Eucalypt* forest. *Hydrological Processes* 12:251-265.

Quinn, P.F., K.J. Beven and R. Lamb. 1995. The ln(a/tanB) index: How to calculate it and how to use it within the TOPMODEL framework. *Hydrologic Processes*. 9:161-82.

Renard, K.G., G.R. Foster, G.A. Weesis, D.K. McCool and D.C. Yoder. 1997. Predicting soil erosion by water: A guide to conservation planning with the revised universal soil loss equation (RUSLE). US Department of Agriculture, Agriculture Handbook No. 703, Washington, DC. 404 pp.

Rice, R.M. 1974. The hydrology of chaparral watersheds. In: *Proc. Symp. on Living with the Chaparral*. Riverside, CA. pp. 27-33.

Rinne, J.N. 1996. Short-term effect of wildfire on fishes and aquatic macroinvertebrates in the southwestern United States. *North American Journal of Fisheries* 16:653-658.

Robichaud, P.R., and T.A. Waldrop. 1994. A comparison of surface runoff and sediment yields from low- and high-intensity site preparation burns. *Water Resources Bulletin* 30(1):27-34.

Rosgen, D. L. 1994. A classification of natural rivers. *Catena* 22:169-199.

Sampson, A.W. 1944. Effect of chaparral burning on soil erosion and soil-moisture relations. *Ecology* 25(2):171-191.

Sampson, R. Neil and Leon F. Neuenschwander. 2000. Characteristics of the Study Area and Data Utilized. In Sampson, R.N., R.D. Atkinson, J.W. Lewis (Eds.), *Mapping Wildfire Hazards and Risks:* Papers from the American Forests scientific workshop, September 29-October 5, 1996, Pingree Park, CO. The Haworth Press, Inc., New York.

Savage, S.M. 1974. Mechanism of fire-induced water repellancy in soil. *Soil Science Society America Proceedings* 38:652-657.

Scott, D.F. 1993. The hydrological effects of fire in South African mountain catchments. *Journal of Hydrology* 150:409-432.

Scott, D.F. and D.B. Van Wyk. 1990. The effects of wildfire on soil wettability and hydrological behavior of an afforested catchment. *Journal of Hydrology* 121: 239-256.

Taskey, R.D., C.L. Curtis and J. Stone. 1989. Wildfire, ryegrass seeding and watershed rehabilitation. In: *Proceedings of the Symposium on Fire and Watershed Management*. USDA Forest Service, Pacific Southwest Forest and Range Experiment Station, General Technical Report PSW-109. pp. 115-124.

Tiedemann, A.R., C.E. Conrand, J.H. Dieterich, J.W. Hornbeck, W.F. Megahan, L.A. Viereck and D.D. Wade. 1979. Effects of fire on water: A state-of-knowledge review. USDA Forest Service General Technical Report WO-10, 27 pp.

Troendle, C.A., M.R. Kaufmann, R.H. Hamre and R.P. Winokur. 1987. Management of subalpine forests: building on 50 years of research. USDA Forest Service General Technical Report RM-149. pp. 68-78.

USFS. 1981. Guide for predicting sediment yields from forested watersheds in the Northern and Intermountain regions. 48 pp.

Water Resources Center. 1989. Fire and watershed management. *California Water*, Number 2, University of California, Riverside, CA. 5 pp.

Wells, W.G. II. 1987. The effects of fire on the generation of debris flows in southern California. In: *Debris flows/avalanches: Processes, Recognition and Mitigation*, J.E. Costa and G.F. Wieczorek, eds. Geol. Soc. Am., Rev. Eng. Geol. VII:105-114.

Wells, W.G. II, P.M. Wohlgemuth, A.G. Campbell and F.H. Weiriche. 1987. Postfire sediment movement by debris flows in the Santa Ynez Mountains, California. In: *Erosion and Sedimentation in the Pacific Rim. Proc. Corvallis Symp*, R.L. Beschta, T. Blinn, G.E. Grant, G.G. Ice and F.J. Swanson, eds., IAHS Publ. no. 165. pp. 275-276.

Whelan, R.J. 1995. *The Ecology of Fire*. Cambridge University Press. 343 pp.

Wohl, E.E. and P.P. Pearthree. 1991. Debris flows as a geomorphic agent in the Huachuca Mountains of southeastern Arizona. *Geomorphology*. 4:273-292.

Wohlgemuth, P.M., S.G. Conrad, C.D. Wakemand and J.L. Beyers. 1996. Postfire hillslope erosion and recovery in chaparral: Variability in responses and effects of postfire rehabilitation treatments. Presented at the *13th Conference on Fire and Forest Meteorology*, 27-31 October, 1996, Lorne, Australia.

Chapter 5

Modeling Biotic Habitat High Risk Areas

Don G. Despain
Paul Beier
Cathy Tate
Bruce M. Durtsche
Tom Stephens

SUMMARY. Fire, especially stand replacing fire, poses a threat to many threatened and endangered species as well as their habitat. On the other hand, fire is important in maintaining a variety of successional stages that can be important for other animals such as elk. Methods are given here on a variety of ways to approach risk assessment to assist in prioritizing areas for allocation of fire mitigation funds. One example looks at assessing risk to the species and biotic communities of concern followed by the Colorado Natural Heritage Program. One looks at the risk to Mexican spotted owls. Another looks at the risk to cutthroat trout, and a fourth considers the general effects of fire and elk. *[Article copies available for a fee from The Haworth Document Delivery Service: 1-800-342-9678. E-mail address: <getinfo@haworthpressinc.com> Website: <http://www.haworthpressinc.com>]*

KEYWORDS. Wildfire, habitat, threatened and endangered species, succession

Don G. Despain is affiliated with Montana State University, Bozeman, MT 59717. Paul Beier is affiliated with Northern Arizona University, Flagstaff, AZ 86011. Cathy Tate is affiliated with U.S. Geological Survey, Denver, CO 80255. Bruce M. Durtsche is affiliated with Bureau of Land Management, Denver, CO 80225. Tom Stephens is affiliated with Colorado Natural Heritage Program, Fort Collins, CO 80523.

[Haworth co-indexing entry note]: "Chapter 5. Modeling Biotic Habitat High Risk Areas." Despain, Don G. et al. Co-published simultaneously in *Journal of Sustainable Forestry* (Food Products Press, an imprint of The Haworth Press, Inc.) Vol. 11, No. 1/2, 2000, pp. 89-117; and: *Mapping Wildfire Hazards and Risks* (ed: R. Neil Sampson, R. Dwight Atkinson, and Joe W. Lewis) Food Products Press, an imprint of The Haworth Press, Inc., 2000, pp. 89-117. Single or multiple copies of this article are available for a fee from The Haworth Document Delivery Service [1-800-342-9678, 9:00 a.m. - 5:00 p.m. (EST). E-mail address: getinfo@haworthpressinc.com].

INTRODUCTION

Fire and other large disturbances often have effects that are detrimental to species of plants and animals that are of special concern to segments of society and therefore to land managers. On the other hand these same disturbances may provide needed or expanded habitat for other species, also of concern. It is important to know where the critical areas are and to have a rational priority process to allocate scarce monetary resources. This is needed to provide the best return on scarce dollars and provide protection in the most critical places.

We have attempted to bring together, on a very broad scale, the information necessary to give an estimate of where fire may pose a risk to habitat for rare and endangered species and for species and biological communities of concern in western Colorado. We have also chosen a group of species for individual consideration. Indicators of habitat were derived from digital maps all of a scale and resolution that precludes exact locations or acreage estimates. We have tried to indicate areas that have a high likelihood of containing significant proportions of each species' habitat. Significant habitat may exist outside these areas and some of the areas we delineated may not have the relevant habitats. Extensive field work would be necessary to validate these maps and estimate their accuracy.

We chose examples from species of birds, fish, big game, and plant communities. The Mexican spotted owl's existence is being threatened by loss of older successional stages through even-aged silvicultural practices and wildfires. These owls nest in pockets of old trees perched on the sides of steep sided canyons and forage in denser late succession and climax forests. Big game and sport fisheries are important to Colorado's economy. Anglers and hunters travel from around the world to fish for trout in the clear streams of the mountainous west and hunt the abundant game animals. The threatened greenback cutthroat trout, *Onchorynchus clarki stomais*, and Colorado River cutthroat trout (*Onchorynchus clarki pleuriticus*) and Rio Grande cutthroat trout (*Onchorynchus clarki virginalis*) are species of special concern in the state of Colorado and candidate species for Federal listing of threatened and endangered species. Elk are a popular game animal. Late successional stages of plant communities and their special habitats are being threatened by fire, timber harvest and real estate developments.

RISK TO SPECIES AND BIOTIC COMMUNITIES
OF SPECIAL CONCERN

We estimated general fire risk to species and biotic communities of special concern by looking at the entire suite of these entities listed by the Colorado

Natural Heritage Program. We determined the risk to these entities by combining their locations with the probability that this location will burn.

Methods

We overlaid the list of locations of species and biotic communities of concern with each drainage and calculated the density of sites for each drainage. Density was defined as the frequency of sites within a drainage divided by its area. Density of fire starts was also determined as an index of probability that the area will burn (Neuenschwander et al. 2000). The ranges of these densities were arbitrarily divided into three approximately equal classes by first obtaining a histogram of each of the densities. The densities were logistically distributed (Figure 5.1 shows the distribution of fire densities) with all but a very few densities falling within contiguous classes.

The range of values from the beginning of the first to the end of the last contiguous classes was divided by three. (Outliers were combined with the third class.) The first third was designated low with a value of one, the middle class was designated moderate with a value of two and the third class was designated high with a value of three. The two density values for each drainage were added together to provide a risk index with values from two to

FIGURE 5.1. Distribution of fire start densities.

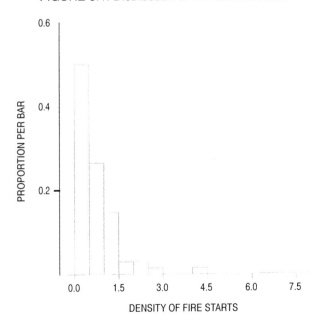

six. Six indicated a very high risk drainage having a high density of both fire starts and locations. Five indicated a high risk drainage having high locations or high fires and a moderate density in the other category. Four indicated a moderate risk drainage having moderate densities in both categories or high in one but low in the other. Three indicated a low risk drainage having low density in one category and moderate in the other. Two indicated a very low risk drainage with both low starts and low location densities.

Results

Two drainages fell into the very high category, four in high, seven in moderate, 21 in low and 35 in very low (Figure 5.2). The two drainages with very high risk and the four with high risk would be obvious candidates for mitigating measures to reduce fire risk to species or communities of special concern. However, wherever a fire starts, the location of any threatened or endangered species habitat or any of the species or biotic communities of concern that could be impacted by the fire should be made known to those responsible for management of the fires.

RISK TO MEXICAN SPOTTED OWLS
FROM STAND-REPLACING FIRES

Although many species of forest birds reach their highest abundance in a particular forest type or seral stage, most avian species in Western forests use a variety of forest types and seral stages. Among the few exceptions with respect to forest type are Williamson's sapsucker, closely associated with aspen forests, especially trees infected with the *Fomes* fungus (Crockett and Hadow 1975), and golden-cheeked warbler, found only in Ashe juniper woodlands (Pulich 1976). With respect to seral stage, more species are highly dependent on old-growth forests than other seral stages (Hejl et al. 1995), but only a handful of species, such as the northern spotted owl, Mexican spotted owl, Vaux's swift, and marbled murrelet, are obligate users of late-succes-sional conifer forests. No forest birds are obligates to mid-seral stages of forests (Hejl et al. 1995). Of the several species that reach their highest abundance in early-successional forests, most would survive on farms and in suburbs in the absence of early seral forests. Perhaps only the black-backed woodpecker is a true obligate for early successional forests, rarely breeding outside stands of dead trees in the first 10 years after stand-replacing fire (Hutto 1995).

We modeled the response to stand-replacing fires for the Mexican spotted owl. Of the other species mentioned above, only Williamson's sapsucker and black-backed woodpeckers occur in the western Colorado study area. We chose not to model fire risks to these species because we could not imagine any realistic fire scenario that would threaten these species. Absolute fire suppression could eventually threaten these two species, but that is so unlikely as to be practically impossible.

Our goal was to suggest the degree to which habitat suitable for Mexican spotted owls in Colorado might be at risk due to wildfires. We chose watersheds as the landscape unit of analysis because Neuenschwander et al. (2000) assessed risk of stand-replacing fires at this spatial scale.

The Mexican spotted owl uses multistoried forest stands, with dense canopy (typically over 65%) and high basal area (especially in large trees and large snags) for roosting and nesting sites (Ganey and Dick 1995). In 1993 the US Fish and Wildlife Service (USFWS) listed the owl as threatened. The Service listed two immediate threats to the species in the southwest: use of even-aged silviculture and risk of stand-replacing wildfires.

The Recovery Plan for the Mexican spotted owl (US Fish and Wildlife Service 1995) identified only 17 owl sites in Colorado during 1990-1993. Three of these are in the southwestern corner of the state and the other 14 lie in the Front Range of the Rocky Mountains. In Colorado, Mexican spotted owls are typically found in rock canyons within the ponderosa pine, Douglas-fir, and mixed-conifer forests (Ganey and Dick 1995). Two canyon settings are used for nesting. Some nests are in small (< 2.5 acres) patches of Douglas-fir in the bottom of, or on perched ledges within, sheer-walled slickrock canyons. Other nests are in steep canyons with exposed bedrock cliffs in several tiers. Owls range outside the nest stand to forage in the forests on perched benches, in the canyon bottoms, and on plateaus above the canyon rims.

The Nature Conservancy provided information from their Biological Conservation Database on elevations of 20 Mexican spotted owl nest sites in Colorado. These ranged from 5820 to 9100 ft, with all but 3 sites lying between 6500 ft and 7800 ft elevation. Colorado has not been systematically surveyed for spotted owls, and they may also breed in higher-elevation forests of Colorado. However, we suspect that the owls are excluded from higher elevations by an inability to tolerate prolonged snows and low temperatures. This speculation is supported by the observations that (1) in the Pacific northwest, northern spotted owls are a low- to mid-elevation species, and (2) Colorado represents the northern limit of the Mexican spotted owl's geographic range.

We did not attempt to predict likelihood of owl occurrence, but rather characterized each 0.8-km^2 cell as potential Mexican spotted owl habitat if it satisfied all 3 of the following conditions (1) Elevation was between 5200 and 9100 feet, (2) Vegetation type was ponderosa pine, mixed conifer, or

Douglas-fir (Loveland et al. 1991, modified as indicated in Sampson and Neuenschwander, 2000), and (3) Slope, as determined by a digital elevation model, exceeded 50% at $\geq 30\%$ of the sample 100 points in that cell. We then ranked each of 60 forested watersheds in Colorado from most to least suitable for owls based on the number of cells within a polygon rated as suitable. We evaluated fire risk to each watershed by computing the percent of suitable vegetation (i.e., of total acres in mixed conifer, ponderosa pine, and Douglas-fir types) that burned during the previous decade (Neuenschwander et al. 2000).

The habitat model indicated that 19 of 60 forested watersheds contained some areas that qualified as suitable habitat (Figure 5.3). Only 10 of the 60 watersheds contained at least 5 km^2 of suitable habitat and might therefore be characterized as significant habitat areas. The model correctly identified as suitable all watersheds in which owls are known to occur. However, it also identified as suitable large areas where Mexican spotted owls have not been documented, such as the San Juan Mountains in southwest Colorado and Grand Mesa in northwest Colorado (Figure 5.3). In the absence of systematic surveys, it is impossible to know how much the model overestimated geographic areas in which the owls occur. Indeed, the model may underestimate owl distribution in some cases. For instance, in 1996 the National Park Service surveyed Dinosaur National Monument (in extreme northwestern Colorado) for owls for the first time (D. Willey, pers. comm.). When owls were detected in the Monument, the northern limit of the species distribution was pushed northward by about 250 miles! Thus, although we suspect that the model made more errors of commission (predicting a watershed as suitable when it is not) than errors of omission (failure to correctly identify suitable watersheds), only future surveys will settle this issue. Within watersheds correctly identified as suitable, the model probably overestimated the amount of suitable habitat. These errors occurred because, with the coarse data available, the model could not consider habitat elements known to be important to owls, such as numbers of large downed logs, large trees, and large snags, and presence of multiple canopy layers (Ganey and Dick 1995).

Most watersheds containing suitable owl habitat had only a small fraction of suitable forest types burned by wildfire during the past decade, but there was considerable variation among watersheds (Table 1, Figure 5.4). For instance, the watershed with the highest value in terms of potential value to owls ranked 30th in terms of historical fire risk (Table 1). Only 1.7% of the mixed-conifer, Douglas-fir, and ponderosa pine forests in that watershed burned during the most recent 10-year period. Similarly, most of the watersheds that had most extensive fires in the mixed-conifer, ponderosa pine, and Douglas-fir forests (the forest types preferred by owls) had low rankings in terms of suitability for spotted owls. However, the watersheds that ranked

FIGURE 5.2. Composite wildfire risk to threatened and endangered species, by watershed, Colorado study area.

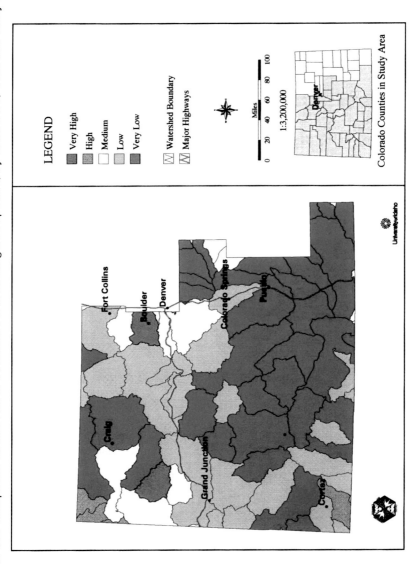

FIGURE 5.3. Areas in Colorado study area predicted to be suitable habitat for Mexican spotted owl, based on elevation, topography and vegetation. Polygons delineate watersheds and dots indicate predicted suitable habitat.

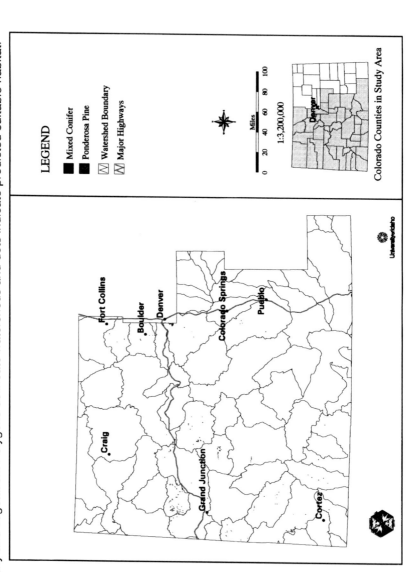

LEGEND

- Mixed Conifer
- Ponderosa Pine
- Watershed Boundary
- Major Highways

1:3,200,000

Miles

0 20 40 60 80 100

Colorado Counties in Study Area

TABLE 1. Cross-ranking of 60 watersheds in western Colorado in terms of (a) value as suitable habitat for Mexican spotted owls, and (b) risk of wildfire. Habitat suitability was ranked in terms of number of 0.8-ha cells in the watershed with suitable vegetation, topography, and elevation. Fire risk was ranked in terms of percent of acres of suitable forest types (Douglas-fir, mixed conifer, and ponderosa pine) that burned during 1986-1995.

Ranking as Owl Habitat	Ranking for Fire Risk	% of Suitable Forest Burned	Ranking for Fire Risk	Ranking as Owl Habitat
1	30	1.7	1	18
2	7	9.4	2	40[b]
3	8	8.6	3	40[b]
4	10	7.1	4	40[b]
5	33	0.9	5	40[b]
6.5	19	3.7	6	40[b]
6.5	23	2.9	7	2
8	34	0.8	8	3
9.5	45[a]	0.0	9	40[b]
9.5	26	2.3	10	4

[a]A rank of "45" for fire risk indicates that this was one of 24 watersheds that had no fires in suitable habitat during the previous decade.
[b]A rank of "40" for owl suitability indicates that this was one of 41 watersheds for which the model identified no suitable spotted owl habitat.

2nd and 3rd in owl value also ranked 7th and 8th in terms of fire risk, with 9.4% and 8.6% of the acreage in key forest types having burned during 1986-1995 (Table 1). For these 2 watersheds, wildfire may represent a significant risk to suitable owl habitat. However, Mexican spotted owls have not been documented to occur in these 2 watersheds.

We believe that recent fire history is a reasonable predictor of future fire risk in the preferred forest types in these watersheds. A plausible counter-argument is that extensive recent fires in suitable habitat would render the area safe from future fires until fuels and fuel ladders can again sustain crown fires. However, of the 19 watersheds with suitable owl habitat, in only one watershed did wildfires burn over 10% of the appropriate forest types in the last 10 years (Figure 5.4). Thus, for most watersheds, past fires apparently have not been extensive enough to "fireproof" the area against future fires.

The sites where owls nest are often in perched benches within canyons. Because steep cliffs and slickrock are natural firebreaks, these sites are relatively fireproof. Thus the model certainly overstates the risk to nest sites. However, Mexican spotted owls forage at some distance from the nest, including the plateaus above their nest sites. Thus our analysis may fairly represent the risk to the owl's foraging habitat.

The Recovery Plan for Mexican spotted owl (US Fish and Wildlife Service 1995) recognized stand-replacing fire as one of the primary threats to the species, and encouraged the use of prescribed fire in and near Protected

FIGURE 5.4. Frequency distribution of watersheds by percent of selected forest types (i.e., mixed conifer, Douglas-fir, and ponderosa pine) that burned during 1986-95, for 32 watersheds in Colorado study area. The 32 watersheds are where our model identified habitat suitable for Mexican spotted owls.

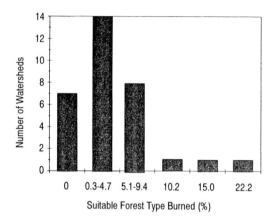

Activity Centers (PACs) to reduce the risk of stand-replacing fire. Our model suggests that there is only a low risk that wildfire will destroy large areas of owl foraging habitat in Colorado (with even less risk to nesting habitat). Thus the degree of threat posed by wildfire may be less in Colorado than in other parts of the species' range.

RISK TO TROUT FROM STAND-REPLACING FIRES

Fire can affect fish populations directly and indirectly by influencing riparian vegetation and the amount of sediment transported to the stream from soil erosion after a fire. Each of the factors will affect different life stages of the fish community. During spawning, trout require gravel nesting areas which are free of silt and sediment to allow flow to carry oxygen to eggs. When trout reach the fry stage they use the edge habitat in shallow areas of the stream which typically have a silt/sand covering, algae and invertebrate food resource, and cover to protect them from predation by adults. Adult trout feed primarily on insects within the stream and from terrestrial (riparian vegetation) sources. Cutthroat trout (*Onchorynchus clarki*) and rainbow trout (*Onchorynchus mykiss*) spawn in the spring whereas brook trout (*Salvelinus fontinalis*) and brown trout (*Salmo trutta*) spawn in fall.

Riparian vegetation provides a buffer zone between the terrestrial and stream environment which provides bank stability, traps sediment to the stream, and is a source of nutrients. Overhanging vegetation serves as cover, is a source of woody debris for maintenance of channel structure and refuge for fish, and provides habitat for invertebrate species that serve as a food resource to adult fish. Fire effects on riparian vegetation can be both beneficial and detrimental. Complete removal of riparian vegetation after a fire reduces its function as a sediment trap and bank stabilizer and reduces stream cover and habitat for invertebrates that serve as fish food resources. A fire that has a moderate effect on riparian vegetation (one that may cause a more diverse vegetation type but maintain its function as sediment filter and bank and channel stabilizer) may be beneficial by increasing algal growth through increased nutrients and sunlight to the stream channel. Algae provide an important food resource to the fry life stage of trout species. Riparian vegetation that remains intact after fire can maintain the same functions as pre-fire conditions.

Soil erosion after fire can increase sediment input to the stream until the land revegetates. Increased sedimentation affect fish populations by filling spawning gravels, covering macro-invertebrate habitats, reducing algal growth, and reducing fish habitat by filling in pools. If the sedimentation effect is short term then it may not have a long-term detrimental effect on the fish population. Removal of vegetation and increased erosion potential can increase peak runoff which may then affect the channel structure and removal of woody debris.

Our goal was to identify watersheds where trout would be at risk after severe fires due to sedimentation. The model used to assess the risk of a severe fire was at the watershed scale.

Model for Trout Species

To assess the risk to trout species from wildfire, a map was produced that combined the wildfire risk model (i.e., ignition frequencies) as described by Neuenschwander et al. (2000), the erosion risk model as described by Mac-Donald et al. (2000), and the distribution of cutthroat trout including the threatened greenback cutthroat trout, and two species defined by the state as being of special concern, the Colorado River cutthroat trout and the Rio Grande cutthroat trout. Five categories of wildfire risk were defined based on ignition frequencies (Figure 5.5). Five categories of erosion risk were also defined as very low (0-30 Erosion Risk), low (31-60 Erosion Risk), medium (61-90 Erosion Risk), high (91-120 Erosion Risk), very high (>120 Erosion Risk).

In general, if the erosion risk is low to very low, then the potential for sedimentation to affect fish in the stream segment is low as well. If the erosion risk is medium, then the potential to affect fish populations in the stream reach may be high immediately after the fire, but then recover within

2-3 years as the watershed recovers. If the erosion risk is high to very high, then the potential to affect fish is greater for a long period of time. Stream segments in this latter category are typically steep and have the capacity for severe soil erosion. This would lead to high sediment deposition in the stream channel over longer time periods (>2 years).

There are limitations of the data sets to interpreting the affects of severe fire on fish populations. One limitation is that the scale was too coarse to accurately categorize the effects of fire on riparian vegetation at any given point. Second, the resolution of stream segments was too coarse to determine exactly where specific fish populations of concern, such as cutthroat trout, would be at risk due to severe fire.

Results

Watersheds with high to very high fire risk in the northwestern (6 watersheds) and southwestern (2 watersheds) regions had no cutthroat trout populations of concern in streams (Figure 5.5). Most cutthroat trout populations were located in watersheds in the northeastern region (Figure 5.6). Two watersheds in this northeastern region with high to very high risk of wildfire had numerous populations of greenback cutthroat trout located in the headwaters (Figure 5.7). Most cutthroat trout populations in these two watersheds, however, were located in streams where low to very low erosion risks on the adjoining lands indicated a low risk of sedimentation affecting the trout population if a severe fire should occur. Two of the 29 known populations of greenback cutthroat trout occurred in stream reaches with both high risk of fire ignition and high to very high erosion risk (Figure 5.7).

Areas of high to very high erosion risks were located primarily in the central and southwest regions in the study area (Figure 5.5). Many of these watersheds have numerous populations of Colorado River and Rio Grande cutthroat trout but have low to very low risk of wildfire (Figure 5.5). Although the wildfire risk is low in these watersheds, these cutthroat trout populations would be at high risk of sedimentation if a severe wildfire occurred. This would also apply to other widely distributed and economically important trout species, such as brook trout, rainbow trout, and brown trout.

One area of concern for trout species is the watershed with very high risk of wildfire in the southwestern corner of the state. This watershed also has areas of high to very high erosion risks. This combination of factors indicate that a severe wildfire has a high potential to affect some trout populations in this watershed.

Discussion

Overall, the risk of wildfire to affect populations of the greenback cutthroat, Rio Grande cutthroat and Colorado River cutthroat trout is relatively

FIGURE 5.5. Distribution of cutthroat trout in relation to ignition frequencies and erosion risks for watersheds in Colorado study area.

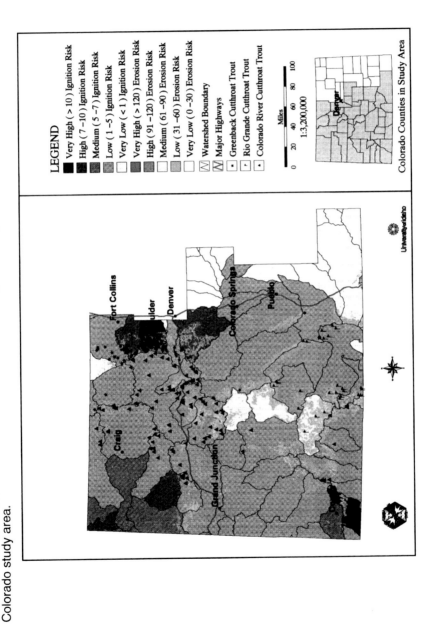

FIGURE 5.6. Distribution of cutthroat trout in relation to ignition frequencies and erosion risk for selected watersheds in northeastern Colorado study area.

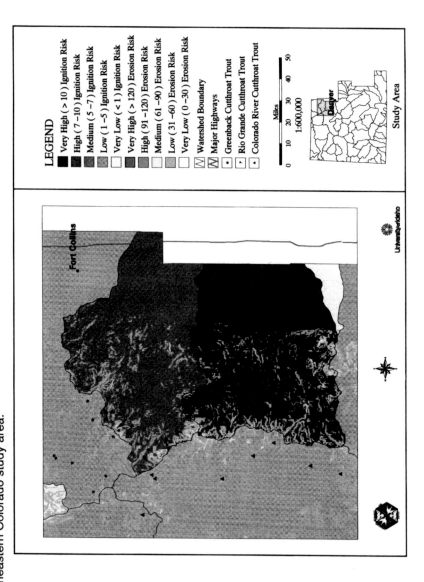

LEGEND

Very High (> 10) Ignition Risk
High (7 –10) Ignition Risk
Medium (5 –7) Ignition Risk
Low (1 –5) Ignition Risk
Very Low (< 1) Ignition Risk
Very High (> 120) Erosion Risk
High (91 –120) Erosion Risk
Medium (61 –90) Erosion Risk
Low (31 –60) Erosion Risk
Very Low (0 –30) Erosion Risk
Watershed Boundary
Major Highways
Greenback Cutthroat Trout
Rio Grande Cutthroat Trout
Colorado River Cutthroat Trout

1:600,000

Miles
0   10   20   30   40   50

Study Area

low (Figure 5.5). This is also true for other trout populations with the exception of the watershed with very high ignition frequencies and erosion risks in the southwestern corner of the state.

The relationship of fish distribution to erosion risk and wildfire risk did not consider how fire affects the riparian vegetation. Thus, low to very low erosion risks have low potential for delivering sediment to stream channel, but if the riparian vegetation was changed to create open patches, a fire may have a beneficial effect on fish populations. If the fire has no effect on riparian vegetation in old growth stands, this may have a negative effect on long-term reproductive sustainability of the trout population because little light may reach the channel to produce algae necessary for the fry life stage. Similarly, if the riparian vegetation was altered due to fire in medium erosion risk streams, the short-term effects may be detrimental due to sedimentation, but in the long-term, the patchiness created may benefit fish populations.

Other factors that affect fish populations that were not discussed include road, mining activities, timber harvest, water quality, flow modifications and current riparian habitat status. Suppression activities may also increase sediment input to streams. Fire lines, whether constructed by hand or bulldozers, must be properly constructed and rehabilitated to minimize sediment production. Future studies related to determining the risk of fish populations to severe fires would need to be conducted at finer scale resolution (i.e., stream channels within these larger watersheds) and consider these other factors to get a better indication of risk to fish populations. Additional factors that would help interpret the effect of wildfire on fish populations would be the probability of storms of high enough magnitude to cause massive soil erosion and changes in channel structure in medium to very high erosion risk streams after a wildfire, and the recovery of soil conditions and vegetation after a wildfire.

Acknowledgments

Cutthroat trout locations were provided by the Colorado Natural Heritage Program's Biological Conservation Database. Special thanks go to Bruce Roseland of the U.S. Fish and Wildlife, Lakewood, Colorado for information and discussions on the Greenback Cutthroat Trout and other trout species in Colorado streams. Thanks go to Alicia Lizarraga for preparations of cutthroat trout distribution maps.

GENERAL FIRE EFFECTS ON ELK HABITAT

The social and economic significance of elk to the citizens of the State of Colorado is extremely high. The elk is a symbol of wilderness, open space, and freedom to both the city dweller and the outdoorsman. The welfare of elk is inextricably tied to management of the land (Thomas and Lyon, 1987). In Colorado, the coniferous forests of ponderosa pine, Douglas-fir, lodgepole pine, subalpine fir, and Engelmann spruce are important to elk. Most elk herds are migratory, moving down from high elevations when snow accumulates in fall or early winter. When the snow depths recede in spring, elk move from their lower elevation winter ranges to higher elevation summer ranges.

Fire is an integral component of the forest ecosystem in Colorado. In varying degrees, all the conifer forests of Colorado have burned sometime in the past. These fires are important to elk because they assure a continued rejuvenation of forest vegetation. Forest recovery following fire proceeds through a succession of plant communities, each with unique values to elk. The repeated burning and recovery of the vegetation is a part of the constant change that produces elk habitat diversity and productivity.

Our goal was to use a geographic information system (GIS) and readily available digital databases to identify and demonstrate the general effects of fire and predicted wildfires on elk forage habitat, hiding cover, and thermal cover on summer (includes spring-fall transition range) and winter ranges at the landscape scale in Colorado.

Methods

Our overall approach involved using GIS to develop a general elk summer and winter range model at the landscape scale that would provide a basis for relating fire effects to elk habitat in western Colorado. Linkage of the elk habitat model outputs to the effects of fire and predicted wildfire occurrence were established by characterizing elk summer and winter forage areas, thermal cover, and hiding cover in terms of successional stages of the plant communities. Further linkage of elk cover and forage areas to fire effects was accomplished by determining the effects of fire and predicted wildfire on plant communities as described in terms of successional stages. Finally, anticipated changes to elk habitat attributable to the occurrence and effect of predicted wildfires were identified by using predicted wildfire occurrence information provided by Neuenschwander et al. (2000).

Probable elk distribution, summer range, and winter range derivation–A digital database of elk overall range and winter range for the Larimer County area was provided by the Colorado Division of Wildlife (CDOW). Those areas of the CDOW Larimer County area database that were within the delineated elk overall range and not part of the delineated winter range were assumed to be elk summer range. The Larimer County area elk summer and winter range map was overlaid with a digital elevation model (DEM) derived elevation database to determine the minimum and maximum elevations for Larimer County area elk summer and winter ranges. Through extrapolation, these derived minimum and maximum elevations were used to delineate elk summer and winter ranges for the larger study area. The first generation probable elk summer and winter range map for the study area was adjusted by excluding Douglas, Elbert, El Paso, and Pueblo Counties because of their relatively high level of urbanization. The resulting elk summer and winter range map displays three categories: (1) areas that may be used solely as elk summer range, (2) areas that may be used solely as elk winter range, and (3) areas that may be used as both elk summer and winter range (Figure 5.8).

Probable elk summer and winter range association with vegetation–The probable elk summer and winter range map was overlaid with a vegetation type map derived from AVHRR imagery to determine what vegetation types are present within probable elk summer, winter, and summer/winter ranges (Loveland et al. 1991, modified as indicated in Sampson and Neuenschwander 2000). A tabular output was produced by the GIS to determine acres of each vegetation type present in each probable elk range category (Table 2).

Vegetation association with successional stages–Information about dominant plant species, general habitat form and vegetation structure (height) at different ages, and the general characteristics of the dominant plant species at different successional stages was compiled and analyzed for each vegetation type found in the elk range categories. In a manner similar to Thomas et al. (1979), five general successional stage categories were characterized for the elk habitat model: (1) grass-forb, (2) shrub-seedling, (3) sapling-pole, (4) young forest, and (5) mature forest. Each vegetation type was assigned to the appropriate successional stage category (Table 3). The combined information about the plant communities and their associated successional stages was the basis for the existing successional stage map for the study area. A tabular output was obtained to determine acres of each existing successional stage category (Table 4).

Successional stage association with elk forage and cover habitat–Our elk habitat model features only foraging areas, thermal cover, and hiding cover considerations for elk. Information about each elk habitat category was analyzed in relation to each successional stage category to determine associations. Each successional stage category was assigned to the appropriate elk habitat category. The combined information about the elk habitat categories and their associated successional stages was the basis for the existing elk habitat map for the study area (Figure 5.8). A tabular output was obtained to determine acres of each existing elk habitat category (Table 4).

Anticipated changes due to fire and predicted wildfire–Information from fire records was compiled and analyzed by Neuenschwander et al. (2000) to determine the number of ignitions (fires) and acres burned in the study area during 1986 through 1995, by vegetation type. The information includes compilation of acres burned. This fire occurrence information was used as a basis for predicting fire ignitions and acres burned for each vegetation type, based on the last decade's experience. Predicted wildfires and acres burned in each vegetation type, general effects on vegetation and elk habitat (Table 5), and proportional changes in elk habitat was the basis for the determination of anticipated general effects on elk habitat from fires of 50 acres or greater size for the next decade (Table 6).

Results and Discussion

Fire influences the relationship between vegetation and wildlife through effects on food, cover, water, and space. Typically, the effects of fire on wildlife are noted as changes in abundance of food and cover. Food and cover are probably the two most visually obvious determinants of habitat suitability. In this model, direct changes in vegetation and the resulting influence on successional stages and elk forage and cover were used as the indicators of fire effects.

Derivation of the minimum and maximum summer and winter range elevations from the Larimer County area elk database allowed extrapolation to the larger study area. The model based this extrapolation on summer range corresponding to 3441 meters and higher, winter range from 1583 to 1729 meters, and summer-winter range from 1729 to 3441 meters in elevation. Vegetation types present in probable elk summer, summer-winter, and winter ranges are presented in Table 2. On this basis, vegetation types most important to elk on summer ranges included aspen, lodgepole pine, mixed conifer, spruce-fir, and alpine tundra. During winter, annual grass, perennial grass, sage, aspen, pinyon-juniper, ponderosa pine, and mixed conifer appear to be most important.

Habitat structure or form follows successional trends in most plant communities. For the most part, pre-burn plant composition and the individual plant species response to fire determines the composition of the post-burn plant community, at least initially. This model provides for these linkages (Table 3). Our characterization of the association between vegetation and successional stages recognizes that some plant habitats support only herbaceous vegetation (i.e., grasslands) while others support woody vegetation (shrublands and forests). Fires in habitats that support only shrubs return the area to a grass-forb stage which returns to a shrub stage within 10-20 years. Fires in forest habitats initiate plant community development through a progression of successional stages that are typified by greater horizontal and vertical diversity and woody species development. Our characterization of successional stages and habitat form in forested habitats assumes the general age of the stands to be 0 to 10 years in the grass-forb successional stage, 10 to 20 years in the shrub-seedling category, 20 to 50 years in the sapling-pole category, 50 to 90 years in the young forest category, and greater than 90

FIGURE 5.7. Cutthroat trout distribution and erosion risk for two high-ignition watersheds in Colorado study area.

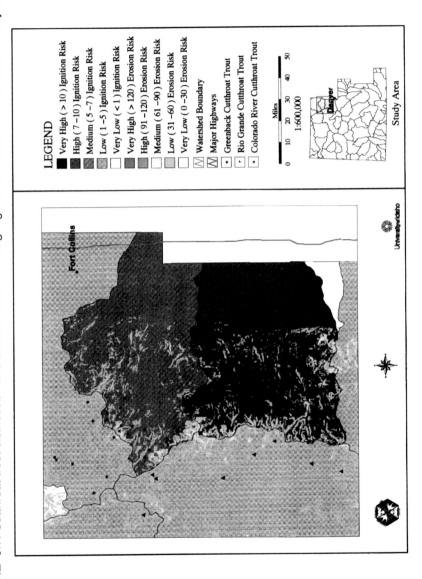

FIGURE 5.8. Probable elk seasonal range and primary use areas, Colorado study area.

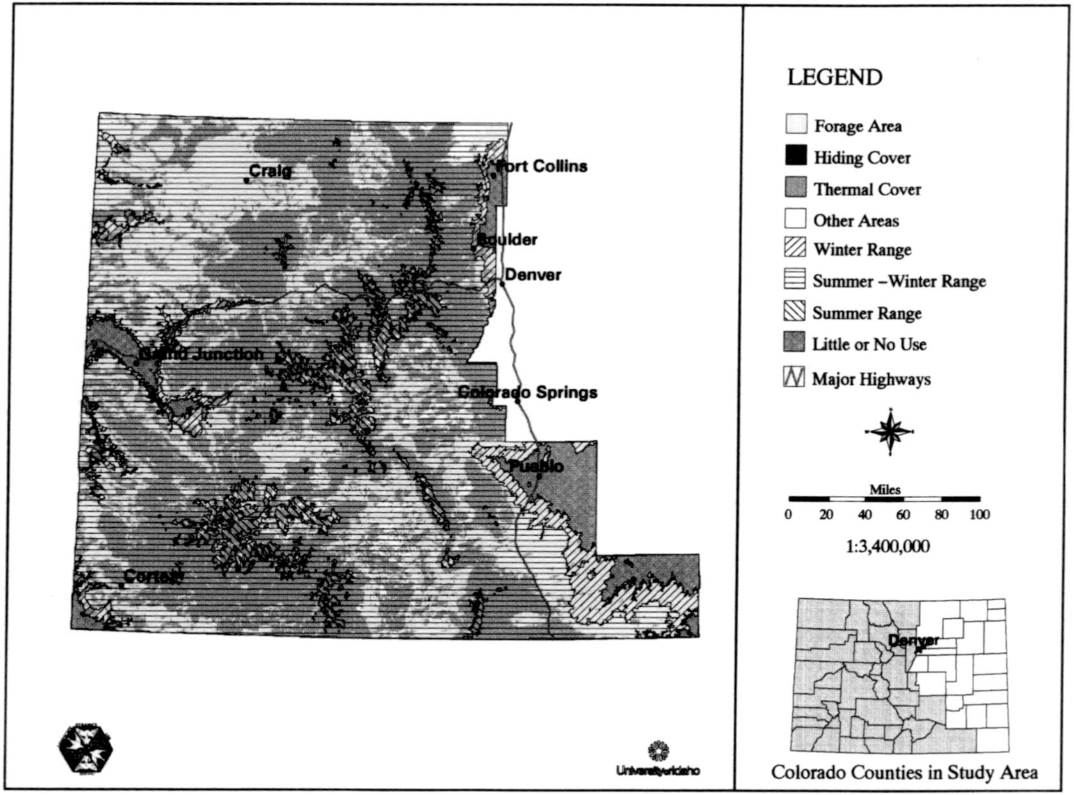
Craig
Fort Collins
Boulder
Denver
Grand Junction
Colorado Springs
Pueblo
Cortez
LEGEND
Forage Area
Hiding Cover
Thermal Cover
Other Areas
Winter Range
Summer –Winter Range
Summer Range
Little or No Use
Major Highways
Miles
0 20 40 60 80 100
1:3,400,000
Denver
Colorado Counties in Study Area

TABLE 2. Vegetation association with probable elk summer, summer-winter, and winter ranges. Acreage rounded to nearest thousand.

Vegetation Type	Elk Range			
	Summer	Summer-Winter	Winter	Total Acres (000's)
Annual Grass	3	2501	145	2649
Perennial Grass	21	6265	1145	7431
Alpine Tundra	61	41	–	102
Sage	1	2148	674	2823
Riparian	–	–	–	–
Chaparral	–	4	–	4
Aspen	24	2834	4	2862
Pinyon-Juniper	10	2537	330	2877
Lodgepole Pine	808	6734	6	7548
Ponderosa Pine	1	3086	146	3233
Mixed Conifer	14	1978	58	2050
Spruce-Fir	1163	4873	26	6062
Total Acres (000's)	2106	33001	2534	37641

TABLE 3. Vegetation association with successional stages. YES indicates a vegetation type that may achieve a habitat form like that characterized by these successional stages, NO indicates a vegetation type that does not.

Vegetation Type	Successional Stage				
	Grass-Forb	Shrub-Seedling	Sapling-Pole	Young	Mature
Annual Grass	YES	NO	NO	NO	NO
Perennial Grass	YES	NO	NO	NO	NO
Alpine Tundra	YES	NO	NO	NO	NO
Sage	YES	YES	NO	NO	NO
Riparian	YES	YES	NO	NO	NO
Chaparral	YES	YES	YES	NO	NO
Aspen	YES	YES	YES	YES	NO
Pinyon-Juniper	YES	YES	YES	YES	NO
Lodgepole Pine	YES	YES	YES	YES	NO
Ponderosa Pine	YES	YES	YES	YES	NO
Mixed Conifer	YES	YES	YES	YES	YES
Spruce-Fir	YES	YES	YES	YES	YES

years in the mature category. It should be noted that habitat structure or form can include dead as well as living components.

When disturbances such as fire occur on the landscape, successional stages are usually reset to the grass-forb stage or converted to a fire maintained climax stage. The characteristics of a fire, whether natural or prescribed, has an influence on the response of the vegetation and ultimately on elk habitat. Fire severity or vegetation consumption by the fire, and mortality

TABLE 4. Elk habitat association with successional stages. YES indicates an elk habitat type that may achieve a form like that characterized by these successional stages, NO indicates an elk habitat type that does not. Acreage rounded to nearest thousand.

Elk Habitat Type	Successional Stage					
	Grass-Forb	Shrub-Seedling	Pole-Sapling	Young	Mature	Acres (000's)
Summer Range	86	1	0	843	1177	2107
Forage	YES	YES	NO	NO	NO	–
Hiding Cover	NO	NO	YES	NO	NO	–
Thermal Cover	NO	NO	YES	YES	YES	–
Winter Range	1290	674	0	487	85	2536
Forage	YES	YES	NO	NO	NO	–
Hiding Cover	NO	NO	YES	NO	NO	–
Thermal Cover	NO	NO	YES	YES	YES	–
Su. & Wi. Range	8807	2148	3	15191	6851	33000
Forage	YES	YES	NO	NO	NO	–
Hiding Cover	NO	NO	YES	NO	NO	–
Thermal Cover	NO	NO	YES	YES	YES	–
Total Ac. (000's)	10183	2823	3	16521	8113	37643

TABLE 5. Estimated general fire effects on vegetation, successional stage, and elk habitat. Acreage rounded to nearest thousand, 0 indicates little or no change, L indicates 0-20%, M indicates 21-80%, H indicates 81-100% consumption of vegetation; NO indicates little or no unburned patches within fire area, YES indicates unburned patches in fire area, NA indicates not applicable, + indicates positive, − indicates negative).

Vegetation Type	Fire Severity	Mortality	Patchy	Forage	Hiding	Thermal
Annual Grass	L	L	NO	+	NA	NA
Perennial Grass	L	L	NO	+	NA	NA
Alpine Tundra	L	L	NO	+	NA	NA
Sage	L	H	YES	+	NA	NA
Riparian	L	L	YES	+	NA	NA
Chaparral	M	H	NO	+	−	NA
Aspen	L	L	YES	+	−	−
Pinyon-Juniper	L	M	YES	+	−	−
Lodgepole Pine	M	H	NO	+	−	−
Ponderosa Pine	M	M	YES	+	0, −	0, −
Mixed Conifer	M	M	YES	+	0, −	0, −
Spruce-Fir	M	M	YES	+	0, −	0, −

TABLE 6. Vegetation, total predicted fires larger than 50 acres and acres burned, and estimated general effects on successional stage, and elk habitat. Acreage rounded to nearest thousand. (G-F indicates grass-forb, S-S indicates shrub-seedling, S-P indicates sapling-pole, Y indicates young, M indicates mature, > indicates increase, < indicates decrease, 0 indicates little or no change, L indicates low, M indicates moderate, H indicates high.)

Vegetation Type	Fires Per 10 Years	000's Acres Burned	Successional Stage	Expected Significance & Effect to Existing Elk Habitat
Annual Grass	54	48	G-F 0, >	H > Winter Forage
Perennial Grass	212	251	G-F 0, >	H > Winter Forage
Alpine Tundra	0	0	G-F 0, >	None
Sage	112	211	S-S 0, >	H > Winter Forage
Riparian	0	0	S-S 0, >	None
Chaparral	0	0	S-P 0, >	None
Aspen	16	10	Y <	H > Forage, < Thermal, > Hiding
Pinyon-Juniper	74	54	Y <	H > Winter Forage, < Winter Hiding & Thermal
Lodgepole Pine	35	13	Y <	L > Summer Forage, < Summer Hiding & Thermal
Ponderosa Pine	62	52	Y <	H > Winter Forage, < Winter Thermal
Mixed Conifer	23	9	M <	L > Forage, < Thermal
Spruce-Fir	29	19	M <	L > Summer Forage, < Summer Thermal

(lethality), an estimate of tree and shrub canopy death, are useful indicators of the direct effect of fire on vegetation (Table 5). Severity and mortality vary dependent on several factors, including the kind of vegetation or fuel present. High severity, high intensity fire can significantly alter plant communities and habitat, sometimes for very long periods of time. In this instance, major adjustment to the existing successional stage may occur, ultimately resulting in changes in elk forage, hiding cover, and thermal cover. Alternatively, fires of low severity and lethality may have little effect on habitat. Variation in the completeness of burning is common within a single fire. Several small fires produce more habitat edge than most single large fires. Patchy or irregular burns increase the diversity of the habitat, particularly in an area with one or a few vegetation types all in the same or similar habitat form or structural condition. The physical mix and patterning of the vegetation, edges, structural components, and seral stages is crucial in determining how beneficial a fire may be to elk. Timing of a fire relative to the phenological stage of the plants is an important factor in determining the post-burn plant community makeup. The size and phenological stage of coniferous tree buds influences their resistance to fire. Large buds such as on ponderosa pine that have already produced scales are more resistant to fire than small and or scaleless buds (Ryan et al. 1988).

Many factors can significantly influence elk use of an area including the presence of roads, fences, and other human activities such as logging operations. Elk avoid heavily traveled forest roads from spring through fall (Edge

1982). Management of a burned area with elk considerations in mind is crucial to ensure benefits to elk are realized. Firelines, roads, and trails may require removal or closure. Burns are an attraction to livestock and depending on terrain and proximity to water may subject the area to over utilization. Appropriate control and management of livestock is essential to avoid competition with elk for forage, reduction in habitat structure, or alteration of expected vegetative response.

The vegetation and the characterization of successional stages as elk forage, hiding cover, and thermal cover is the basis for the elk habitat map. This model associates elk foraging habitat with the habitat form present in the grass-forb and shrub-seedling successional stage category, hiding cover with sapling-pole, and thermal cover with the young forest and mature forest successional stage categories (Table 5). Site preference studies show that elk usually prefer to graze on burned as opposed to unburned sites (Canon 1985, Canon et al. 1987, Leege 1979, Lowe 1975, Lowe et al. 1978, Rowland 1983). Elk clearly exhibit a strong affinity to burned areas. Fire improves the quality of forage under aspen stands (Canon 1985, Canon et al. 1987, Debyle et al. 1989, Gruell and Loope 1974), regenerates decadent aspen stands, opens the understory, increases forbs and grasses by reducing shrubs, and increases suckering. Elk are capable of making excessive use of burned areas that are insufficient in size. Following fire, most preferred elk forage species are enhanced by an increase in nutrients (Asherin 1973, Debyle et al. 1989, Leege 1968, Rowland 1983). However, many studies conclude that an increase in forage quantity is more significant than an increase in forage quality (Bartos and Mueggler 1979, Canon 1985, Davis 1977, Gruell and Loope 1974, Jourdonnais et al. 1990, Leege 1968, Leege 1979). Prescribed fire is used routinely to enhance elk foraging habitat in many Western states and can be used to encourage early spring green up of grasslands by reducing litter, slow or prevent conifer dominance in important foraging areas, increase palatability of forage, reduce the height of browse species, and stimulate regeneration through sprouting or heat scarification of seed (Jourdonnais and Bedunah 1990, Leege 1979, Weaver 1987). In Glacier National Park fires increased carrying capacity on winter range by creating a mosaic of thermal and hiding cover and forage areas (Martinka 1976). Elk more readily use burned areas if their cover requirements are met in close proximity to the burned area. Kramp and others (Kramp et al. 1983) reported that elk prefer burns less than 8.6 acres, and use of burns decreased with an increase in distance to cover. Standing dead trees may provide adequate cover within burns (Davis 1977). If structural conditions or habitat form of vegetation in a successional stage category are at or near the perceived optimum for a particular kind of elk habitat and in healthy condition prior to burning, fire would not be of benefit.

Outputs of this model indicate an overall positive influence on elk forage, hiding cover, and thermal cover will occur as a result of the 50 acre and larger fires predicted for the next decade (Table 6). Overall, these fires will increase the amount of vegetation types associated with earlier successional stages on the landscape. Based on the location of the fires and their relative significance to the amount of existing elk habitat, a significant increase in winter forage is expected. Reductions in winter thermal cover should be expected to occur, but will likely not be significant. Vegetation in burned areas will progress toward later successional stages over time. This will likely result in an increase in both elk summer and winter hiding cover 20 to 30 years after predicted fires occur.

This elk habitat and fire effects model successfully demonstrates the use of GIS and use of available digital and other information. Constraints were such that only elk habitat requirements for feeding, hiding cover, and thermal cover could be considered. Future modeling efforts could include additional important elk management considerations such as reproductive cover and the influence of roads. Limitations of the vegetation database did not allow detailed analysis of important considerations such as fire intensity and severity. These same limitations did not allow detailed analysis of mosaic, juxtaposition, interspersion, size, or diversity of vegetation, successional stages, or elk habitat parameters. The use of higher resolution and accurate thematic mapper derived vegetation maps could prove to be of great value in enhancing the resolution of model outputs. More accurate information on elk seasonal use areas will also be of value. The use of finer resolution data will allow expansion and adjustment of our "prototype" model. Field testing and verification would be appropriate at that time. It should be noted that the primary sources of fire effects information were the NWCG Fire Effects Guide and the USFS computerized Fire Effects Information System.

Conclusions

The evaluation of the effects of fire varies depending on the perspective. Catastrophic fire is a term that has meaning only to the person using it. In the case of elk, fire should be viewed as an overall positive influence on their habitat. Managers must not be so concerned with the short term effects that they lose sight of the future needs of a species (Bunting et al. 1987) and this is truly the case with elk. Given our constraints and need to protect human life and property from fire, whether natural or prescribed, we must strive to achieve goals and objectives related to elk and other wildlife. This can be accomplished. A strong commitment by all involved to give careful consideration and accommodation to the needs of elk, when integrated "up front" into fire, land, and resource decision making, can minimize conflict and avoid missed opportunities while achieving a "win-win" situation for people and elk.

RISK TO LATE SUCCESSIONAL FORESTS

Although we could not obtain appropriate digitized data for this exercise, we feel it is important to retain (or if necessary restore) substantial tracts of late successional forests of ponderosa pine, Douglas-fir, and other forest communities in Colorado. No avian species (with the possible exception of the Mexican Spotted Owl in Douglas-fir) are obligates for such forests, and few avian species reach their maximum densities there. Nonetheless, such forests are valuable because they provide important reference sites against which to assess conditions in more intensely managed forests. If large enough (and buffered from external disturbances long enough) such forests could answer or at least shed light on questions such as: What is the natural fire regime of an undisturbed forest in the ponderosa pine type? To what extent does the presumed 5-15 fire interval reflect burning by Native Americans? What bird and mammal species reach highest abundance (or highest reproductive success) in late-successional forests? Do any plants or insects depend on such forest conditions?

Conclusion

We have provided examples of both fire and lack of fire on a few species. The effects of fire is different for each species and would take an inordinate amount of time and space to fully elucidate and is far beyond the capacity of the workshop that provided this report. However, there is value in looking at where there may be concentrations of habitat for known species of concern. Fires in areas where more species may be affected would be more likely to affect a species of concern. We hope that we have provided examples from a broad enough range of species that land managers can use similar techniques to assess fire effects potential for other species and habitats.

REFERENCES

Anderson, L. 1994. Terrestrial Wildlife and Habitat. In: Ch. VII, Fire Effects Guide. National Wildfire Coordination Group. National Interagency Fire Center. Boise, Idaho. NFES# 2394. 15 p.

Asherin, D. A. 1973. Prescribed burning effects on nutrition, production, and big game use of key northern Idaho browse species. Unpublished Dissertation. University of Idaho. Moscow. 96 p.

Bartos, D. L., and W. F. Mueggler. 1979. Influence of fire on vegetation production in the aspen ecosystem in western Wyoming. In: pp. 75-78. Boyce, Mark S., and Larry D. Hayden-Wing, Larry D., eds. North American Elk, Ecology Behavior and Management. University of Wyoming. Laramie, WY.

Bunting, S. C., L. Neuenschwander, and G. Gruell. 1987. Guidelines for prescribed burning sagebrush-grass rangelands in the Northern Great Basin. General Technical Report INT-186.USDA, Forest Service, Intermountain Forest and Range Experiment Station, Ogden, UT.

Canon, S. K., P. J. Urness, and N. V. Debyle. 1987. Habitat selection, foraging behavior, and dietary nutrition of elk in burned aspen forest. *Journal of Range Management.* 40(5): 443-438.

Canon, S. K. 1985. Habitat selection, foraging behavior, and dietary nutrition of elk in burned vs. unburned aspen forest. Unpublished thesis. Utah State University, Department of Range Sciences, Logan. 110 p.

Crockett, A. B., and H. H. Hadow. 1975. Nest site selection by Williamson's and Red-naped sapsuckers. *Condor.* 77:365-368.

Davis, P. R. 1977. Cervid response to forest fire and clearcutting in southeastern Wyoming. Final Report Cooperative Agreements Nos. 160391-CA and 16-464-CA, U.S. Department of Agriculture, Forest Service and University of Wyoming. Laramie, WY: University of Wyoming, Department of Zoology and Physiology. 94 p.

Debyle, N. V., P. J. Urness, and D. L. Blank. 1989. Forage quality in burned aspen communities. Research Paper INT-404. U.S. Department of Agriculture, Forest Service, Intermountain Research Station. Ogden, UT. 8 p.

Edge, D. 1982. Distribution, habitat use, and movements of elk in relation to roads and human disturbance in western Montana. Unpublished thesis. University of Montana, School of Forestry and Wildlife Biology, Missoula. 98 p.

Ganey, J. L., and J. L. Dick, Jr. 1995. Habitat relationships. US Fish and Wildlife Service. Mexican spotted owl recovery plan, Volume II. 42 p.

Gruell, G. E., and L. L. Loope. 1974. Relationships among aspen, fire, and ungulate browsing in Jackson Hole, Wyoming. Lakewood, Co: U.S. Department of the Interior, National Park Service, Rocky Mountain Region. 33 p. In cooperation with U. S. Department of Agriculture, Forest Service, Intermountain Region.

Hejl, S. J., R. L. Hutto, C. R. Preston, and D. M. Finch. 1995. Effects of silvicultural treatments in the Rocky Mountains. In: T. E. Martin and D. M. Finch (eds.). Ecology and Management of Neotropical Migratory Birds. Oxford University Press, New York.

Hutto, R. L. 1995. Composition of bird communities following stand-replacement fires in northern Rocky Mountain conifer forests. *Conservation Biology.* 9:1041-1058.

Jourdonnais, C. S., and D. J. Bedunah. 1990. Prescribed fire and cattle grazing on elk winter range in Montana. *Wildfire Society Bulletin.* 18(3): 232-240.

Kramp, B. A., D. R. Patton, and W. W. Brady. 1983. The effects of fire on wildlife habitat and species. USDA, Forest Service. Wildlife Unit Technical Report. Southeastern Region, Albuquerque, NM. 29 p.

Leege, T. A. 1968. Prescribed burning for elk in northern Idaho. In: pp. 235-253.

Proceedings Annual Tall Timbers Fire Ecology Conference; 1968 March 14-15; Tallahassee, FL: Tall Timbers Research Station.

Leege, T. A. 1979. Effects of repeated prescribed burns on northern Idaho elk browse. *Northwest Science.* 53(2): 107-113.

Loveland, T. R., J. M. Merchant, D. O. Ohlen, and J. F. Brown. 1991. Development of a land-cover characteristics database for the conterminous U.S. *Photogrammetric Engineering and Remote Sensing* 57(11): 1453-1463.

Lowe, P. O., P. F. Folliott, J. H. Dieterich, and D. R. Patton. 1978. Determining potential wildlife benefits from wildlife in Arizona ponderosa pine forests. USDA Forest Service General Technical Report RM-52. Rocky Mountain Forest and Range Experiment Station. Fort Collins, CO. 12 p.

Lowe, P. O. 1975. Potential wildlife benefits of fire in ponderosa pine forests. Unpublished Masters of Science thesis. University of Arizona, Department of Renewable Natural Resources, Tucson. 131 p.

Martinka, C. J. 1976. Fire and elk in Glacier National Park. In: pp. 377-389. anonymous. Proceedings, Montana Tall Timbers Fire Ecology Conference and Intermountain Fire Research Council Fire & Land Management Symposium; 1974 October 8-10; Missoula, MT. Tallahassee, FL: Tall Timbers Research Station: 377-389.

MacDonald, Lee H., Robert Sampson, Don Brady, Leah Juarros, and Deborah Martin. 2000. Predicting Erosion and Sedimentation Risk from Wildfires: A Case Study from Western Colorado. In Sampson, R.N., R.D. Atkinson, J.W. Lewis (Eds.), Mapping Wildfire Hazards and Risks. Papers from the American Forests scientific workshop, September 29-October 5, 1996, Pingree Park, CO. The Haworth Press, Inc., New York.

Neuenschwander, Leon F., James P. Menakis, Melanie Miller, R. Neil Sampson, Colin Hardy, Bob Averill and Roy Mask. 2000. Indexing Colorado Watersheds to Risk of Wildfire. In Sampson, R.N., R.D. Atkinson, J.W. Lewis (Eds.), Mapping Wildfire Hazards and Risks. Papers from the American Forests scientific workshop, September 29-October 5, 1996, Pingree Park, CO. The Haworth Press, Inc., New York.

Pulich, W. M. 1976. The golden-cheeked warbler. Texas Parks and Wildlife Department, Austin. US Fish and Wildlife Service. 1995. Recovery plan for the Mexican spotted owl, Volume I. Albuquerque New Mexico. 172 p.

Rowland, M. M. 1983. A fire for winter elk. *New Mexico Wildlife Magazine.* 28(6): 2-5.

Ryan, K. C., D. L. Peterson, and E. D. Reinhardt. 1988. Modeling long-term fire-caused mortality of Douglas-fir. *Forest Science* 34(1):190-199.

Sampson, R. Neil and Leon Neuenschwander. 2000. Characteristics of the Study Area and Data Utilized. In Sampson, R.N., R.D. Atkinson, J.W. Lewis (Eds.), Mapping Wildfire Hazards and Risks. Papers from the American Forests scientific workshop, September 29-October 5, 1996, Pingree Park, CO. The Haworth Press, Inc., New York.

Thomas, J. W., and L. J. Lyon. 1987. Elk & Rocky Mountain Majesty. In: pp. 145-149. Kallman, H. (ed), Restoring America's Wildlife 1937-1987. USDI, Fish and Wildlife Service.

Thomas, J. W., R. J. Miller, C. Maser [and others]. 1979. Plant Communities and Successional Stages. In: Thomas, J. W., ed. Wildlife Habitats in Managed Forests–The Blue Mountains of Oregon and Washington. Agric. Handbook 553. Washington, DC: USDA Forest Service. pp. 22-39.

US Fish and Wildlife Service. 1995. Recovery Plan for the Mexican Spotted Owl, Volume I. Albuquerque, New Mexico.

Weaver, S. M. 1987. Fire & elk: summer prescription burning on elk winter range, a new direction in habitat management on the Nez Perce National Forest. *Bugle: The Quarterly Journal of the Rocky Mountain Elk Foundation.* 4(2): 41-42.

Chapter 6

A Screening Method
for Identifying Potential Air Quality Risks
from Extreme Wildfire Events

Helen Getz Rigg
Roger Stocker
Coleen Campbell
Bruce Polkowsky
Tracey Woodruff
Pete Lahm

SUMMARY. Historic land use policies and fire suppression practices have resulted in vegetation conditions and fuel loadings on public lands which increase the potential for catastrophic wildfire events. Air quality is an important consideration during wildfire events and in implementing the fire or smoke management practices which reduce the likelihood and/or impacts of such events. Regulations exist at the national, state, and in some cases local level, to protect public health and visibili-

Helen Getz Rigg is affiliated with Idaho Division of Environmental Quality, Boise, ID 83706. Roger Stocker is affiliated with Western States Air Resources Council, Portland, OR 97204. Coleen Campbell is affiliated with Colorado Air Pollution Control Division, Denver, CO 80246. Bruce Polkowsky is affiliated with US Environmental Protection Agency, Research Triangle Park, NC 27711. Tracey Woodruff is affiliated with US Environmental Protection Agency, Washington, DC 20460. Pete Lahm is affiliated with USDA Forest Service, Phoenix, AZ 85012.

[Haworth co-indexing entry note]: "Chapter 6. A Screening Method for Identifying Potential Air Quality Risks from Extreme Wildfire Events." Rigg, Helen Getz et al. Co-published simultaneously in *Journal of Sustainable Forestry* (Food Products Press, an imprint of The Haworth Press, Inc.) Vol. 11, No. 1/2, 2000, pp. 119-157; and: *Mapping Wildfire Hazards and Risks* (ed: R. Neil Sampson, R. Dwight Atkinson, and Joe W. Lewis) Food Products Press, an imprint of The Haworth Press, Inc., 2000, pp. 119-157. Single or multiple copies of this article are available for a fee from The Haworth Document Delivery Service [1-800-342-9678, 9:00 a.m. - 5:00 p.m. (EST). E-mail address: getinfo@haworthpressinc.com].

ty/scenic values from the air pollution from many sources, including fire. Air regulators can make informed decisions, and better set priorities in support of implementing wildfire prevention actions and management alternatives, if they can identify and rank areas susceptible to wildfire events relative to the potential risk to air quality values. This information can also be used by land managers to better incorporate air quality considerations into their overall analysis of the cumulative socio-environmental effects of potential wildfire events. This paper describes a screening method which utilizes readily available data sources and Geographic Information System (GIS) tools in conjunction with an air quality dispersion model to rank risks from potential wildfire events relative to public health and visibility air quality values. The screening method is piloted using information from the State of Colorado. The results can be used to indicate and communicate the potential risk to air quality values for those areas which are susceptible to wildfire events. *[Article copies available for a fee from The Haworth Document Delivery Service: 1-800-342-9678. E-mail address: <getinfo@haworthpressinc.com> Website: <http://www.haworthpressinc.com>]*

KEYWORDS. Air pollution, wildfire, smoke, particulate matter, visibility

INTRODUCTION

In recent years an increase in extreme wildfires has been noted in the United States. A 1994 article in American Forests magazine observed, "Change agents and processes such as drought, pests, and wildfires are normal components of forest environments. Unfortunately, the current conditions in many Inland West forests allow these normal situations to become catastrophic events" (Sampson, et al., 1994, p.15). The National Commission on Wildfire Disasters noted in its' 1994 report, that "millions of acres of forest, grasslands and deserts in the United States face abnormally high risks of wildfire due to altered species composition, excessive fuel buildup and increased ignition opportunities" and predicts that in the future wildfires will be larger, more intense and far more difficult to control (NCWD 1994).

Scientists, land managers, and interest groups have expressed concern at the range of potential environmental, economic, health and social impacts of extreme wildfire events. The air quality issue is one of many that must be considered in addressing the problem and identifying management alternatives. Air quality considerations include the regulatory, health, safety and scenic value issues of both the wildfire problem and possible solutions. Tools are needed for strategic decision making relative to wildfire emission management. The ability to conduct relative comparisons of the air quality risk in various land management units will support decision makers in making reasoned decisions about where to prioritize resources and direct management attention.

Extreme wildfires create air emissions that can threaten human health and impair visibility. This paper addresses a method to screen the relative, potential impacts on human health and visibility resulting from wildfires occurring in various land management units identified as at risk from extreme wildfire. The focus is on particulate matter emissions because of their large contribution to public health and visibility impacts. The basic premise of this approach is that wildfires in different locations will have different air quality impacts because of factors such as differences in vegetation type, density and condition resulting in different emissions; variations in climatology and geography effecting air dispersion and transport patterns; and differences in the types and number of impacted health and visibility receptors. The screening methodology can be used as a tool by decision makers in prioritizing land management units for wildfire prevention or mitigation efforts on the basis of air quality protection. Air quality results should ultimately be considered in conjunction with other environmental and societal impacts for integrated decision making. Geographic Information System (GIS) technology can be a valuable tool in both analyzing and visually presenting the results of complex spatial impacts, like air quality, as well as for considering the combined impacts of multiple factors.

APPROACH

The screening method was developed by the Air Quality/Health Risk Working Group, a cross functional team of air quality scientists and policy implementors, formed in support of the *Harnessing Data and Technology for Forest Health Decisionmaking Workshop*, Pingree Park, Colorado in the fall of 1996. This workshop was sponsored by the American Forests' Forest Policy Center in cooperation with the Environmental Protection Agency; USDA Forest Service, Bureau of Land Management; National Park Service, Colorado Forest Service and Colorado State University. Other teams from this workshop addressed other environmental and societal impacts. Each team used the western half of Colorado as a pilot area for applying their work. The work conducted during and subsequent to this workshop is referred to simply as the "Pingree Project" in this paper.

The air quality team identified public health and visibility values as two distinct categories of air quality impacts from extreme wildfires. Separate characteristic values were identified in each of these two categories based on accepted standards and consideration of the kinds of data which would be expected to be readily available to a public agency and easily utilized in GIS format. Information on the 11 most likely wildfires in the study area was obtained from the Wildfire and Pest Hazard Working Group. This group provided both the locations for likely outbreaks of large, intense wildfires and emission factors for particular fuel categories (Neuenschwander et al. 2000). The characteristics and emission factors for these wildfires were entered into a prototype air quality dispersion model under development by the Western States Air Resources Council (WESTAR). This model was loaded with climatological, geographic and base emissions information for the study area. The results indicated the levels of particulate in the most likely travel path of the plume from each fire. The ranking scheme involved overlaying the output from the dispersion model with base demographic information and the corresponding air quality receptors for public health and visibility categories. Summing the points assigned to each category resulted in a number which could be used to rank wildfires which pose the greatest risk to public health and/or visibility values.

DEFINING AIR QUALITY IMPACTS FROM WILDFIRE

It is well documented that extreme wildfires can significantly impact air quality. Both gaseous and particulate emissions occur during the combustion of forest fuels. The emission rates (the amount of emissions produced per unit of time) can vary significantly depending on a variety of factors including fuel types, amount, condition and combustion characteristics. For example, smoldering is indicative of incomplete combustion and results in significantly more emissions. The impacts of emissions on public health and visibility will depend not only on the emission amount but on meteorology, topography and general dispersion conditions.

Particulate impacts are used as the basis for prioritizing risk in this paper. Particulates are of key interest because of their chemistry and size.

> This means that most wood smoke particulates (over 90%) are small enough to enter the human lung and a significant fraction (those under 2.5 μm) are respirable, and that their average size is sufficiently near the wave length of visible light that they can scatter sunlight and reduce visibility dramatically. Since toxic organic chemicals often attach themselves to particles, respiration of particles increases their toxicity. (Pyne, et al. 1996)

The federal Clean Air Act (CAA) and state regulations have established regulatory standards for protection of air quality from particulate pollution on the basis of public health and visibility values. Smoke from wildfires can both impair visibility and produce levels of particulate matter which are correlated to increased respiratory illnesses. For this project, the visibility and health effects of particulate air quality impacts from wildfires are evaluated separately. This is consistent with the separate federal (and in some cases state) standards which exist for public health and visibility.

The workgroup considered a variety of receptors for both the health and visibility risk determinations used in this paper. The final selections were narrowed to those receptors for which information should generally be available in GIS format through state or federal air quality agencies or through the U.S. census for any state or region being evaluated. Health receptors were selected to reflect general population exposed, sensitive population groups and populations considered for chronic exposure potential. Visibility receptors were selected to reflect areas where a high public value is placed on visibility and areas where public safety may be adversely impacted. The particulate levels at which these receptors would be impacted were identified based on current and proposed federal standards and related support materials.

PUBLIC HEALTH RELATED RISKS

The potential for public health impacts from air pollution should be a key consideration in prioritizing areas where public land management actions may reduce or increase the potential of extreme wildfires. Wildfires can pose a significant health risk to those exposed to air pollution emitted from the fires. Wildfires can produce high emissions of particulates which may result in significant concentrations of particulate matter (PM) in the air, posing serious health risks. Air quality monitoring conducted in a number of western communities during actual wildfire incidents have indicated particulate at levels which are considered a threat to public health standards (WESTAR 1995).

Much work exists which documents that exposure to elevated levels of particulate matter has been associated with a number of adverse health effects ranging from morbidity to mortality effects. Studies of the effects of excessive air pollution episodes, such as that in London in 1852, have found that elevated levels of PM cause significant morbidity and mortality effects (USEPA 1996a). Studies of lower, more typical, ambient concentrations of PM have also found similar results (USEPA 1996b). The studies of mortality and PM indicate the deaths are primarily due to cardiovascular and respiratory related causes. In addition, elevated particulate matter levels are associated with a number of morbidity effects (USEPA 1996a and 1996b). These include decreased lung function, respiratory impacts and hospital admissions for cardiovascular and respiratory effects. The evidence also indicates certain populations are more susceptible to health effects from exposure to particulate matter. Susceptible groups include those with certain preexisting health conditions and the elderly, children and infants. In addition, long-term studies of exposure to PM show that chronic exposure can result in a number of respiratory and cardiovascular related effects and increased risk of mortality (USEPA 1996a and 1996b; Pope et al. 1995; Dockery et al. 1993; Woodruff et al. 1977). Both susceptible groups and those who experience higher chronic exposure to PM need to be considered when evaluating wildfire risks because they may suffer adverse health outcomes at lower levels of PM than the general population.

Under the Clean Air Act, the Environmental Protection Agency (EPA) is responsible for reviewing the literature on particulate matter and setting a national standard which is intended to protect public health. Information from these reviews and standard setting can be used to evaluate potential health effects from PM emissions from wildfires. At the time of the initiation of the Pingree Project, EPA had a standard for particles less than 10 microns in size (PM_{10}) based on its 1987 review of the literature. During the course of the Pingree Project, EPA completed a review of this standard and concluded there is strong evidence for health effects occurring at lower levels of particulate matter than had been observed in the past (EPA 1996a and 1996b). In addition, they concluded that there is substantial quantitative and qualitative information indicating that fine particles, those less than 2.5 microns ($PM_{2.5}$), are a better surrogate for the particulate matter associated with the observed morbidity and mortality effects (Schwartz et al. 1996). As a result of this review, EPA set a new standard for $PM_{2.5}$. In addition, when EPA considered the particle standard they recognized that a new $PM_{2.5}$ standard would not protect against the effects of elevated coarse particle levels. While there is less epidemiological and toxicological evidence suggesting adverse health effects from low levels of coarse particles, there is still concern for adverse effects because of their ability to penetrate the thoracic region. Thus, EPA recommended that the current annual PM_{10} standard be retained as a surrogate for coarse particle measures. This paper relies heavily on the PM_{10} standard, which was the only particulate standard in effect at time of the initiation of the Pingree Project. Although limited $PM_{2.5}$ emissions and data were available during the course of this project, an attempt was made to incorporate some consideration of the more recently adopted $PM_{2.5}$ standard into its' analysis.

Given the potential health risks posed from exposure to particulate matter from fires, prioritization of the air quality health related risks from varying wildfires needs to consider several factors which influence the risk to the population exposure to particulate matter:

1. Population Exposed;
2. Exposure Window; and
3. Level of Exposure.

Population Exposed

General population density in the vicinity of potential wildfires is a good basis for comparing the total numbers of people potentially exposed to unhealthful levels. Although the workgroup felt it was important to consider the total number of people that may be impacted by a wildfire in any risk based analysis, this should not be misinterpreted to mean that individuals living in lower density areas are of any less concern. Because it is important to look beyond just numbers, the workgroup also emphasized sensitive populations and populations at risk of chronic exposure in the evaluation methodology.

Sensitive populations for which data can readily be extracted from the census were defined as older persons (those older than 55 years of age) and children (those under 12 years of age). As discussed earlier, these groups have been found to be particularly sensitive to PM exposure. The workgroup was especially interested in individuals of all ages with existing cardiovascular or respiratory problems. Unfortunately, information on these populations is not uniformly and readily available. As a result, it was not included in the screening methodology and subsequent analysis. It is still indicated on the summary table because of its importance and should be considered in future work when available.

The workgroup identified population groups that would be at risk from chronic exposure due to lifestyle or environment. This included people living in identified particulate nonattainment areas. Since, the $PM_{2.5}$ standard is new and thus determinations of $PM_{2.5}$ attainment status had not been made at the time this paper was developed, this paper relies on the well established data on designated PM_{10} nonattainment areas. People living in nonattainment areas already experience unacceptable and unhealthful PM_{10} levels in the area they live and any additional impacts are considered significant. Populations which were defined as below the poverty line in the census were included in the at risk group because they might be expected to have lifestyle, living and or working conditions which would make them more susceptible to PM effects. Additionally, they would be less financially capable of considering options of escaping the air pollution exposure from a wildfire event through temporary relocation. Finally, families using wood heat as their sole heating source were identified as potentially having a high chronic exposure to particulates from their residential heating related wood burning.

Exposure Window

Exposure window is the time period over which the population is exposed to particulate matter. Both long-term and short-term exposure to particulate matter has been associated with an increased risk of mortality and morbidity effects. Exposure from fires generally represents an acute health risk, since levels are elevated over a few days and then decline. Thus, it is most appropriate to consider levels of particles over a 24-hour period since this is the period often considered in planning to insure that the population is protected against acute health effects.

Levels of Exposure

As discussed above, EPA recently reviewed and set new particulate standards which are intended to protect public health. The rationale and proposed levels for both the current PM_{10} and the (at the time) expected $PM_{2.5}$ standards were used as a basis for choosing particle matter levels and populations at risk from air quality impacts of potential wildfires.

As part of EPA's risk management strategy, EPA has chosen to focus on the annual $PM_{2.5}$ standard as the controlling standard between the 24-hour and the annual standard. A daily level was chosen that protects against excessive excursions of fine particles, since meeting a low annual standard is likely to insure that the overall distribution of fine particle levels does not result in excessive 24-hour PM levels. Since there are few epidemiology studies which use directly measured $PM_{2.5}$, consideration of the 24-hour level relies on the extensive data from the acute mortality study conducted in six cities in the eastern United States (Six Cities Study). This study includes air quality data over 7-9 years for 6 cities in the United States (Schwartz et al. 1996). The analysis found statistically significant or marginally statistically significant increases in mortality associated with increases in fine particles in 5 of the 6 cities. The 98th percentile of daily values observed in the studies ranged from 34 to 90 $\mu g/m^3$ with most of the values being around 40 $\mu g/m^3$. Given the data from the six-city study and the proposed $PM_{2.5}$ standard at the time this paper was under development, it is reasonable to pick 50 $\mu g/m^3$ of $PM_{2.5}$ as a level of concern when prioritizing wildfires for their impacts on the general population.

However, given that certain segments of the population are more susceptible to exposure to wildfires, it is appropriate to pick a lower 24-hour level when considering effects on children, the elderly and those with preexisting conditions. Again, considering the data from the Six City Study, a level of 35 $\mu g/m^3$, which is the lower end of the 98th percentile in the study, would be an appropriate level for prioritization of wildfires for children, elderly, those with preexisting conditions and people who live in PM nonattainment areas.

As discussed above, it is still important to protect against the health effects from excessive exposure to coarse particles. EPA retained the annual PM_{10} standard as a surrogate for coarse particle measures. Similarly, when considering coarse particle matter exposure levels to be used in identifying areas where there is a potential acute exposure to coarse particles, it would be appropriate to use the PM_{10} standard. Thus, a 24-hour PM_{10} level of the standard of 150 $\mu g/m^3$ is a level of concern for the general population. Similar to using two different criteria for prioritizing $PM_{2.5}$ levels based on susceptible populations, consideration should be given to these populations for a secondary level of concern from PM_{10}. Thus, a 24-hour PM_{10} levels over 100 $\mu g/m^3$ for children and elderly and people in PM_{10} nonattainment areas would be an appropriate level of concern for prioritization of wildfires.

PUBLIC HEALTH RECEPTORS

Public health value tables were developed using the public health exposure factors identified (population exposed, exposure window, levels of exposure). Table 1 summarizes the workgroup's recommendations on appropriate public health receptors to consider in ranking wildfires occurring in different land management units for their potential to impact public health. Table 2 indicates the levels at which PM_{10} and $PM_{2.5}$ are of concern. By using these tables in combination, the population groups identified at risk from wildfire emissions and the documented PM levels at which that impact would be expected to occur are identified. A visual representation of population density and the location of PM_{10} nonattainment areas is shown in Figure 6.1 following the tables.

VISIBILITY RELATED VALUES

Poor visibility impacts scenic values and vistas as well as public safety. It has been well documented that fire is a major source of visibility impairment

TABLE 1. Public Health Receptors

Population Groups	Categories at Risk	Wildfire Emission Levels Code
General Population	Population Density (especially high)	E, C, H, F
Sensitive Populations	Persons > 55 Years of Age	A, C, F, G
	Persons < 12 Years of Age	A, G, C, H
	Persons with Existing Cardiovascular-Respiratory Conditions	A, B, D, G
Chronic Exposure Populations	Wood as Sole Source of Heat	A, D, H, C, F
	Poverty Level	A, D, H, C, F
	PM_{10} Nonattainment Area Resident	A, C, H, C, F

TABLE 2. Public Health Indices

	$\mu g/m^3$	Averaging Time	Code		$\mu g/m^3$	Averaging Time	Code
PM_{10}	100	4 hr	A	$PM_{2.5}$	340	1 hr	F
	150		B				
	400	1 hr	C		35	24 hr	G
	100	24 hr	D		50	24 hr	H
	150		E				

A, F, C, H–most important   B&G not used

(Pyne et al. 1996). Visibility is a serious air quality concern and is protected under the Clean Air Act under prevention of significant deterioration (PSD) provisions. Particular protections are provided to those areas classified as Class 1. This includes national parks, wilderness areas and some Indian reservations. The national goal in these areas is to restrict man-made visibility impairment. Not only are these public lands at risk from wildfire themselves, but they are frequently located near other large tracts of public lands, such as national forests, which may experience wildfires. States and local jurisdictions sometimes may have even more stringent protections for visibility as a result of state or local values. In the state of Colorado, these are frequently areas which have spectacular views of distant mountains recognized as having significant value. Jurisdictions which have a record of recognizing visibility values or regulating visibility protection were considered in determining the receptors for ranking visibility impacts from wildfire. Public safety considerations were also taken into account.

Five visibility values were identified as providing decision makers adequate information to assess the visibility impacts of wildfire potential statewide and to prioritize prescribed fire projects. Some of these values, such as cities with visibility standards and the protection of visibility in specified Class II areas, are applicable in Colorado but are not necessarily so in other states. When performing a statewide assessment, the air quality programs and regulations of the particular state should be considered in determining the visibility values to evaluate. This project assesses the impact on visibility of smoke from wildfires based on the following five value categories:

1. Federal Class I areas;
2. Cities with visibility standards;
3. Class II areas with visibility protection (Class II +);
4. Airports; and
5. Ground transportation.

Federal Class I Areas

On August 7, 1977, Congress designated large national parks and wilderness areas as Class I areas. These areas are afforded visibility protection under the Clean Air Act. The Federal 1977 Clean Air Act Amendment in Section 169A, set as a national goal,

> . . . the prevention of any future and remedying any existing, impairment of visibility in mandatory class I Federal areas which impairment results from man-made pollution. (Clean Air Act of 1977)

This protection is important as a criteria for ranking the relative impacts of extreme wildfires. There are twelve federally designated Class I areas in Colorado. Designated Class I areas in Colorado are:

- Black Canyon of the Gunnison Wilderness;
- Eagles Nest Wilderness;
- Flat Tops Wilderness;
- Great Sand Dunes Wilderness;
- La Garita Wilderness;
- Maroon Bells-Snowmass Wilderness;
- Mesa Verde National Park;
- Mount Zirkel Wilderness;
- Rawah Wilderness;
- Rocky Mountain National Park;
- Weminuche Wilderness; and
- West Elk Wilderness.

Cities with Visibility Standards

Several communities in Colorado have adopted a standard for visibility. These communities have determined a Standard Visual Range (SVR) of less than 32 miles is visually unacceptable. When visibility conditions are measured or predicted to occur which would be worse than the standard, mandatory and voluntary emission reduction programs are triggered. Since these communities are implementing measures to protect visibility in their communities and have determined visibility is an important value to them, these areas were included in the ranking matrix. The visibility standard applies to all communities within the Automobile Inspection and Readjustment (AIR) program region. These include:

- Denver metropolitan area;
- Fort Collins; and
- Colorado Springs.

Class II Areas with Visibility Protection (Class II+)

In addition to the Class I areas, there are eight wildland areas in Colorado which have been designated by Colorado regulation to be given visibility protection at the same level as the federally designated areas. In these areas, the increase in sulfur dioxide emissions from a major stationary source cannot exceed the allowable increase limit set for the Class I Federal Areas. Although sulfur dioxide is not a significant component of wood smoke, because the regulation was adopted to provide protection of visibility, the impacts on visibility from wildland fires are considered in this assessment. The following Colorado scenic areas are classified as Class II+ in this assessment:

- Florissant Fossil Beds National Monument;
- Colorado National Monument;
- Dinosaur National Monument;
- Black Canyon on the Gunnison National Monument (those portions not already included as Federal Class I area);
- Great Sand Dunes National Monument (those portions not already included as Federal Class I area);
- Uncompahgre Mountain Forest Service Primitive Area;
- Wilson Mountain Forest Service Primitive Area; and
- Gunnison Gorge Recreation Area (lands administered by BLM as of October 27, 1977).

Air and Ground Transportation

Another layer in the air quality assessment is the impact on visibility, and as a result public safety, at airports. Location coordinates are readily available for airports servicing scheduled commercial flights. It is important to note that a few airports, such as military airfields, might be missed using these data. For the statewide assessment, the commercial airport should cover the general vicinity of the noncommercial airports.

Major roadways were considered as an additional layer. For the state level assessment, it was determined the PM_{10} National Ambient Air Quality Standards (NAAQS) will provide sufficient protection of ground transportation. If priorities for prescribed fire were determined within a more localized area, the effect of the smoke on specific roadways would be an important value to assess. The NAAQS is to protect the general public welfare and on a statewide basis would provide adequate visibility for safe roadways.

VISIBILITY RECEPTORS

Visibility value tables were developed using the visibility exposure factors identified. Tables 3 and 4 summarize the recommendations on visibility value receptors to be considered in ranking wildfires occurring in different land management units for their potential to impact visibility. In Table 3, a code indicates which measurement(s) provide assessment data for each chosen visibility value. Table 4 summarizes the criteria and appropriate measurement(s) for evaluating each of the visibility values. By using these tables in combination, the visibility values identified at risk from wildfire emissions and the documented PM levels at which that impact would be expected to occur are identified. The location of these receptors is shown in Figure 6.2.

There is strong technical support for the individual measurements selected to be used in evaluating the smoke impacts on visibility from wildfire emissions. During the permitting process for major stationary sources of air pollutants, the visibility impacts due to the source's emissions have historically needed to be estimated to cause less than 5% change in the visibility conditions at federal Class I areas. If the impact of the source's emissions is greater than 5%, mitigation efforts will be a permit condition (Blett et al. 1993).

TABLE 3. Visibility Receptors

Visibility Values	Wildfire Emission Levels Code
Class I Area	A
Cities with Visibility Standards	B
Class II Areas with Class I Visibility Protection	A
Airports (Scheduled Passenger Flights & Military)	C
General Public Welfare (Population Density)	D, E

TABLE 4. Visibility Indices

Description	Code
5% Extinction Reduction Based on th 90th Percentile Seasonal Standard Visual Range (1 Deciview)	A
Standard Visual Range < 32 Miles (During Daylight Hours and RH < 70%)	B
Standard Visual Range < 5 Miles	C
Secondary: 150 μg/m^3 PM$_{10}$ (24 hr avg)	D
Primary: 50 μg/m^3 PM$_{2.5}$ (24 hr avg)	E

Change in visibility can be expressed as extinction or deciview. A change in visibility condition of one deciview is perceptible to the average person. Since this is the criterion used for stationary sources of air pollutants, it is used to evaluate the wildfire smoke impacts on visibility impacts at the federal Class I areas. Views with little visibility impairment are more sensitive to visibility degradation than hazier views. To protect visibility in the federal Class I areas, the cleanest days are of concern and the acceptable change is compared to the average Standard Visual Range (SVR) on the cleanest ten percent of the days monitored. For this assessment, the seasonal SVR of the cleanest summer days will be used for comparisons of acceptable change. If wildland fire activity in other seasons is assessed, it is recommended that one compare visibility changes to the average SVR of the cleanest days during the corresponding season.

Standard Visual Range (SVR) indicates how clearly distant objects are visible. If one can distinctly see the color and features of scenery which is 30 miles away, the SVR is 30 miles or 48 km. The higher the SVR value, the clearer the view. The Visibility Standard is based on the results of a photo series study involving over 200 Colorado Front Range residents. When visibility was less than 32 miles, Colorado residents consistently indicated the visibility impairment was unacceptable. The standard is in effect in a number of Colorado communities. For the purpose of this assessment, the SVR value of the standard is used as a surrogate for unacceptable visibility impairment from smoke in both cities with visibility standards and Colorado Class II+ areas. With more research or actual fire activity evaluation, another value may be considered in the future as more appropriate for use with smoke specific visibility impacts.

For the purposes of the protection of air or ground transportation, a SVR of 5 miles is considered sufficient viewing range to allow for safe driving or flying conditions. Federal Aviation Administration rules require flights to switch to instrument controls when visibility is less than 5 miles.

BACKGROUND CONDITIONS AND VALUES

In order to properly evaluate the air quality impacts from wildfire on various receptors in the state, the estimated smoke impacts of the wildland fire are added to a background value for each receptor category. The resulting impacts are used to evaluate and compare the smoke impacts. Background values for the purpose of this project are estimates of the total expected contribution from emission sources other than wildfire, or the expected air quality without the impact of wildfire, during the period being modeled. This is considered the baseline which the emissions or air quality impacts from a particular wildfire are added to in order to calculate the total emissions or air

FIGURE 6.1. Location of sensitive receptors for population density and PM₁₀ nonattainment areas.

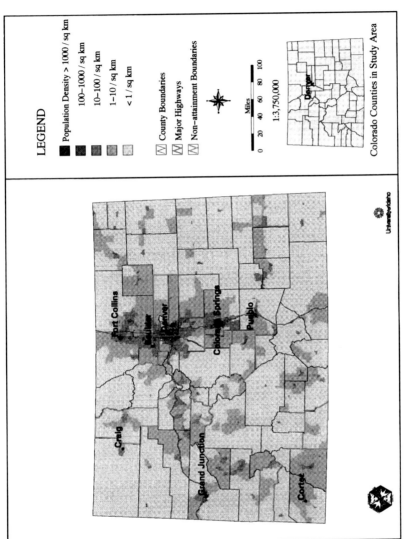

133

FIGURE 6.2. Key receptors used in the evaluation of the visibility effects from the 11 simulated wildfires.

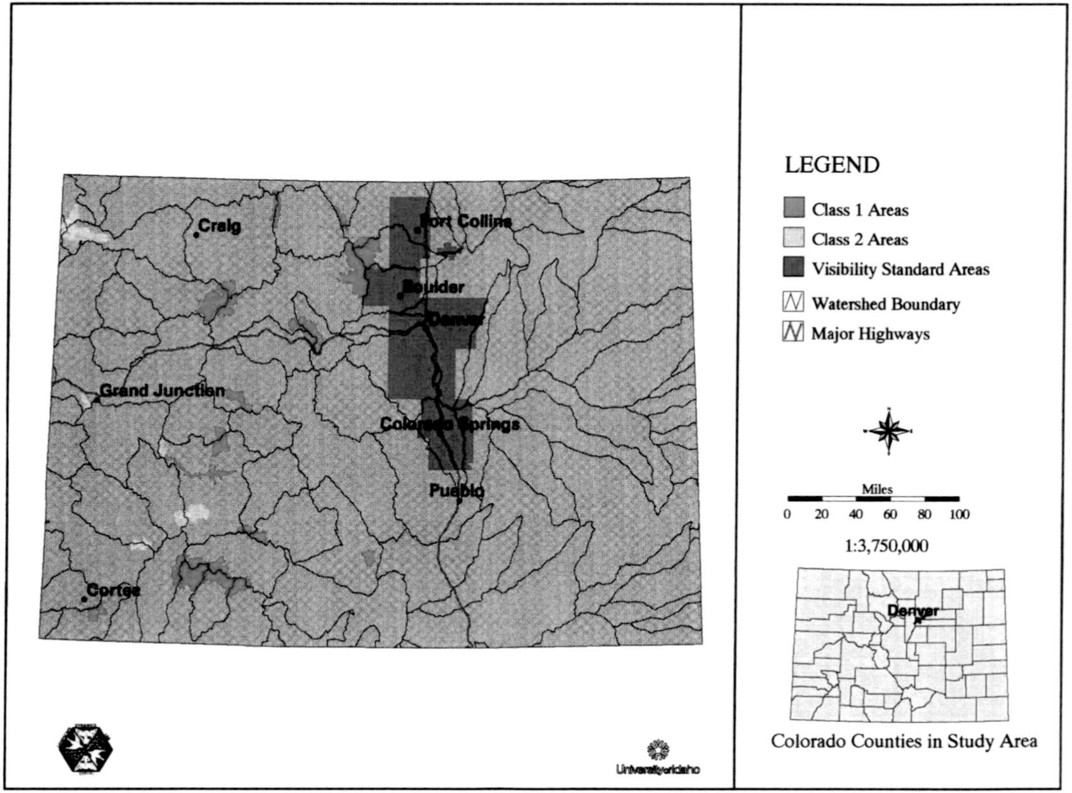

quality impact. For the air quality assessment, background values were included as model parameters for ambient PM_{10} concentrations and extinction values.

Background PM_{10} Concentrations

Average summertime PM_{10} concentrations for Colorado were determined from the Colorado Air Pollution Control Division's statewide PM_{10} monitoring network. These data are reported to the EPA Aerometric Information Retrieval System (AIRS). The arithmetic mean of the 24-hour ambient PM_{10} concentrations during June, July, and August for the years 1990 to 1996 were calculated. Statewide this system consisted of 59 sites in communities around Colorado. Ambient PM_{10} concentrations have been monitored in all of these communities, except Cripple Creek, for at least five years. At times, a monitoring site must be relocated and the data over the six year period may have been collected from more than one monitoring sites. In many of the Colorado communities, several sites are monitored on a routine basis.

The concentrations were assigned to four categories (blue, green, yellow, and red). Any of the communities with a yellow or red plot within their boundary were given a "modeling domain" and the average monitored concentration assigned to the domain. These domains include a number of smaller Colorado communities which are PM_{10} nonattainment areas (the NAAQS for PM_{10} has been exceeded) or have consistently monitored elevated PM_{10} concentrations. The remainder of these areas, are the higher population communities of the state. One of the areas, encompassing the front range from Fort Collins to Denver, contains plots in the red and yellow categories and is the most densely populated part of the state.

As mentioned previously, five years of monitoring data are not available for Cripple Creek. This small mountain town is surrounded on three sides by a bowl which rises steeply around the town. On the surrounding hillsides are a number of unpaved roadways. Legalized gambling has increased the number of mobile sources in the area in the last several years. Ambient monitoring was initiated in 1995 due to anecdotal reports of 'very hazy' days. Within the first month or two of monitoring, the NAAQS for PM_{10} was exceeded. The town has not been designated as a PM_{10} nonattainment area, but a memorandum of understanding (MOU) between the town and the U.S. EPA is being negotiated. This is one of the first times an MOU has been used to reduce ambient PM_{10} concentrations. Through the MOU the town of Cripple Creek has agreed to institute control strategies. Even though the community is not an official non attainment area, it was included in the analysis for impact to nonattainment communities.

For the remaining communities in the state, an average PM_{10} concentration was estimated using the average of the monitored concentrations in the lowest category (blue plots). This value was estimated to be 19 $\mu g/m^3$. For the remainder of the state, except the Class I Wilderness areas, this should be an adequate value for this assessment. For a statewide prioritization system, the states' ambient particulate monitoring systems should provide a broad base to use for background values. This assessment is of the impact of wildfires. In Colorado wildfires occur most frequently in July, June and August. If the assessment was expanded to evaluate wildland fires or wildfires for an alternate or expanded area, the background values for the corresponding months should be used.

Background Visibility

Extinction (the amount of haze) has been monitored in Denver since 1990 with a transmissometer. This data is processed and reported by the Colorado Air Pollution Control Division. The median and 90th percentile of the standard relevant averages during June, July and August from 1990-1992 were calculated. Extinction is monitored for comparison to the State's Visibility standard during daylight hours and when relative humidity is less than 70%. During most years the transmissometer has not been operational during a large portion of June. The transmissometer is removed annually in May and/or June for annual cleaning, maintenance, and calibration. For the purposes of a statewide assessment, the extinction value measured should be adequate for the entire summer season.

The National Park Service (Binkley et al. 1997; Maniero 1997) and the U.S. Forest Service (Copeland 1997) calculated extinction values for various Class I areas. These extinction values are calculated from the IMPROVE particulate monitoring sites. The USFS values are in draft form. The values are to be recalculated and the cleanest days are expected to be about 10% cleaner at Mt. Zirkel and Weminuche Wilderness Areas. The USFS does not have the seasonal values calculated at this time. The annual median and 90th percentile extinction values were used for the USFS wilderness areas. It is known the summertime visibility will be poorer than the annual value. The extinction data from the NPS wilderness areas could be used to develop a correction factor to estimate the seasonal values from the annual data. It was decided the extra work would not significantly improve the assessment. To increase the accuracy of the assessment, the final, seasonal data set should be utilized when it is available.

Background extinction values for most of the remaining Class I and Class II wilderness areas and parks were estimated from the NPS and USFS Class I areas with monitored extinction values. Generally an area was assigned a background extinction value from the closest monitored Class I area of similar elevation.

The extinction measured at Juniper Mountain from June through August,

1995 for the Mount Zirkel Wilderness Area Reasonable Attribution Study of Visibility Impairment (Watson and Blumenthal 1996) was used as the background extinction for Dinosaur National Monument. Juniper Mountain is 40 km east of the Monument and the elevation is similar to parts of the Monument. The Monument and Juniper Mountain are situated between two coal fired generation stations, Deseret to the west and Tri State Generating Station to the east. Although on individual days visibility at the two sites most likely varies, the seasonal average should be similar and provides a locally monitored value for the state wide assessment.

Because of the close proximity to the Great Sand Dunes National Monument (GSDNM) and the similar climates, the background extinction measured for the GSDNM was used for the Florissant Fossil Beds National Monument.

METHODS FOR EVALUATION OF AIR QUALITY IMPACTS OF WILDFIRES IN COLORADO

This section contains the methods used in the air quality impacts screening analysis. The information used to conduct this analysis came primarily from other work products produced at the workshop at Pingree Park, Colorado during the fall of 1996 and available meteorological information. The purpose of this analysis is to evaluate the impact of Colorado wildfires from an air quality perspective to be used in determining those areas which:

1. Have a high probability of having a wildfire present, and
2. Incur smoke impacts at "sensitive" areas (receptors) downwind of the fire plume.

The two criteria suggested above involve both spatial and temporal inhomogeneity issues. Therefore, the first step in evaluating the impact of Colorado wildfires is to choose a time period which is probable to produce wildfires throughout Colorado. A handbook published by the USDA Forest Service (Schroeder and Buck 1970) split the United States into 15 broad zones based on geographic and climatic factors which were determined to give each area a distinctive synoptic wildfire signature. Two of these zones, Great Basin and Southern Rockies, cover the state of Colorado and therefore, suggest typical synoptic meteorological conditions associated with the wildfires in this State. One of the synoptic patterns common to both areas is a dry cold front which is found east of the Rockies with a cold core high centered on the Great Basin region. This pattern is especially common in wildfires which occur in the eastern foothills of the Rocky Mountains since it also produces chinook winds which serve to produce dry conditions with high wind speeds. This in no way exhausts the research used to identify meteorological variables common in wildfire events, but provides an easy way of examining readily available regional information to determine wildfire starts. This is accomplished using National Weather Service Daily weather maps which contain both surface and upper air information needed to determine what synoptic patterns existed on any given day.

For this Colorado analysis, the year 1992 was chosen to extract a time period to fit the mentioned synoptic pattern. This year was chosen primarily due to recent work by the Grand Canyon Visibility Transport Commission (GCVTC) which has done extensive analysis on conditions in the west for this year. The time period which most closely matches the synoptic conditions mentioned above in the summer of 1992 is a four day period from 1-5 August. During this period, a stationary front was evident along the eastern foothills of the Rocky Mountains providing light rainfall to the eastern plains of Colorado. Historical fire activity for this time suggested that a number of large wildfires occurred in the northwestern portion of Colorado around Dinosaur National Monument during this time with these fires reaching sizes of approximately 500 acres in size and lasting 2-3 days (BLM lands, National Interagency Fire Center).

Emissions Inventory and Source Locations

The emissions inventory used in any modeling study provides crucial information regarding the amount of pollution generated by any given source. In order for an evaluation to be done on the impact of this pollution, it is important that accurate information be obtained related not only to meteorology but also an emissions inventory. The inventory used in this analysis was obtained by a group of experts attending the Colorado meeting and is outlined in (Neuenschwander et al. 2000) and so only a brief overview and the application of the inventory to the sources will be discussed here. This inventory contained estimates of emissions for PM_{10} on a grid which covered the entire state of Colorado at a 1km pixel resolution. The base information used in deriving these emissions estimates was a vegetation map of the same resolution which allowed for the conversion of vegetation into emissions estimates.

The site locations for the potential wildfires were also chosen by the emissions experts and included two steps. The first step was to identify a number of watershed regions throughout Colorado which were likely to experience a wildfire in the future. These were chosen based on historical information on wildfire starts. Eleven watersheds were selected for consideration in the Pingree Project. A specific site within each watershed was chosen to represent the wildfire origin. This was chosen based on considerations such as vegetation type, aspect and elevation. The first two columns in Table 1 show the origin latitude/longitude pairings of each of these fires. It was then necessary to obtain the emissions at each of these locations. The size of each of these wildfires was chosen to be 25,000 acres to represent catastrophic worst case conditions and to match the meteorological modeling domain which used a grid with 10km spacing (25,000 acres = 100 km^2). In order to obtain the total emission for a 25,000 acre fire at any of the locations, it was necessary to total the emissions estimates of the 100 cells surrounding the origin location of and individual wildfire. (The origin represented the center of any 100 km^2 grid box.) The next step involved the allocation of these total emissions estimates to some wildfire emissions profile to obtain an hourly emissions rate for a fire. This was done using a scheme outlined in Stocker (2000) in which the total emissions are mapped to an emissions profile generated by an emissions production model run for another modeling study, the Interior Columbia River Basin study (ICRB) (Scire 1996).

Emissions were based on a vegetation layer developed by the Fire Emissions Project of the Grand Canyon Visibility Transport Commission (GCVTC). This vegetation layer was then correlated with Interior Columbia River Basin Assessment (ICRB) vegetation types. Emissions had already been calculated for the ICRB effort using a method described by Hardy (2000). The pounds of PM_{10} per acre for the identified CRB vegetation types were then supplied for the air quality analysis herein.

The emissions production model run in the ICRB produced emissions profiles for four common vegetation types for a number of fire sizes and types. The four types of vegetation which were included in the analysis were grass, shrub, ponderosa pine and mixed conifer. Therefore, in order to match one of these ICRB profiles to the Colorado vegetation it was necessary to obtain the vegetation information for each of the 11 source locations. These vegetation classes were obtained from the base map used in estimating the emissions and are included in Table 1 along with the ICRB vegetation profile class used to determine the emissions profile. It is important to notice that in very few cells were the vegetation classes defined by a single class. Therefore, the final vegetation class was mostly chosen to represent the dominant vegetation in a cell. This is true for all but the last source mentioned in which ponderosa pine was chosen as the vegetation emissions class. This exception was made because although grass was the dominant vegetation class, the high percentage of ponderosa pine coupled with the high emissions for this 100 km^2 cell suggest that a large woody fuel profile is more appropriate for this source. The small diameter woody fuel profiles (shrub and grass) suggest that a 25000 acre fire duration is on the order of 12 hours; whereas, the large fuels can last on the order of days. Due to the limitation of the modeling period, the large fuel profiles (ponderosa pine and mixed conifer) were restricted to three days in order to allow for plume advection, dispersion and deposition before the end of the meteorological simulation.

Meteorology MM5 Simulation

A mesoscale, prognostic, meteorological model, MM5, was run for this analysis to obtain information pertaining to meteorological conditions input for transport of pollution from a source location to some receptor. This model allows for the determination of not only large synoptic scale circulation patterns but also more localized mesoscale transport patterns import for de-termining impacts local to a wildfire origin. This model also has the ability to reproduce precipitation patterns which are important for understanding wet deposition of a plume along the transport path. A more detailed description of this model can be found in Seaman and Anthes (1981).

MM5 for this analysis was run for the entire test period, 1-5 August 1992, producing wind, temperature, turbulence and moisture information necessary as an input for the air quality model which in used to determine source impacts to particular receptors. Figure 6.3 shows a plan view of the surface vector winds for a time 30 hours after the start of the meteorological simula-tion. This time period (11 p.m. MST on 1 August 1992) indicates the "snap-shot" flow patterns which were active in transporting emissions from all of the 11 fires. (Wildfires were started randomly between the hours of 11 a.m. and 4 p.m. on 1 August 1992.) Figure 6.3 suggests that at this time, flow patterns suggest divergence over the continental divide very common to nocturnal drainage flows in complex terrain. On the east side of the moun-tains, where a stationary front lies along the foothills, a strong surface wind is blowing from north to south with strong winds of 30 mph around the Colora-do/Wyoming border. Transport patterns west of the divide can be shown to exhibit two major patterns. The first is a anticyclonic motion which tends to indicate transport from northwestern Colorado will be advected towards the south and west. The second pattern can be seen in the southwestern corner of the state and indicates cyclonic motion with transport primarily towards the west. This southern pattern contains the weaker winds with many areas of calm conditions indicating stagnation at the surface.

Air Quality (REMSAD) Simulation

REMSAD is an Eulerian grid model capable of handling advection/diffu-sion, chemical transformation and deposition for both PM and toxics issues. Its formulation closely follows that of the EPA regulatory ozone regulatory model, UAM-V with both being developed at Systems Applications Interna-tional. Important features of this model as they pertain to this study include such things as secondary aerosol production, wet and dry deposition and the ability of this model to utilize, in part, the meteorological fields produced by the MM5 model. Further explanation of this model can be found in the users guide for this model (SAI, 1996).

PRIORITIZATION OF AREAS BASED ON RANKING METHODOLOGY

The results of the modeling analysis will produce concentration estimates associated with each of the 11 critical watersheds discussed in a previous section. These concentration estimates provide the basic information necessary for evaluating both air quality related health and visibility impacts to areas in Colorado. It was determined by the workgroup that visibility and health would be treated separately and not merged to come up with an overall air quality ranking. This was because both values are independent since they are not measure in the same way and have different impacts. Therefore, a ranking was developed for the 11 watersheds for visibility and health.

Prioritization Methodology for Public Health

The basic concept involved in the creation of a methodology for prioritizing risk on the basis of public health impact was associated with determining the relative risk which the modeled wildfires posed to the identified public health values. Seven public health values were recommended as described earlier in this paper. They are as follows:

1. General population density.
2. Population density of persons > 65 in age.
3. Population densities of persons < 12 in age.
4. Population densities of persons having a propensity for Cardiovascular-respiratory ailments.
5. Population densities of persons using wood as a sole heat source.
6. Populations densities of people below the poverty level.
7. PM_{10} nonattainment areas.

Table 1 lists these variables as the categories of persons at risk. Of these seven elements, the fourth one dealing with respiratory ailments was the only variable not included in the ranking scheme due to lack of available information. The other variables were obtained from census information obtained from the 1990 national census (Case et al. 2000). Table 2 contains a list of health indices (A-E) which are used to measure exceedences for the seven elements listed above. The second column of Table 1 gives the dependant indices associated with each of the seven elements.

Equations 1 and 2 define the actual formula developed to evaluate the health impacts for the study. This is a *receptor* based formula which means that exceedences are determined based on the total concentration estimates which are "observed" at a receptor. (A receptor is defined as any 1 km^2 model cell within the analysis domain. Receptors for the health analysis were further broken out into those inside and outside PM_{10} nonattainment areas [NAA]. There were approximately 300,000 grid cells used for the Colorado domain.) The total concentration used in this analysis was the sum of concentration impacts from all sources to a particular receptor plus a summertime background. An exceedence was given a value of 1 for any receptor above any threshold value applicable to an associated risk category and a value of 0 was given to a non-exceedence at a particular receptor at a given time. Prior to testing for exceedences, an initial test was performed on the grid cells based on population assumptions. For any of the five terms located on the right-hand-side (RHS) of equation 1, the test included determining if for the PD variable, there was even one person which permanently resided in a grid cell ,and for the other four variables whether the population in a grid cell contained higher than average numbers of the particular population groups (poor, youth, etc.) being tested for (measured by determining if the population density in a grid cell was at least one standard deviation above the average for the State of Colorado). For those locations passing through the initial test, the five terms on the RHS of equation 1 for a time (t), the value of a term was given a value between 0 and 1 (e.g., the PD term has four components or tests, if all four tests were to exceed the allowable limits (see Table 2), a value of 1 would be assigned to this term; if only one of the four tests showed an exceedence, a value of 0.25 would be assigned. This effectively normalizes the variables in this equation). These values were then summed over the length of the five day simulation to obtain an estimation of the health impact over the entire time period at each receptor (RR[x,y]). A two tiered test was then performed on the receptor locations which were based on the assumption that any impact to a PM_{10} NAA should be given more weight than an area outside these regions. Equation 2 shows the formulation for this two tiered approach with a constant value of $RR_{(max)}$ being added to any term located within a NAA to make sure any impacts to these locations are given the most weight. ($RR_{(max)}$ is determined as the maximum receptor ranking for any of the cells located outside a NAA.) The gridded total ranking score is then partitioned back to the individual wildfire watersheds based on the proportion of the total concentration due a wildfire located in the individual watershed. These watershed numbers were then tallied for each hour of the simulation to come up with final wildfire impact rankings based on health considerations.

Equation 1:

$$RR(x,y) = \sum_{t=1}^{T} \left(\sum_{m=1}^{4} PD_m + \sum_{n=1}^{4} P55_n + \sum_{o=1}^{4} P12_o + \sum_{q=1}^{5} WS_q + \sum_{r=1}^{5} Pq_r \right)_t \quad (1)$$

where:
RR = receptor ranking based on the impact from fires affecting this receptor at time t,
PD = Population density impact function based on the four elements defined in Table 1,
P55 = Population density impact function for persons over the age of 55 for the four elements outlined in Table 1,
P12 = Population density impact function for persons under age 12 for the four elements outlined in Table 1,
WS = Population density impact function for persons with wood stoves as their sole source of heat for the five elements outlined in Table 1, and
Pq = Population density impact function for persons living below the poverty level for the four elements outlined in Table 1.

Equation 2:

$$PS_j = \frac{[PM_{10}]_j}{[PM_{10}]_{total}} \sum_{l=1}^{R} F(l) \begin{cases} F(l) = RR_l + RR_{(\max_l, l \notin PM_{10}NAA)}, l \in PM_{10}NAA \\ F(l) = RR_l, \; l \notin PM_{10}NAA \end{cases} \quad (2)$$

where:
PS = Ranking of the source j (wildfire/watershed),
T = time,
R = receptor location (all grid cells),
$[PM_{10}]_j$ = estimated concentration of PM_{10} from receptor (l) associated with source j, and
$[PM_{10}]_{total}$ = estimated concentration of PM_{10} at receptor (l) from all receptors

Prioritization Methodology for Visibility

The primary emphasis for the evaluation of impact from wildfires on visibility was not population as was the case in the previous section but was on sensitive geographical locations which have either been given special treatment at a federal or local level.

According to the listing of ranking variables outlined in Table 3, there are five visibility values which need to be considered in the visibility ranking formula. The five values are each associated with one of the five visibility indices indicated in Table 4 as follows:

1. Class I = 5% extinction change
2. Class II = 10% extinction change
3. Areas with visibility standards = 76 Mm^{-1}
4. Airports = SVR < 5 miles
5. General Public welfare = exceedance of 1 hour values for $PM_{2.5}$ as outlined in Table 1.

Equation 3 defines the ranking scheme for visibility that was used in this study. This formula is a *source* oriented formula that evaluates fire impacts individually and never attempts to estimate what the combined impact from multiple fires would be. Therefore, the "trigger" level is based only on the impact from an individual fire; thus, individual fires will need to produce exceedences of the visibility indices at particular receptors in order to be counted in the ranking formula. (Concentration estimates are defined for a given receptor as the concentration estimate "observed" at the receptor from a particular source plus a background estimate from this receptor.)

$$\checkmark_i = \sum_{j=1}^{T} \sum_{k=1}^{R} [E_{ijk}^1 \times 10 + \frac{BEXT_{ijk} - BEXT_{thresh}}{0.05 \times BEXT_{thresh}} + E_{ijk}^2 \times 5 +$$

$$\frac{BEXT_{ijk} - BEXT_{thresh}}{0.10 \times BEXT_{thresh}} + E_{ijk}^A \times 5 + \sum_{m=1}^{3} E_{ijk}^{2.5} \times C_m + \sum_{n=1}^{3} E_{ijk}^{10} \times C_n \quad (3)$$

where:
RV$_i$ = Visibility ranking for source i (fire/watershed),
T = time,
R = Receptor location (e.g., Class I, Class II, etc.),
E^1 = number of exceedences in Class I areas,
E^2 = number of exceedences in Class II areas and cities with visibility standards,
BEXT$_{ijk}$ = extinction based on PM_{10} "observed" concentrations from source i at time j and receptor k,
BEXT$_{thresh}$ = threshold PM_{10} extinction based on the values outlined in Table 4,
E^A = number of exceedences at major airports (currently only Denver International is considered),

$E^{2.5}$ = number of exceedences of the 1 hour $PM_{2.5}$ one hour criteria from Table 4,
E^{10} = number of exceedences of the 1 hour PM_{10} one hour criteria from Table 4, and
C_n, C_m = factors based on population density
$\qquad$ C_1 = 3 (if population density greater than $1000/1$ km^2 cell)
$\qquad$ C_2 = 2 (if population density greater 100 but less than $1000/1$ km^2 cell)
$\qquad$ C_3 = 1 (if population density less than $100/1$ km^2 cell).
note − 'cell' refers to a modeling domain cell with boundaries locations defined by the model.

As an example of how this formula would be evaluated, consider the following example for the impact of a single source on multiple receptors at a given time. In this example, Fire A impacts one Class I area, two cities which have visibility standards, no airports, and exceeds the health standards in three high population areas, one medium population center and 2 lower population centers for $PM_{2.5}$. Let us further assume that the impact in both the Class I and cities were 10% above BEXT$_{thresh}$. This scenario would then produce the following numbers from equation 3:

$$RV = 1*10 + .1/.05 + 2*5 + .1/.1 + 0 + 3*3 + 1*2 + 2*1 + 0 + 0 + 0 = 36$$

This example suggests the impact only at a single time. To get a total ranking for a source, the RV at every time would need to be evaluated and totaled to come up with a complete ranking for a given source.

RESULTS

The results of the simulated wildfires for the 11 critical watersheds are discussed in this section. As discussed previously, meteorological fields for a five day event in August 1992 were input into the REMSAD air quality model to simulate transport, diffusion and transformation processes for the 11 wildfires which were all 25,000 acres in size. Figure 6.4 shows the maximum 24 hour surface PM_{10} concentrations for the two longest duration (three days) wildfires that were simulated. Both of these wildfires began on August 1, 1992 sometime between 12:00 and 5:00 p.m. based on a random start time as discussed earlier. Figure 6.4 presents impacts from only two of the eleven wildfires that were simulated but serve as good examples of the types of plume impacts to be expected from a wildfire in this region.
As was shown in Figure 6.3, the circulation patterns were quite different depending on which side of the continental divide the fire originated. The wildfire located in southwestern CO around Cortez shows evidence of very

localized impact with the highest concentration areas located near the origin of the fire; in fact, only three of the modeled grid cells were significantly impacted by this wildfire (significant impact is defined as PM_{10} concentrations > 10 $\mu g/m^3$) with the largest value being 71 $\mu g/m^3$. (This was most likely a result of the light winds near the surface in this area which suggested stagnant conditions over this area during this time.) The portion of the wildfire plume which impacted the surface farther to the west had much lower 24 hour values which did not exceed 10 $\mu g/m^3$ with the plume tending to spread towards the west in association with the anticyclonic circulation as a result of the subtropical high centered on the Nevada/Utah border during this time.

The eastern foothills wildfire originating south of Denver was more regional in its impact with a larger area experiencing concentrations above 26 $\mu g/m^3$. Wildfires which were originated east of the continental divide during these times were influenced by the stationary front which was present over the area and produced transport winds which moved plumes from north to south as evidenced by the plume concentration pattern. The maximum PM_{10} concentration for this wildfire was 52 $\mu g/m^{3.}$

Health Evaluation

The purpose of the health evaluation in this study was to prioritize the 11 critical watershed areas in Colorado based on the potential health impact from PM_{10} concentrations generated from a wildfire initialized in each of the 11 critical watersheds. One key element in this scheme involves the concentration estimates from each of 11 wildfires simulated in the modeling that was completed (Table 5). However, in order to get a clear understanding of the health impact to an area it is also important that some background concentration be added to these wildfire impacts in order to get an idea of the health impact into any given area. Not only is it important to understand the air quality impact of a given wildfire, but it is also important to have an understanding of what areas in a state need to be given the highest level of protection. In the formula description section of this paper, the evaluation for health included two main criteria, population density and PM_{10} nonattainment areas.

Figure 6.1 indicated the population density and PM_{10} nonattainment area boundaries for Colorado. This figure shows that for the evaluation of health risk, there is one area in Colorado which weighs heavily in the evaluation of the health formula. This area is known as the front range of the Colorado Rocky mountains and can be generally identified as the area between Colorado Springs and Fort Collins. This area contains the majority of the nonattainment areas and the largest area of high population density (as defined by population density > 1000 km^{-2}).

Once the impacts from the potential wildfires has been determined along

FIGURE 6.3. MM5 surface wind speed and direction on 1 August 1992 at 11:00 p.m. MST. (Full barbs represent speeds of 5 ms^{-1}. Circles represent calm winds. [1 ms^{-1} = 2.2 mph])

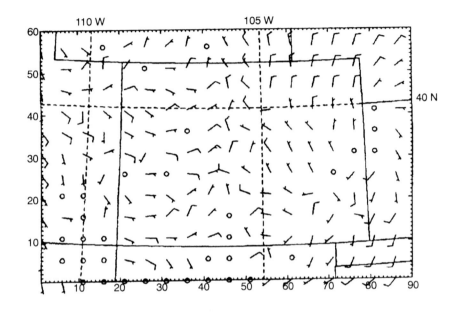

BARB VECTORS: FULL BARB = 5 m s^{-1}

FIGURE 6.4. Maximum 24 hour PM$_{10}$ concentration from two of the eleven simulated wildfires.

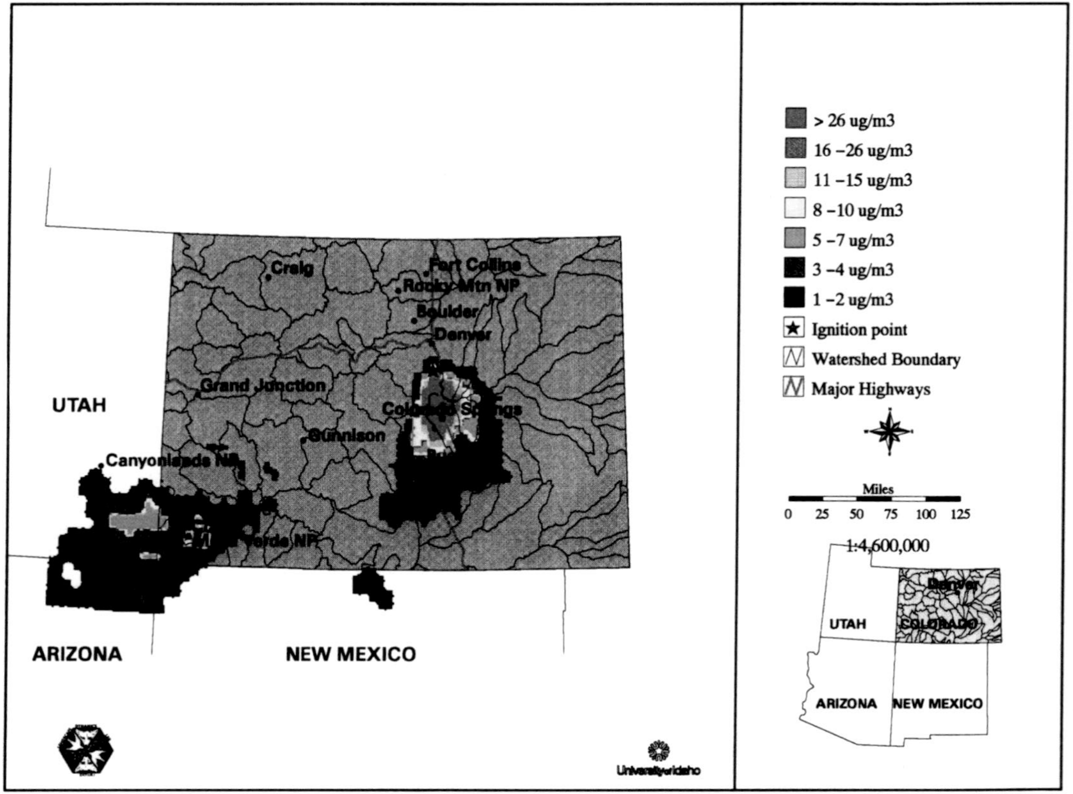

TABLE 5. Emission and Source Characteristics of Modeled Wildfires

Location (°longitude)	Location (°latitude)	Emissions (lbs) PM_{10}	Vegetation Distribution	Vegetation Emission Class
108.63	40.64	1.11E9	96% sage; 4% pinyon	shrub
108.70	40.63	9.60E8	57% sage; 30% pinyon; 13% grass	shrub
109.01	40.57	5.84E8	66% grass; 22% pinyon; 12% sage	grass
108.77	40.53	8.27E8	57% grass; 34% pinyon; 9% sage	grass
108.97	40.25	9.66E8	42% pinyon; 30% sage; 28% grass	shrub
105.09	40.34	2.99E8	79% grass; 18% cottonwood; 3% ponderosa	grass
108.19	40.03	6.81E8	66% grass; 22% pinyon; 12% sage	grass
105.11	40.15	6.56E8	74% grass; 11% cottonwood; 8% mix conifer; 7% other	grass
104.97	39.44	3.75E9	86% ponderosa; 14% grass	ponderosa pine
108.88	37.25	1.41E9	41% grass; 21% sage; 17% pinyon; 17% ponderosa; 4% mix conifer	grass
108.50	37.20	1.88E9	55% grass; 33% ponderosa; 12% sage	ponderosa pine

with the determination of where sensitive receptors are located within a state, it becomes a simple exercise to rank the 11 watershed regions based on the formula discussed earlier. Table 6 shows the actual numbers for each of the five variables in the health equation along with a column which indicates whether a wildfire had an impact to any PM_{10} nonattainment area for the state of Colorado.

As discussed previously, the health evaluation is a two tiered system involving two tests associated with population density and PM_{10} nonattainment criteria. In order to evaluate the health impact, it was determined that all PM_{10} nonattainment areas would be given the highest priority in the evaluation. This was accomplished by guaranteeing that the sum of the columns foar general population, youth, elderly, poor, and wood heating in Table 6 for areas associated with nonattainment impacts (last column, Table 6) would never be below the highest total from a wildfire that didn't impact a nonattainment area. To do this, a constant of 38 was added to the three wildfires which impacted nonattainment areas. (The value 38 was determined by adding up all wildfire numbers from each of the eight wildfires which did not affect a nonattainment area and taking the maximum of these eight and adding 1 to make sure that all nonattainment impacts wildfires would be

TABLE 6. Health values associated with each of the five terms in the health equation mapped to each of the 11 critical wildfires which were simulated. The final column gives an indication of whether a wildfire impacted a PM_{10} nonattainment area with a value of 1 indicating an impact and a value of 0, no impact.

Latitude	Longitude	General	Youth	Elderly	Poor	Wood	NAA
108.62	40.64	.02	18.50	0	0	0	0
108.70	40.63	.01	15.80	0	0	0	0
109.01	40.57	.09	29.78	0	0	0	0
108.77	40.53	.09	33.84	0	0	0	0
108.97	40.25	.44	14.44	0	0	0	0
105.09	40.34	1.84	4.20	5.65	0	0	1
108.19	40.03	0	36.76	0	0	0	0
105.11	40.15	14.92	21.34	21.25	.14	.14	1
104.97	39.44	128.49	4.46	17.10	4.66	4.66	1
108.88	37.25	0	0	0	0	0	0
108.50	37.20	11.09	6.64	0	0	0	0

given the highest priority.) It should be noted that this weighting scheme would need to be recalculated if used at another location, since it was developed specific to this analysis.

Table 6 indicates that for the Colorado wildfire data, the three most crucial elements of the evaluation are nonattainment, general population density and youth population density distributions. It is important to note that initial testing of the impact data on the sensitive areas with population density severed only to indicate that an area would be accounted for in the evaluation of a given wildfire. The actual number in Table 6 represents whether an area experienced impacts that exceeded the concentration values outlined in the criteria tests shown in Table 2. Therefore, while it would be impossible for there to be an impact to a youth population and not the general population in a given area as outlined in this scheme, it is possible that even though both areas indicate an impact that according to the criteria outlined in Table 2, youth population would receive a contribution from this scheme while general population would not. This is indeed the case for this Colorado data set for the wildfire located at (108.19°W/40.03°N) with a relatively strong influence from the youth criteria and no influence from general population.

Figure 6.5 shows the results of the health evaluation for the 11 critical wildfires simulated in this case study analysis. It not surprising to see that the areas that are determined to be the highest potential damaging from a health perspective are those located on the eastern slope of the Rocky Mountains along the front range of Colorado. This area has the most concentrated population centers with most of this area being identified as PM_{10} nonattainment.

The reasons why the most southern of these three areas produced the highest ranking was due to the high fuel loading in this grid cell along with the fact that the wildfire was allowed to burn for a longer duration than either of the other two. (Wildfires lasting only 12 hours initiated in the afternoon would have little surface impact to areas downwind of a wildfire due to the stable conditions during the night which inhibits mixing of the plumes from these short duration wildfires). Fires in the western portion of Colorado did not have much impact when considering health evaluations due to the plume transport towards the West away from sensitive receptors in the State.

Visibility Evaluation

The evaluation of visibility for the 11 wildfires used in this case study were given separate distinction from the health evaluation done in the last section due to the separate issues associated with visibility and health. While the health evaluation dealt primarily with areas associated with high population density, visibility is more concerned with keeping pristine environments free of pollutants which degrade scenic vistas in recreational areas where visibility is an important characteristic (i.e., Class I and II areas). This does not mean to say that there is not some overlap between these two schemes with regards to visibility in cities which have visibility standards or those areas in high population density locations which have major road arteries and/or airports but that the primary focus is different.

Figure 6.2 illustrated the critical areas in Colorado that are defined as sensitive receptors when addressing visibility concerns. As has been mentioned previously, Colorado is unique in that it has developed state standards for visibility along the front range area of Colorado covering most of the cities in this region. This geographical area is similar to the area outlined by general high population in the health section. However, the health and visibility schemes developed for this evaluation treat these two areas very differently with the visibility scheme weighting the influence from this region much lower than Class I regions where the maintenance visibility is a priority. Most of the Class I and II regions located in Colorado are not in the western half of the State but are located in the Rocky Mountains and western slope with the exception of Rocky Mountain National Park located in west of Fort Collins and Boulder in the northern portion of the State.

Table 7 provides the values obtained when the simulated wildfire concentrations from the 11 sources where evaluated using the visibility formula described earlier in this paper. It is obvious that the impact to the visibility formula is captured primarily with the first three terms in the equation. This seems appropriate as there is only one airport that is considered in this analysis (Denver International) and the population density terms were included in this evaluation only to make sure that the primary standards for

PM_{10} were not being ignored in this analysis. Of the first three terms, wild-fires originating in the northeastern portion of the State exhibit a tendency towards impacting Class II areas, specifically Dinosaur National Monument. Wildfires located in the southwestern portion of Colorado have more of a tendency to impact Class I areas due there close proximity to Mesa Verde National Park. The fires on the eastern slope of the Rocky Mountains once again show the strongest correlation with large urban areas and specifically those cities with visibility standards. Figure 6.6 shows the spatial distribution of the visibility rankings for the 11 wildfires which were simulated for this case study. There are significant differences in the map for visibility when compared to the health map. The highest area for health ranking is the southern most critical watershed along the front range of the Rocky mountains. This area for visibility is defined as a "low" priority area. This drop in priority is mostly due to the lack of any Class I or Class II areas to the south of the ignition point indicated by a star in the associated watershed in the Figure. This drop is also due to the origin of the fire being farther south and with the plume movement to the south, the impact to areas within the boundaries of cities with visibility standards is less (see Table 7).

The largest value obtained in the visibility criteria is a value of 195 located in the watershed containing the city of Boulder. While some of the impact is due to direct contact with Rocky Mountain Park located to the west of Boulder, the majority of the impact is from the variable associated with cities with visibility standards. This is also the case with the watershed to the north of this very high region which was only given a medium ranking. The difference between these two adjacent areas can be explained by the higher emissions from the wildfire located around Boulder (see Table 6) and also the origin of the wildfire in the northern most cell being more east then the other producing a plume which does not impact the Rocky Mountain National Park due to the southerly travel of plumes located in this flow regime (see Figure 6.3).

The second highest visibility ranking value is associated with a ponderosa pine burn located close to Mesa Verde National Park. As Table 3 indicates, all of the 177 points associated with this burn can be linked to PM_{10} impacts to Class I areas both in Utah (Canyonlands which was included in this analysis) and Mesa Verde National Park. It is interesting to note that while this southern most fire produced the second highest ranking, the wildfire located about 30 km west suggested no impact to the visibility formula. Once again, the reasons for this difference can be linked to the location of the wildfire in the northern adjacent cell being west of Mesa Verde with a transport wind blowing east to west, a higher emissions for the southern wildfire and the fact that the northern wildfire was completed in 12 hours during stable conditions and the southern wildfire took three days to complete thereby incurring strong mixing of the wildfire plume during the daytime hours. The wildfires located

FIGURE 6.5. Ranking of "critical" watersheds in Colorado using health-based criteria.

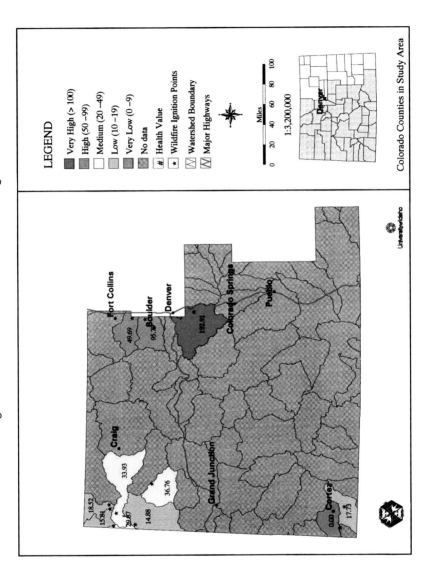

FIGURE 6.6. Ranking of "critical" watersheds in Colorado using visibility-based criteria.

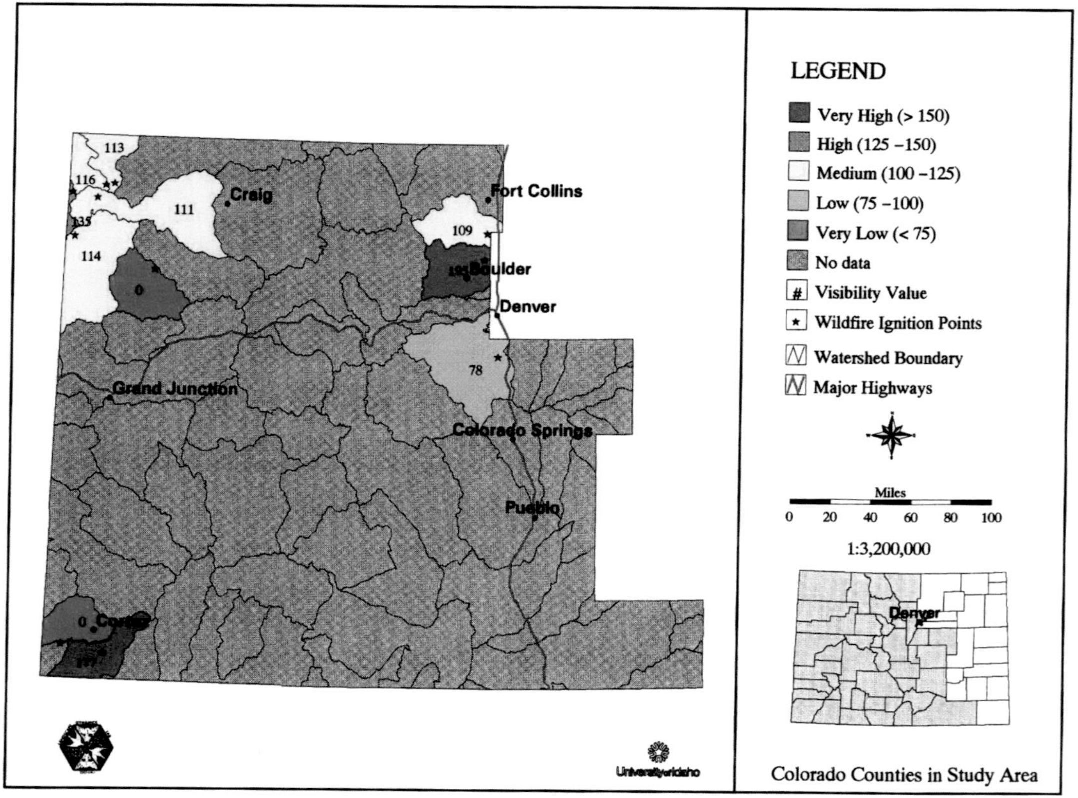

TABLE 7. Visibility values associated with each of the seven terms in the visibility equation.

Latitude	Longitude	Class I	Class II	Cities	Airports	Population (>10,000)	Population (>1000)	Population (<100)
108.62	40.64	11	102	0	0	0	0	0
108.70	40.63	22	94	0	0	0	0	0
109.01	40.57	88	47	0	0	0	0	0
108.77	40.53	0	111	0	0	0	0	0
108.97	40.25	102	12	0	0	0	0	0
105.09	40.34	0	0	109	0	0	0	0
108.19	40.03	0	0	0	0	0	0	0
105.11	40.15	59	0	132	0	0	4	0
104.97	39.44	22	0	56	0	0	0	0
108.88	37.25	0	0	0	0	0	0	0
108.50	37.20	177	0	0	0	0	0	0

in the northwestern portion of Colorado were all grassland/shrub fires lasting 12 hours and impacting both Class I and Class II areas due to the southwesterly movement of the plumes from these wildfires. Figure 6.6 reflects the overall visibility rating by watershed.

CONCLUSIONS

Air quality is an important consideration during wildfire events, and in determining fire or smoke management or prevention strategies. Air quality visibility and public health values are important for a variety of reasons including public health, aesthetics, safety and regulatory compliance issues. The analysis described in this paper demonstrates that different wildfires will have different air quality impacts. This is a result of factors such as differences in vegetation type, density and conditions resulting in different emissions, variations in climatology and geography effecting air dispersion and transport patterns, and differences in the types and number of impacted health and visibility receptors. Data were selected to reflect health and visibility values which are generally available. GIS analysis can utilize these data in conjunction with wildfire data and air quality modeling results to assess which potential wildfire events may pose the greatest risk to public health and/or visibility air quality values.

The results can be used to both indicate and visually communicate the relative potential risk to air quality values at various locations which are susceptible to wildfire events. By using prescribed fire emission data, rather than wildfire data, and making the necessary adjustments for factors such as seasonality, the same type of GIS analysis could be used to identify values at risk during prescribed burning. Improved availability of quality wildfire data and simplified air modeling methods would make this type of analysis more accessible and improve the quality of the results.

Agencies with GIS capabilities, good fire and meteorological data and air quality modeling capabilities may find this type of analysis a useful tool for identifying areas where wildfires may pose the greatest risk to air quality values. This information can be used in combination with other environmental and societal impacts as a basis for communicating risk, reasoned decision making and prioritizing resources for either prevention or response planning.

REFERENCES

Binkley, D., C. Giardina, and I. Dockersmith. 1997. *Status of Air Quality and Related Values in Class I National Parks and Monuments of the Colorado Plateau.* Denver. 10-8, 11-12.

Blett, T., L. Dobson, M. Edwards, K. Foster, D. Hacklow, and K. Wolff. 1993. Manageing Air Resources In the Rocky Mountain Region (6.2-6.3). Denver: U.S. Department of Agriculture, Forest Service, Region 2.

Clean Air Act, 42 U.S.C.A., para. 7491 (a)(1).

Case, Pamela, Brian Banks, Eric Butler, and Ronald Gosnell. 2000. Assessing Potential Wildfire Effects on People. In Sampson, R.N., R.D. Atkinson, J.W. Lewis (Eds.), *Mapping Wildfire Hazards and Risks.* Papers from the American Forests scientific workshop, September 29-October 5, 1996, Pingree Park, CO. The Haworth Press, Inc. New York.

Copeland, S. 1997. *Standard Visual Range Maps, Draft.* US Forest Service Rocky Mountain Forest and Range Experiment Station work product. Fort Collins. Median and 90th Percentile Maps.

Dockery, D.W., C.A. Pope III, X. Xu, J. Spengler, J.H. Ware, M. Fa, B. Ferris, and F.F. Speizer. 1993. An association between air pollution and mortality in six U. S. cities. *New England Journal of Medicine* 329:1753-1759.

Hardy, C.C., R.E. Burgan, and R.D. Ottmar. 2000. A database for spatial assessment of fire characteristics, fuel profiles, and PN10 emissions. In Sampson, R.N., R.D. Atkinson, J.W. Lewis (Eds.), *Mapping Wildfire Hazards and Risks.* Papers from the American Forests scientific workshop, September 29-October 5, 1996, Pingree Park, CO. The Haworth Press, Inc. New York.

Maniero, T. 1997. *Summary Statistics for Rocky Mountain National Park, 3/1/93-2/29/96.* National Park Service Air Resources Division Table. Summer Season Table.

NCWD. 1994. *Report of the National Commission on Wildfire Disasters*, R. Neil Sampson, Chair. Washington, DC: American Forests, April 1994. 29 pp.

Neuenschwander, Leon F., James P. Menakis, Melanie Miller, R. Neil Sampson, Colin Hardy, Bob Averill and Roy Mask. 2000. Indexing Colorado Watersheds to Risk of Wildfire. In Sampson, R.N., R.D. Atkinson, J.W. Lewis (Eds.), *Mapping Wildfire Hazards and Risks*. Papers from the American Forests scientific workshop, September 29-October 5, 1996, Pingree Park, CO. The Haworth Press, Inc., New York.

Pope III, C., M. Thun, M. Namboodiri, D. Dockery, J. Evans, F. Speizer, and C. Health. 1995. Particulate air pollution as a predictor of mortality in a prospective study of U.S. adults. *American Journal of Respiratory and Critical Care Medicine* 151: 669-674.

Pyne, Stephen J., P. Andreas, and R. Laven. 1996. *Introduction to Wildland Fire*, 2nd Ed. NY: John Wiley and Sons Inc.

Sampson, R. Neil, D. Adams, S. Hamilton, S. Mealey, R. Steele, and D. Van De Graaff. 1994. Assessing Forest Ecosystem Health in the Inland West. American Forests. March-April: 13-16.

Schroeder, M.J. and C.C. Buck. 1970. *Fire weather . . . a guide for application of meteorological information to forest fire control operations*. U.S. Department of Agriculture Forest Service, Agricultural Handbook 360.

Schwartz, J; D.W. Dockery, and L.M. Neas. 1996. Is Daily Mortality Associated Specifically with Fine Particles? J. Air Waste Manage. Assoc.: accepted.

Scire J.S. and V.R. Tino. 1996. Modeling of wildfire and prescribed burn scenarios in the Columbia River Basin. Final report. USDA Forest Service report no. 1459-01. March 1996.

Seaman, N.L. and R.A. Anthes. 1981. A mesoscale semi-implicit numerical model. Quart. *J. Roy. Meteor. Soc.* 107, 167-190.

Stocker, R; and Core, J. 2000. Methodology for determining wildfire and prescribed fire air quality impacts on areas in the western United States. In Sampson, R.N., R.D. Atkinson, J.W. Lewis (Eds.), *Mapping Wildfire Hazards and Risks*. Papers from the American Forests scientific workshop, September 29-October 5, 1996, Pingree Park, CO. The Haworth Press, Inc., New York.

Systems Application International. 1996. *Users guide for the Regulatory Modeling System for Aerosols and Deposition (REMSAD)*. SYSAPP-96/42. September, 1996.

U.S. Environmental Protection Agency. 1996a. Air Quality Criteria for Particulate Matter. Research Triangle Park, NC: National Center for Environmental Assessment. Office of Research and Development. April 12, 1996.

U.S. Environmental Protection Agency. 1996b. Review of the National Ambient Air Quality Standards for Particulate Matter: Policy Assessment of Scientific and Technical Information–OAQPS Staff Paper. Office of Air Quality Planning and Standards., Office of Air and Radiation. July 1996.

Watson, J.G. and D. Blumenthal. 1996. *Mt. Zirkel Wilderness Area Resonable Attribution Study of Visibility Impairment*. Denver. Volume II: Results of Data Analysis and Modeling, Part 2 of 2- Appendices to Final Report, Appendix E, Figure E.4.

Western States Air Resources Council (WESTAR). 1995. *Wildfire Emergency Action Plan Implementation Guideline*. Unpublished Draft.

Woodruff, T.S., J. Grillo, and K. Schoendorf. 1977. The relationship between selected causes of postneonatal infant mortality and particulate air pollution in the United States. *Environmental Health Perspectives* 105:608-612.

Chapter 7

Assessing Potential Wildfire Effects on People

Pamela Case
Brian Banks
Eric Butler
Ronald Gosnell

SUMMARY. This chapter describes how people, communities and human artifacts can be represented in a spatial information system so that we can anticipate the social effects of wildfire. The chapter describes pertinent social and economic components of a geographic information system for the State of Colorado. It shows how this kind of information system can be used to estimate the physical risk to people and houses, and to draw useful inferences about the nature of secondary, less tangible, and less predictable social effects resulting from wildfire.

Two types of wildfire episodes are employed in this chapter to illustrate this process. One simulation scenario describes an episode in which seven fires in excess of 25,000 acres in size have ignited. The second scenario presents a more realistic episode for the State of Colorado, given the present hazards of it's forests. This episode consists of five large and 28 small wildfires, all assumed to occur in the same time period. For purposes of social analysis, we are interested in exploring

Pamela Case, Brian Banks, and Eric Butler are affiliated with USDA Forest Service, Lakewood, CO 80225. Ronald Gosnell is affiliated with Colorado State Forest Service, Lyons, CO 80540.

[Haworth co-indexing entry note]: "Chapter 7. Assessing Potential Wildfire Effects on People." Case, Pamela et al. Co-published simultaneously in *Journal of Sustainable Forestry* (Food Products Press, an imprint of The Haworth Press, Inc.) Vol. 11, No. 1/2, 2000, pp. 159-175; and: *Mapping Wildfire Hazards and Risks* (ed: R. Neil Sampson, R. Dwight Atkinson, and Joe W. Lewis) Food Products Press, an imprint of The Haworth Press, Inc., 2000, pp. 159-175. Single or multiple copies of this article are available for a fee from The Haworth Document Delivery Service [1-800-342-9678, 9:00 a.m. - 5:00 p.m. (EST). E-mail address: getinfo@haworthpressinc.com].

the cumulative effect of all 33 wildfires, and in noting the differences between the two scenarios. We also discuss the utility of modeling exercises such as this in dealing with issues such as expected future residential development and population growth. *[Article copies available for a fee from The Haworth Document Delivery Service: 1-800-342-9678. E-mail address: <getinfo@haworthpressinc.com> Website: <http://www.haworthpressinc. com>]*

KEYWORDS. Wildfire, population, community

BACKGROUND

, Colorado is experiencing a substantial fifteen-year long buildup of population, mostly in the form of immigration from the cities of Southern California, Texas and rural areas of the Northern Great Plains. Colorado's population is expected to double during the ten years between 1995 and 2005. This transformation of a semi-rural state to an urban state appears to be akin to the change which took place in Southern California between 1940 and 1955. Unlike Southern California, much of Colorado is forested. Many of these forests are at particularly high risk for wildfire. They also are the preferred residential location of many of Colorado's new citizens.

About 88 percent of the current population is concentrated along the eastern face of the Colorado Front Range. A significant portion of the remaining residents are located along the I-70 corridor through the mountains, and the small remainder are scattered in the rural areas of the West Slope, the Eastern Plains, and the San Luis Valley. Figure 7.1 shows the current distribution of population densities within the State as a whole.

Approximately 80 percent of the newcomers to the State are expected to settle in the Colorado Front Range area. Current residential development patterns indicate that most of this buildup will take place in the suburbs between the cities, and in the forested and non-forested areas of the foothills. The remaining 20 percent of the newcomers are now settling, or expected to settle in formerly rural areas of the State, especially on the Western Slope. Most of these newcomers are likely to construct houses, and even some new towns in forested areas.

MODELING FIRES AND THE SOCIAL LANDSCAPE

The construction of fire probability and propagation models for this exercise have been discussed in some detail in Neuenschwander et al. (2000).

Nine large wildfires were modeled in the watersheds characterized by the highest ignition probability (Figure 3.5, Neuenschwander et al. 2000). Two of these large fires were modeled in terms of both fire and smoke impact (Figure 6.2, Rigg et al. 2000). We tested these nine large wildfires, burning in the same day, as one wildfire scenario on which to estimate impacts on the social landscape.

A typical fire event day in Colorado is characterized by many large and small fires. After examining the ten-year fire history, we created a second, reasonably realistic scenario for a single fire day. This scenario contains five large fires and twenty-eight smaller ones. The large fires exceed 25,000 acres in size: the smaller fires range between 263 and 550 acres in size.

Both wildfire ignition scenarios are based on probable locations of fires in Colorado, given a ten-year fire history. The ignition model indicates the probable locations of ignition points, and for each ignition point, a likelihood of ignition.

Given the location of a set of probable fire ignitions, we constructed a "fire propagation and suppression surface" to determine the most likely patterns of fire propagation within one small segment of the State's landscapes (Figure 7.2). The propagation and suppression surface is based the effects vegetative (fuel) patterns, summer precipitation patterns, slope and aspect have in encouraging a fire to spread or decline. For example, fires ignited in an area of heavy timber are more likely to spread into larger areas than fires ignited in grassy areas: the timber provides more fuel than the grass. Fires within Colorado's summer monsoon pattern will tend to burn into larger patterns when they are free of the typical precipitation areas. Fires tend to run further when they move upslope and on drier southern aspects.

The propagation and suppression surface is constructed by compositing the patterns of vegetation, summer precipitation, slope and aspect. The end result forms a two-dimensional "stencil" for wildfire propagation. Just as paint can spread within the cut-out area of a stencil, a fire, once ignited, can "spread" within the "cut-out" and low impedance areas of the propagation and suppression surface.

In this particular "stencil," fire ignitions are assumed to be able to propagate rather freely in the green areas where they have timber for fuel, are free of characteristic summer precipitation, and can run upslope and on southern aspects. Wildfires can move into the light blue areas which have shrubs for fuel, experience variable summer precipitation conditions, and have moderate slopes and lateral aspects, but will tend to be more easily suppressed. Wildfires moving into the yellow areas which have grass, or very small plant forms for fuel, lie within the typical summer monsoon pattern, or their slopes face north and downward would be expected to be even more quickly suppressed.

Not unexpectedly, there is a strong correlation between the probable fire ignition points, and the areas of the fire propagation and suppression surface where fires can readily spread. For the second scenario, we used the probability model to identify 33 likely simultaneous ignition points, and the suppression surface to determine how the fires would propagate. This led to the scenario we used to represent a typical summer fire event day.

The US Bureau of the Census has developed several data structures which can be assembled into a geographic information system useful for describing the layout of people and houses on a landscape. The Bureau publishes a set of geographically referenced lines describing the boundaries of counties, places (any city, town or village), metropolitan areas, census tracts, block groups and blocks, Congressional Districts, and some Native American census areas. These lines, called the TIGER-Line files,[1] are NOT assembled into polygons by the Census Bureau, but a number of software products contain tools or functions for creating polygons from them. The lines are delineated at the 1:100,000 scale.

The Census Bureau also publishes a wide variety of population counts and statistics from the Decennial Census. Data published in the STF3A series of files[2] can be attributed to polygons constructed from the TIGER-Line files.

The USDA Forest Service has constructed a complete demographic geographical information system from these components and other sources of data. The information system is called The Common Social Unit Geographic Information System. Coverage exists for most of the Western United States, including Alaska and Hawaii.

In this system, the smallest statistically reliable unit of human geography is the "block group." Block groups range in size from about four city blocks to several hundred acres, and occasionally to several thousand acres in remotely populated rural areas. The block group delineations follow the lines and contours of terrain very closely in rural, mountainous and riverine areas: they are particularly well-suited to the study of physical phenomena such as wildfire. In more populated areas, the block groups follow road, county and even some railroad lines. The Common Social Unit information system also contains layers of information for communities ("places"), for counties, and in New Mexico and Alaska, for Native American census units.

The Forest Service extracted about 750 variables from the 1990 Decennial Census, re-organized these into five basic tables, and attributed these data to the block group, the community and the county polygons. About 100 demographic variables commonly used by social scientists were extracted from the

---

1. TIGER-Line files, US Bureau of the Census, Department of Commerce, Washington, D.C.

2. STF3A series of files, US Bureau of the Census, Department of Commerce, Washington, D.C.

FIGURE 7.1. General population density in Colorado, 1990 Census.

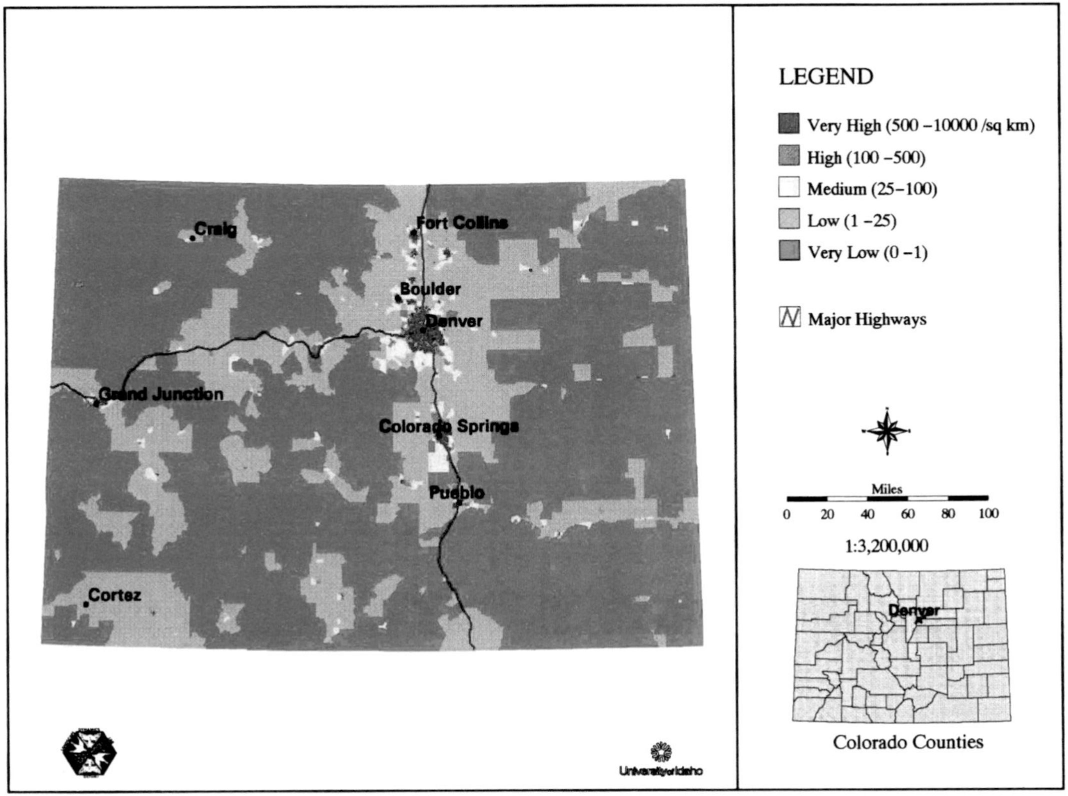

FIGURE 7.2. Fire spread in a high ignition frequency watershed and vegetation types.

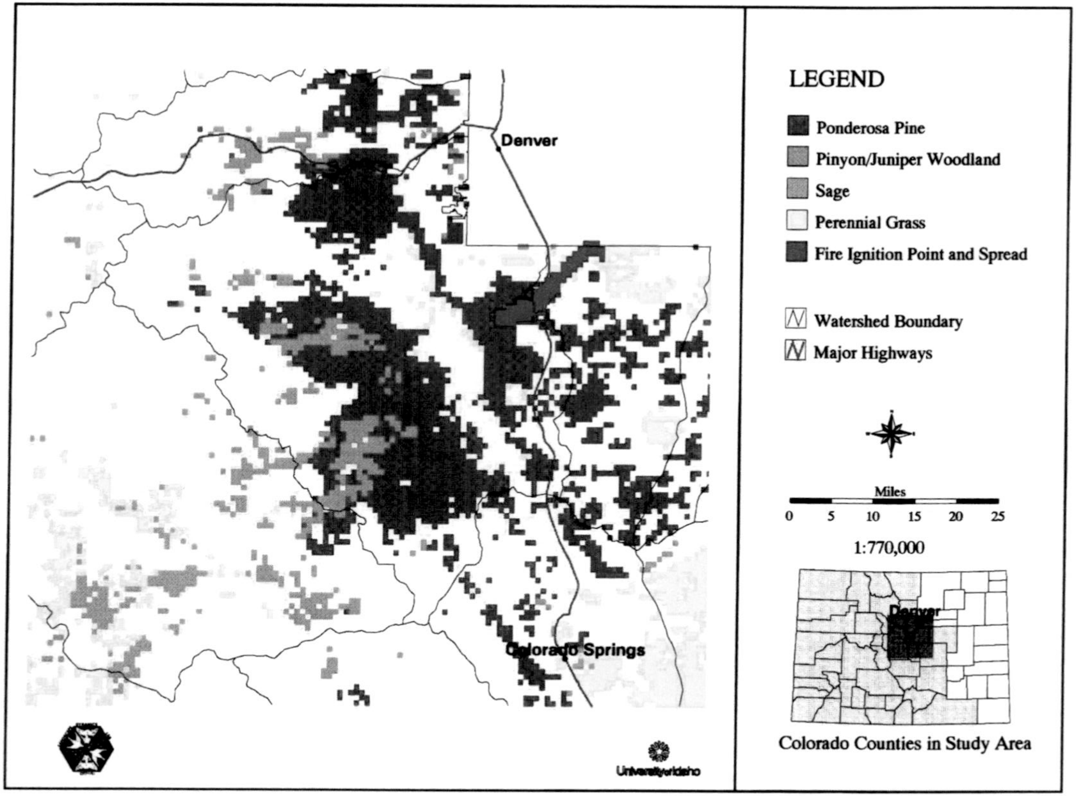

LEGEND

- ■ Ponderosa Pine
- ▨ Pinyon/Juniper Woodland
- ▨ Sage
- ☐ Perennial Grass
- ■ Fire Ignition Point and Spread

- ▨ Watershed Boundary
- ▨ Major Highways

Miles

1:770,000

0 5 10 15 20 25

Colorado Counties in Study Area

Denver

Colorado Springs

164

1980 Decennial Census and attributed to the community and county layers of data. These and other data, such as building permit applications, allow us to trace and predict patterns of residential change.

For our purposes, the first five tables of information provide material useful to this modeling effort. In general, these tables contain the following kinds of information.

Demography Table–Number of persons, number of households, number of families (living in rural and living in urban areas), sex, age, ethnicity (formerly called "race"), education, number of persons per household, native language, country of origin, place of birth if within the US.

Individual Income, Work Table–Employment status, occupation, industry worked within, location of workplace, travel time to work, type or organization working for, income, sources of income, aggregate sources of income, number of workers within the family.

Households Table–Language spoken at home, age of people in the family or in the household, household status, family and household income, family status above and below the poverty line, head of household gender.

Houses Table–Median value of the house, type of structure (trailer, single-family house, condominium, etc.), house value classes, number of housing units, location inside and outside urban areas, number of rooms in the house, presence of plumbing facilities, water sources, type of sewage system, sources of heat.

Residential Development Table–Location of the respondent's resident in 1985, year this structure was constructed, median ages of structures, year respondent moved into this structure.

Current social geographic information systems provide a tremendous amount of useful information for assessments of risk and hazard. But they also are limited to relatively high scales of data aggregation and resolution. At the 1:100,000 scale, these social data sets are not detailed enough to allow us to quantify the complete suite of social effects we know can occur. Additionally, some crucial social effects depend very much on the exact nature of the circumstances surrounding each fire, or upon problematic factors such as the exercise of leadership in emergency conditions. These cannot be adequately described by this data or in any known model. As a result, any analysis of wildfire risk to people is necessarily "strategic," that is, relatively broad in nature. The strengths and limitations of wildfire assessment through spatial data modeling are pointed out in the relevant sections of this analysis.

PHYSICAL EFFECTS OF WILDFIRE

The primary social effects of wildfires are physical: houses and other kinds of human structures burn, people may be injured or die. We can model three of these basic properties of people (*individuals*, the physical structures of *houses*, and the physical structures and facilities of *communities*) relatively easily in the geographic information system.

Houses

Since houses and other fixed human artifacts cannot escape the path of a wildfire as easily as their more mobile builders, we've arbitrarily began our analysis by looking at the risk to housing structures. Besides providing shelter and a place to accumulate material goods and objects of personal or family importance, houses have transferable property value, and sometimes, take on symbolic significance. For example, if people live in relatively close proximity, they often live in neighborhoods organized along economic, cultural or ethnic lines. Therefore, when houses burn, they typically result in a crucial loss of shelter and material possessions, they sometimes result in a loss of material value, and they sometimes lead to a loss or change in economic, cultural or ethnic status or condition.

Some "houses" have characteristics which put them at special risk to wildfire if compensating mechanisms are not employed by the homeowner. For example, many houses in rural areas cannot be provided with water supplies through public water facilities. These draw their water from wells, ponds or other storage devices, typically in amounts sufficient only to the domestic needs of the household. It is much harder to fight fires in these areas because of the lack of large amounts of water. Some houses are heated with fuelwood, and may rely upon fuelwood for cooking and water heating. While these houses are not necessarily at greater risk to wildfire than houses heated with other materials, they may have large amounts of flammable fuelwood stored in close proximity. Other houses are relatively remote, making it harder for fire suppression teams to obtain access to these areas, or they may have difficult ingress and egress routes. Houses with these characteristics can be identified in the geographic information system. Houses with these characteristics can be considered to be special house populations, for purposes of wildfire risk.

Homeowners for this special population may employ compensating mechanisms which reduce the risk of house loss to average or low risk. A home may be constructed of relatively inflammable materials, have "defensible spaces", or have fire-fighting facilities built into them (especially as passive systems such as gravity-fed water tanks and foam retardant systems). The Census Bureau has not collected information about these features of houses in the past, so we are unable to identify houses in this category in this geographic information system. However, such information can be collected rather inexpensively and incorporated into this kind of modeling system.

Houses have different economic value. Some of these values have social significance or provide us with a means of reasoning about potentially important social effects. For example, some houses belong to people of modest income. The property value of the houses may represent a significant amount, or the only form of wealth possessed by such people, or it may be the only possible source of shelter. Loss of shelter to people of modest income can lead to temporary loss of work time, wages, or even to loss of employment. Sometimes, family members are able to remain together only because they have a house, a source of common shelter. If this house is destroyed in a wildfire, the members of the family must separate to obtain alternate shelter for all individuals. The density of low income families, shown in Figure 7.3, can be used as a surrogate indicator of this factor.

A special class of "houses" we are interested in are ranches and farms. These "houses" have additional value as economic production units. If wildfires destroy these structures, families are without housing, but they are also without an immediate means of livelihood. Ranches and farms, of course, are not randomly distributed over the landscape.

In Colorado, housing structures are distributed much as people are distributed, but the forms "houses" take (single-family houses, trailers, multi-family dwellings, group quarters, second and unoccupied homes), their importance to the household, and their economic value are not evenly distributed, and they don't occur randomly over the landscape. Therefore, the location of a wildfire which burns houses will affect different kinds of people in different ways. In the modeling exercise, the first scenario (nine wildfires) results in the following patterns and kinds of effects.

There are limits to the ability to use census data to quantify the effects of wildfire on people and houses. Most wildfires, even fairly large ones, can occur entirely within one unit of social geography. That is, the smallest unit of human geography can be larger than the predicted area of wildfire, particularly within lightly-populated rural areas. Since the Census Bureau delineations do not allow us to locate individual houses within a block group, the best estimate we can make of the social effects of wildfire is to state that a certain number of houses or people of a certain kind may be at risk of destruction by the fire. The more densely populated the area, the more probable the risk.

For Wildfire Scenario 1, we determined that houses and people in 29 areas of the State are likely to be directly affected by the nine fires. The houses at risk in these 29 areas have the characteristics shown in Table 1.

In contrast, the wildfires in Scenario 2 affect 61 areas of the State. Estimated effects of this wildfire pattern are shown in Table 2.

Undoubtedly Scenario 2 involves a greater number of houses than Scenario 1 because there are more fires. But the number of structures involved is lower than might be expected because of the lower overall density of housing

TABLE 1. Houses Likely to Be Affected by the Scenario 1 Wildfire Pattern

House Characteristics	Total Units Potentially Affected	Mean at Any Given Location	Maximum Number at the Highest Concentration Location
Total Structures	13,890	478.96	
Well water supplied houses	2,542	88	266
Other water supply	1,117	38.51	152
Heated with fuelwood	1,615	55.68	173
Farms	583	20.10	55
Very inexpensive houses	85	3	24
Very expensive houses	68	2	40
Approximate dollar amount of the affected houses (sum)	$397,872	$92,742	$292,989

TABLE 2. Houses Likely to Be Affected by the Scenario 2 Wildfire Pattern

House Characteristics	Total Units Potentially Affected	Mean at Any Given Location	Maximum Number at the Highest Concentration Location
Total Structures	29,961	491.16	1,034
Well water supplied houses	12,963	213	902
Other water supply	3,039	49.81	292
Heated with fuelwood	4,351	71.32	261
Farms	872	14.29	55
Very inexpensive houses	115	2	24
Very expensive houses	80	1	34
Approximate dollar value of the affected houses (sum)	$535,801	$87,844	$500,001

structures in rural areas. Unlike Scenario 1, no large fire occurs in close proximity to an urban area. As might be expected, more houses in the special population are at risk. Interestingly, more "very expensive" houses also are at risk in the second, larger fire–more rural scenario.

INDIVIDUALS, FAMILIES, HOUSEHOLDS AND PEOPLE LIVING IN GROUP QUARTERS

The second step of our example analysis, focuses on that portion of *the total population* at risk of injury and death due to wildfire. However, it should be noted that death and injury depend upon unique sets of circumstances surrounding each fire event: these cannot be modeled here. In a geographic information system, we can only describe the total number of people at potential risk, and the number of people having special characteristics which place them in jeopardy.

Some individuals belong to special populations because they have characteristics which lead them to suffer effects disproportionate to the general population. Special populations include families with very young children, the elderly, and households whose members have incomes below the poverty line (Figure 7.3). Risks to these populations come in the form of greater difficulty of successful evacuation, greater susceptibility to health impairment, loss of possessions (which could be irreplaceable for people in two of these special groups), and loss of jobs and income (the likelihood that some people may be employed in relatively marginal jobs where it is impossible to "take time off" to deal with the aftermath of a fire without losing the job).

These special populations are not evenly distributed on the landscape of Colorado, nor are they even distributed within the general populace. Therefore, the location of wildfires can have more or less effect on special sub-populations than it does on the population as a whole. In one sense, the purpose of modeling exercises such as these is to help us evaluate risks to groups of people who might be disproportionately disadvantaged by wildfires. Tables 3 and 4 describe the population groups placed in physical risk by the two wildfire scenarios.

More people are at risk in the second, more numerous fire scenario. As with houses, lower population density in the rural areas implies that the number of people at risk are not as great as might otherwise be expected, given the number of fires. But the Scenario 2 population appears to have disproportionately larger numbers of children and elderly people, and larger

TABLE 3. Population Groups at Risk as a Result of Scenario 1

Populations at Risk	Total People	Mean at Any Given Location	Maximum Number at the Highest Populated Location
Total Population	32,819		
Children	7,912	273	796
Elderly	2,657	92	28
People with incomes below the poverty line in 1989	1,879	65	183

TABLE 4. Population Groups at Risk as a Result of Scenario 2

Populations at Risk	Total People	Mean at Any Given Location	Maximum Number at the Highest Populated Location
Total Population	40,638		
Children	8,177	134	396
Elderly	4,073	67	282
People with incomes below the poverty line in 1989	3,292	54	183

numbers of relatively poor people than the population affected by the fires of Scenario 1.

COMMUNITIES

Communities are defined here as any city, town or village. These have physical structures which can burn in a wildfire. These are buildings, transportation and communication facilities (roads, airports, railroad lines, telephone lines and switch stations), water storage and processing plants, power plants, grocery stores, etc. Any of these features can be identified in advance, in a modeling exercise such as this one. When collecting this information, it is important to remember that not all physical structures belonging to communities are necessarily associated with the community location. (A good example of a disassociated structure is a municipal watershed.)

The physical impact to a given community depends on a set of unique circumstances surrounding each fire event. We cannot model those circumstances, but we can provide representative examples of the kinds of facilities at risk.

Scenario 1 directly threatens the City of Longmont, and the Parker suburb of the Denver metropolitan area. The towns of Cortez, Naturita, Glenwood Springs and Canon City are indirectly affected by the large wildfires.

Scenario 2 does not directly threaten any city or town, although lightly populated high elevation areas of the Colorado Front Range are affected. The towns of Cortez, Rifle, Canon City and Cripple Creek, Jefferson, Shawnee, Bailey and Pine, Lyons, Longmont and Berthod are indirectly affected by the fires.

SECONDARY EFFECTS

Secondary social effects are things associated with people and communities which are not as tangible as individual persons, house structures or city facilities. Social scientists can anticipate and describe several suites of probable effects on people, beginning with effects on perceptions, values and attitudes of individual people and families. The social geographic information system used here allows us to draw some inferences about the nature of these probable effects although the nature of these inferences is necessarily limited. The following discussion illustrates some of these possibilities.

Economic Institutions (Industries)

People tend to live in neighborhoods adjacent to their place of work. It is possible for wildfires to have an impact on some industries simply because a large fire can destroy the homes of a significant fraction of an industry's workforce. For example, a fire near the town of Golden could affect the Coors plant, a fire in Waterton Canyon could affect the aerospace industry in the Denver metropolitan area, and a fire near Boulder could affect the University of Colorado and the federal agencies concentrated there. These industries account for a sizable fraction of the economy of the Denver metropolitan area.

The Census data allows us to identify people by occupation and industry, and therefore to spot unusual concentrations of people working in a single industry. In Scenario 1, there are no unusual concentrations of people working in a particular industry except for the people located near or within the Parker suburb of the Denver metropolitan area. The demographic information recorded by the Census Bureau indicates that about 12 percent of the population are employed in the aerospace industry.

It also is possible to consider potential effects of wildfire on people according to their economic circumstances. (Information which might be useful here is provided by portions of the Census dealing with aggregate income by source.) For example, the wildfire pattern of Scenario 2 primarily affects rural areas. In 1989, about 38 percent of ranchers and farmers suffered a net loss in farm income. About 40 percent of the ranchers and farmers in these areas (those who've already suffered a financial loss as a result of a series of bad crop years) would be affected by Scenario 2 wildfires, somewhat disproportionately to other ranchers and farmers in the State.

Emergency Response Institutions

Communities have a variety of emergency response organizations and institutions intended to protect their citizens against catastrophic events. An important question, given the status of Colorado's forests is "Are there potential cumulative wildfire conditions which could tax the ability of emergency response organizations?" This kind of modeling exercise can help us construct answers to questions like this, or help us identify potential weaknesses in our emergency response configuration relative to the pattern of hazard existing in the State's forests.

Social Service Institutions

Individuals and households directly impacted by a wildfire through death, injury, loss of possessions, job loss or loss of housing often turn to social service institutions for help in recovery. Each community, county, and state provides an array of social services to help people in need, as does the federal government. We can use the modeling capability displayed here to look at the potential cumulative impact on social service organizations, or even upon land stewardship organizations, just as we can use this to consider the possibility of overtaxing emergency response organizations.

SOCIAL RISK ASSESSMENT

To this point, we have discussed the primary ways in which wildfires impact houses and people, and those factors which place some kinds of houses and people in greater risk than other factors, all things being equal. In brief, houses which are dependent upon domestic water supplies and fuelwood, which lie in remote areas, and which have difficult routes of ingress and egress are at greater risk of damage or destruction in the event of wildfire. People who are very young or very old are more likely to experience adverse health consequences in the event of wildfire, and people who have very low incomes are more likely to suffer irreversible economic and social consequences. When these groups of people suffer from wildfires, they have less individual means of recovery, and require more help from the communities they live within.

These observations suggest that a strategic approach to minimizing some of the social impacts of wildfire might focus on concentrations of houses at higher risk and, within those housing areas, households composed of people likely to suffer greater individual and social consequences if they suffer a wildfire. These terms can be defined in several possible ways, of course. Here, we are able to express them as houses more likely to be damaged or destroyed, and people whose age and economic condition makes them more susceptible to the ill effects of wildfire.

Technically, we could approach the problem of strategic risk assessment in three sequential steps. First, we could identify the locations of houses dependent on domestic water supplies and fuelwood, and/or which are in relatively remote locations and have difficult ingress and egress routes. Second, within this special population (for which we can develop a geographic information system "coverage"), we can identify the general locations of households which have higher concentrations of young children and older people, or which have lower income levels. Third, we then could construct a density map of these special populations, reasoning that proportionately greater risk occurs where people and houses with these particular characteristics occur in greater numbers. In effect, we are stating that agencies and institutions responsible for wildfire management might reduce the net social impact of wildfire in forested residential areas by focusing their efforts on these kinds of areas. Figures 7.1 through 7.3 illustrate this method. Figure 7.4 illustrates a "net effect" derived from these 6 indices.

Other social scientists may suggest other types of variables for consider-

FIGURE 7.3. Concentration of low-income people in Colorado, 1990 Census.

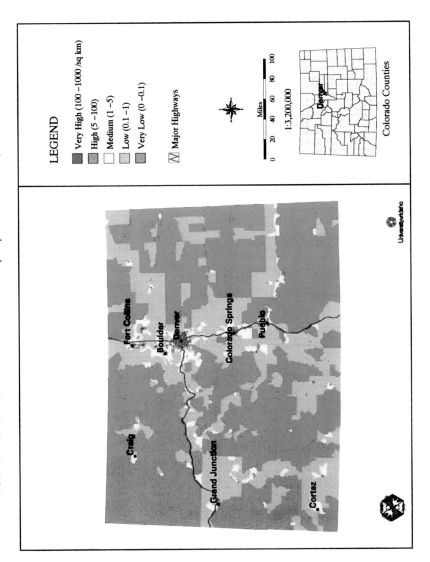

FIGURE 7.4. Composite wildfire risk to human populations, Colorado.

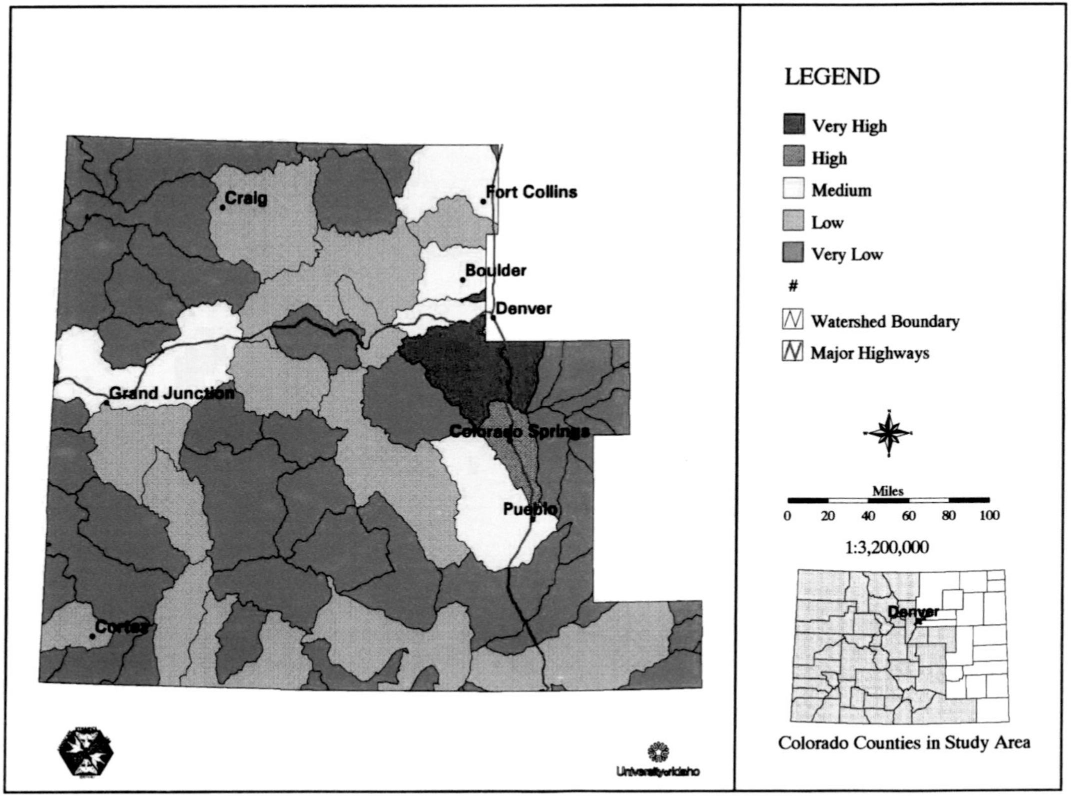

ation. We do not wish to suggest that the approach taken here exhausts the subject, nor that houses and people with the special characteristics described here have greater social or intrinsic value than houses and people with other characteristics. We are simply suggesting that the risk of social impacts due to wildfire can be reduced by reasoning about the way in which people experience the physical effects of wildfire, and the way in which these impacts are transmitted into communities or other aspects of larger social systems. We also suggest that we have some ability to model this line of reasoning. Given that this is so, it is possible for agencies and institutions responsible for wildfire management to identify the locations of populations at greatest risk, assess the magnitude of the risk "problem," and then take a strategic approach to minimizing social risk.

Known Wildfire Hazards in Areas of Expected Population Growth

The most important contribution this kind of modeling capability can make to wildfire assessment may lie in helping us to deal with the future. Past census data collections, in combination with population projections models, land ownership patterns, building permit records, and analysis of water, transportation and communication systems allow us to make reasonably accurate predictions about where and how people will build residences on the land during the next ten years. We can compare these geographical projections with our knowledge of wildfire risk and hazard to assess probable fire hazards in future growth areas. If we can anticipate the nature of probable wildfire risk and hazards in these areas, we have the possibility of deciding how we wish to deal with these risks before they are upon us.

REFERENCES

Neuenschwander, Leon F., James P. Menakis, Melanie Miller, R. Neil Sampson, Colin Hardy, Bob Averill and Roy Mask. 2000. Indexing Colorado Watersheds to Risk of Wildfire. In Sampson, R.N., R.D. Atkinson, J.W. Lewis (Eds.), *Mapping Wildfire Hazards and Risks*. Papers from the American Forests scientific workshop, September 29-October 5, 1996, Pingree Park, CO. The Haworth Press, Inc., New York.

Rigg, Helen Getz, Roger Stocker, Coleen Campbell, Bruce Polkowsky, Tracey Woodruff and Pete Lahm. 2000. A Screening Method for Identifying Potential Air Quality Risks from Extreme Wildfires. In Sampson, R.N., R.D. Atkinson, J.W. Lewis (Eds.), *Mapping Wildfire Hazards and Risks*. Papers from the American Forests scientific workshop, September 29-October 5, 1996, Pingree Park, CO. The Haworth Press, Inc., New York.

Chapter 8

Assessing the Impacts of Severe Fire on Forest Ecosystem Recovery

Kermit Cromack, Jr.
Johanna D. Landsberg
Richard L. Everett
Ronald Zeleny
Christian P. Giardina
Eva K. Strand
Tom D. Anderson
Robert Averill
Rose Smyrski

SUMMARY. The potential for severe impacts from a wildland fire, with or without a subsequent severe precipitation event, was evaluated

Kermit Cromak, Jr., is affiliated with Department of Forest Science, Oregon State University, Corvallis, OR 97331. Johanna D. Landsberg and Tom D. Anderson are both affiliated with USDA Forest Service, Forestry Sciences Laboratory, Wenatchee, WA 98801. Richard L. Everett (Retired) was affiliated with USDA Forest Service, Forestry Sciences Laboratory, Wenatchee, WA 98801. Ronald Zeleny (Retired) was affiliated with Colorado State Forest Service, Fort Collins, CO 80523. Christian P. Giardina is affiliated with Department of Forest Sciences, Colorado State University, Fort Collins, CO 80523. Eva K. Strand is affiliated with Landscape Dynamics Laboratory, Fish and Wildlife Cooperative Unit, University of Idaho, Moscow, ID 83844. Robert Averill is affiliated with USDA Forest Service, Lake Wood, CO 80225. Rose Smyrski is affiliated with Forest Products Laboratory, USDA Forest Service, Madison, WI 53705.

[Haworth co-indexing entry note]: "Chapter 8. Assessing the Impacts of Severe Fire on Forest Ecosystem Recovery." Cromack, Kermit, Jr. et al. Co-published simultaneously in *Journal of Sustainable Forestry* (Food Products Press, an imprint of The Haworth Press, Inc.) Vol. 11, No. 1/2, 2000, pp. 177-228; and: *Mapping Wildfire Hazards and Risks* (ed: R. Neil Sampson, R. Dwight Atkinson, and Joe W. Lewis) Food Products Press, an imprint of The Haworth Press, Inc., 2000, pp. 177-228. Single or multiple copies of this article are available for a fee from The Haworth Document Delivery Service [1-800-342-9678, 9:00 a.m. - 5:00 p.m. (EST). E-mail address: getinfo@haworthpressinc.com].

in western and central Colorado and in a case-study area, the former Rio Grande National Forest (NF) of the San Juan–Rio Grande NF. The evaluation involved identifying the factors that are conducive to vulnerability and, additionally, the factors that impede or enhance recovery. Forest ecosystem characteristics that can increase vulnerability to the effects of a severe fire, with or without a subsequent severe precipitation event, include: relative flammability of forest floor materials, the location of nutrient storage (above or below ground), depth of soil, site quality, steepness of slopes, propensity to produce hydrophobic soils, and likelihood of mass movement, among others. Ponderosa pine forests appear more vulnerable, with natural regeneration more difficult than for some other types because of the relatively higher flammability of the foliar material in the forest floor, storage of nutrients above ground, and the episodic and infrequent years for good natural regeneration in drier areas.

In western and central Colorado, only two subbasins–the subbasin in which Boulder is located and the subbasin south of Denver–had both a high fire-ignition frequency and a high hydrophobicity composite risk index placing them at risk for both fire and erosional overland flow. Construction of a vegetation map showing the forest types most likely to be damaged by a severe wildfire and erosion, particularly on steep slopes, was done using GIS.

The forest landscape structure and composition existing today appear as the result of many stochastic factors operating over decades to centuries. Not all of these historic conditions and processes can be duplicated today. Potential losses of organic matter and nutrients from severe fire and subsequent intense precipitation events can be substantial. *[Article copies available for a fee from The Haworth Document Delivery Service: 1-800-342-9678. E-mail address: <getinfo@haworthpressinc.com> Website: <http://www.haworthpressinc.com>]*

KEYWORDS. Fire, wildfire, nutrients, erosion, soil, GIS, forest ecosystems, thunderstorm

INTRODUCTION

Recent severe fires and subsequent intense precipitation events in western and central Colorado have triggered concern over the likelihood of similar events in the future and their potential impacts on the ecosystems affected. We assume that a severe fire followed by a severe precipitation event will happen again in central and western Colorado and ask these questions: Where is it likely to occur? What makes one area more vulnerable than another? What types of ecosystem impacts will occur? And, which types of areas will take longer to recover? We explore the use of a geographic information system (GIS) to evaluate risk and potential effects of a severe fire in a case-study area–the former Rio Grande National Forest (NF) portion of the San Juan–Rio Grande NF and for western and central Colorado as a whole. We review the primary effects of severe fire on site nutrient capital and vegetation and the effects of a subsequent severe precipitation event.

Fire Severity

We refer to fire severity as a relative, descriptive, qualitative term indicating the conditions found after a fire. Fire severity may be low or light, medium or moderate, high or severe, or anywhere in-between. A light or low-severity fire usually produces a mosaic, removing the forest floor in some locations, damaging some understory vegetation, scorching some trees, and perhaps torching-out a few. A medium or moderate-severity fire may produce complete forest floor consumption in some areas, no consumption in others, and limited consumption elsewhere. It scorches most of the understory vegetation and may partially consume some shrubs; many tree crowns are scorched and a few are consumed.

A severe or high-severity fire consumes the forest floor either by flaming or glowing combustion and leaves mineral soil exposed over most of the site; understory vegetation is consumed, leaving skeletons; and the overstory is reduced to standing, charred remnants–a ghost of the former stand. Because the quantitative terms–intensity, fire intensity, and fire-line intensity–describe heat release during the fire and, accordingly, require measurements of flame length, or flame length and rate of spread, we are using the qualitative terms. Fire severity can be related to specific effects on forest ecosystems, such as amount of fuels consumed, nutrients lost, soil hydrophobicity, or tree mortality (DeBano et al. 1998, Brown and DeByle 1987).

Forest fire intensity and severity have been researched and discussed from a number of different aspects, including, for example, effects on: forest regeneration and secondary succession of vegetation; changes in soil nutrient availability and ecosystem nutrient losses; soil hydrophobicity and soil erosion; and impact on vegetative succession (Raison et al. 1985, Agee 1993, DeBano et al. 1998). Although wildfires can vary greatly in initial intensity, it is the total consumption of forest floor and woody fuels which delineate fire severity and subsequent tree mortality (Brown and DeByle 1987, Ryan and Reinhardt 1988). One effect of severe fire is to heat the soil, with resultant effects on losses of soil organic matter and nutrients, together with an increased risk of soil erosion (Belillas and Feller 1998, DeBano et al. 1998). Where severe wildfires also crown out, there is the least total cover remaining for soil surface protection. Erosion can be severe, particularly where soils are subject to formation of a hydrophobic surface, and there are one or more subsequent intense precipitation events, or other hydrologic events, such as rapid melting of the accumulated snowpack, on a wildfire site before substantial recovery of vegetation occurs (DeBano 1981, De Bano et al. 1998). Wildfires can affect a number of forest ecosystem components, not always negatively (Landsberg 1997, DeBano et al. 1998). However, there are increased risks to normal rates of ecosystem recovery from increasingly severe wildfires, which are being attributed, in part, to past fire suppression, attendant fuel source accumulation (Neuenschwander et al. 2000), and increased risk of watershed level erosion events (MacDonald et al. 2000). Fire severity also affects resprouting of shrubs and the stored seed-bank (Morgan and Neuenschwander 1988); fire intensity and severity influence the degree of soil heating and the effects on soil biotic components (Agee 1993, Valette et al. 1994).

Changes in Ecosystems and Fire Potential

Fire managers have reported an apparent increase in area burned, fire numbers, and fire severity in recent years. The Interior Columbia Basin Ecosystem Management Project reported a general upward trend in fuel loading, potential fire behavior, crown fire potential, and smoke production during the past 30 to 60 years (Ottmar et al. 1998). They found that fuel loading is trending upward and may be the most important factor causing the increase in wildfire impacts across the Interior Columbia Basin. They relate this upward trend to natural succession without fire disturbance, i.e., wilderness areas with a policy of fire exclusion, and non-wilderness areas with a policy of fire exclusion and no alternative fuels management program.

The fire-free, or fire-return, interval of an area is generally determined by measuring the years between consecutive fire scars on trees within that area. Trees with many fire scars are normally chosen to produce a representative measure of the fire-free interval. Such trees often have a pith dating from several hundred years ago, often during the Little Ice Age, which ended about 1870. Accordingly, many of the scar-producing fires occurred during that cooler period. Fire scars dating from the 17th to 19th centuries may underestimate the current potential fire frequency. Present climatic conditions may be expected to produce more fires and more severe fires than what was recorded on old fire-scarred trees. Additionally, fire-free intervals have increased in many forested types of the intermountain west because of a very successful fire suppression program. Without frequent fire, fuel loads have increased, setting the stage for hotter, more severe fires (Waring and Running 1998).

Dramatic changes in fire frequency, fire size, and fire type began in the 18th and 19th centuries with the settlement of these areas by people of European origin. In northern Arizona, prior to European settlement, fires occurred in ponderosa pine forests at frequent intervals, perhaps every 2 to 12 years (Weaver 1951, Cooper 1960, Dieterich 1980); fire scars suggest the average burn covered about 3,000 ac (Swetnam and Dieterich 1985, Swetnam 1990); and that prior to the 1950s, crown fires were extremely rare or nonexistent (Cooper 1960). Together, these characteristics indicate more frequent, less severe, and larger fires than occur now (Covington et al. 1994). When northern Arizona was settled, people of European origin introduced roads, agriculture, and intense grazing–changes which led to reductions in fire size and frequency (Covington and Moore 1994). When people of European origin settled central and western Colorado, they too made similar introductions which very likely produced similar results: reductions in fire size and frequency. In the San Luis Valley of Colorado, near the Rio Grande NF, Spanish settlers introduced sheep grazing in the early 1700s.[1] Over decades, grazing–primarily by sheep but also by larger livestock–reduced the understory fuels which, in turn, decreased the potential for flashy, fast fires. Analysis of fire scars from the Chuska Mountains of Arizona reveals that the early regime of light, frequent fire experienced a sudden and persistent drop in frequency around 1830, when grazing animals caused a reduction in fine fuels (Savage 1991). Later, the introduction of weedy species, such as cheat grass (*Bromus tectorum*), created a new source of flashy fuels. Eventually, settlers established fire-suppression forces to protect people, private property, and natural resources, with major effects on grazing patterns (Frank et al. 1998). This combination of factors has resulted in a longer interval between fires in several important forested types found in Colorado and the intermountain west (Table 1).

During this era of settlement by people of European origin, ecosystem composition and structure were slowly being altered by an unseen process. The Little Ice Age ended about 1870, after which the warmer climate both enhanced and inhibited recruitment, albeit on different sites. Jakubos and Romme (1993) found bands of small lodgepole pine trees (*Pinus contorta* var *latifolia*) that are consistently younger than adjacent forest stands of obvious fire origin. These lodgepole pine bands originate at the edges of dry meadows, and their presence is attributed to the warmer and wetter growing seasons since the end of the Little Ice Age. This change in climatic conditions allows lodgepole pine to move into what formerly were meadow areas.

Conversely, on xeric sites, where ponderosa pine once grew at the edge of its range, the warmer climate since the 1870s may have contributed to a lack of reforestation. Sites on the eastern edge of the Deschutes NF, near Bend, Oregon, once supported scattered huge ponderosa pine trees; nonetheless, repeated planting efforts have failed to establish a new stand since the previous stand was harvested in the 1920s.[2] Further evidence for changes in climate that affect potential recovery after a disturbance event comes from Yellowstone, where the examination of a large amount of lake sediments for

TABLE 1. Changes in fire-free interval from late 19th Century or early 20th Century to the 20th Century in forests with species compositions similar to those of south central Colorado; alphabetical by reference.

Species	Geographic location	Historic mean fire-free interval, years	Current mean fire-free interval, years	Reference
P. ponderosa	Bitterroot National Forest (NF), MT valley edge	6 to 20[1] before 1910 50- to 100-acre basis	fire scars very infrequent after 1910	Arno and Petersen 1983
Pseudotsuga menziesii	montane slope	12 to 20[2]		
Abies grandis	moist canyon	18		
P. menziesii– A. lasiocarpa	lower subalpine slopes	22 to 34[3]		
A. lasiocarpa– P. albicaulis	upper subalpine slopes	57 to 61		
P. ponderosa– P. menziesii	Lolo NF, MT "Lolo-1" ~5,000' elev. on 45-50% south-facing slopes, dry site	32 before 1900[4]	two fires since 1900[4]	Arno et al. 1995
	"Lolo-2" same as Lolo-1	31 before 1900[4]	one fire since 1900[4]	
	"Lolo-3" same as Lolo-1	26 before 1900[4]	no fires since 1897[4]	
	Bitterroot NF, MT "Bitterroot-1" ~5,400-5900' elev., south-facing slopes, dry site	47 before 1900[4]	no fires since 1889[4]	
	Flathead NF, MT "Flathead-1" ~3,600' elev., <8% slopes, moist site	31 before 1900[4]	no fires since 1850[1]	
P. ponderosa– Larix occidentalis	Lolo NF, MT "Lolo-4" moist site	27 before 1900[4]	one fire in 97 years since 1990[4]	Arno et al. 1997
L. occidentalis	"Lolo-5" very moist site	24 before 1900[4]	no fires since 1859	
Larix occidentalis P. contorta	North Fork Flathead Valley, Glacier National Park, MT	5 before 1935	50 after 1935	Barrett et al. 1991
	McDonald Creek-Agpar Mountains, and Middle Fork Flathead area, Glacier National Park, MT	20 to 30+ before 1935	no change from historic	

Species	Geographic location	Historic mean fire-free interval, years	Current mean fire-free interval, years	Reference
P. contorta Dougl. var. *latifolia*	Absaroka Mountains, Yellowstone National Park, WY	stand replacement fires, 200	no change due to long fire-return interval	Barrett 1994
P. albicaulis Engelm.	same as above	stand replacement fires, >350	"virtual absence of such fires. . . after 1880."	
P. menziesii	same as above, lower elevation	non-lethal underburns		
P. ponderosa var. *scopolorum*– *P. menziesii*	Colorado Front Range: Fourmile Canyon, Boulder County, CO	31.8 on a 1500-ac basis	28 on a 1500-ac basis	Goldblum and Veblen 1992
Same as above	Wintersteen Park, Roosevelt NF, CO (53 mi. north of Fourmile Canyon)	66 on a 125-ac basis	27.3 on a 125-ac basis	Laven et al. 1980
P. contorta	Rocky Mountain National Park, CO (25 mi. northwest of Fourmile Canyon)	19.8	6.2	Skinner and Laven 1983
Various pines: *P. lieophylla* (Chihuahua pine) *P. engelmannii* (Apache pine) *P. ponderosa* *P. strobiformus* (southwestern white pine)	17 sites in the Madrean Province Borderlands: AZ, NM; Sonora and Chihuahua, Mexico	surface fires: maximum interval: 8 to 25 years from 1700 to 1900 (no surface area basis given); minimum interval: one year	no fires since 1900 except for two sites with fires in 1972 (no surface area basis given)	Swetnam and Baisan 1996

[1]Range of grand means from three study areas.
[2]Grand mean of one study area.
[3]Range of grand means from two study areas.
[4]On a 2.2 acre basis.

charcoal suggests that fire frequency has varied on millennial time scales in accordance with changes in the intensity of summer drought (Whitlock 1998). Fire intervals of the last 14,000 years have ranged from 50 to 500 years, depending on the climate and vegetation of particular periods (Millspaugh and Whitlock 1995, Whitlock and Millspaugh 1996). This long temporal view shows the close tie between fire and climate; as climate changes, so, too does the importance of fire. Whitlock points out that the current set of climatic conditions is unique on both centennial and millennial time scales. "We cannot reconstruct the forests of the Little Ice Age, Medieval Warm Period [500 to 1000 years ago], or the early Holocene, because the climate and disturbance regime are now different than then" (Whitlock 1998). Accordingly, ecosystem structure and composition which now exists on a site may not be obtainable in the future if the site is altered by a severe disturbance. Changes produced by a severe fire alone or by a severe fire and subsequent sudden, severe thunderstorm event(s) together with substantial erosion may lead to markedly increased time necessary for recovery of previous ecosystem productivity and biodiversity.

An ecosystem must have the climatological conditions necessary to recover from a severe fire and precipitation event. Dendroclimatological analysis has shown that natural regeneration and establishment of ponderosa pine in the southwestern United States is historically episodic and infrequent (Savage et al. 1996). During the 20th century, only two years have met the criteria established for germination and establishment of ponderosa pine: 1919 and 1992. This suggests that ponderosa pine need to germinate early enough to be robust in the ensuing winter frosts and the next spring's drought. "Ponderosa pine requires a warm and wet last few weeks of May, together with an above-average, well-distributed water supply throughout the year," (Savage et al. 1996). With only two such years in the southwestern US during the preponderance of the 20th century, the likelihood of having a good establishment year immediately after a severe disturbance event is problematic. The more moist conditions in Colorado would promote increased recruitment of ponderosa pine and other species, such as Douglas-fir, especially on sites where fire has been excluded. Future forest regeneration may be affected by changes in climate, or by the physiological effects of increased atmospheric CO_2 (Waring and Running 1998).

These factors–the introduction of practices by people of European origin that either increased or decreased fuels, or the effects of the Little Ice Age that either increased or decreased recruitment or stymied reforestation–have altered the vegetation and fuel components from their pre-1870s conditions. Fire-free intervals have increased in many forest types of the intermountain west (Table 1) since the introduction of large-scale fire suppression. Without frequent fires, fuel loads have increased, setting the stage for hotter, more severe fires (Mutch 1994). The acknowledged increase in area burned, numbers of fires, and fire severity, apparently due to increasing fuel loads resulting from successful fire suppression, auger for a more severe fire scenario in the future–a scenario which may make ecosystem recovery more problematic.

While the potential for severe fire may be increasing, an increased focus concurrently has been developing on the importance of inherent disturbance, including fire, in managing forest ecosystems (Attiwill 1994) and maintaining forest productivity (Kimmins 1996). Fire is necessary for maintaining some species on a site, and is the most rapid means of nutrient turnover. In degraded landscapes, reestablishment of forests and soil fertility, together with soil animal and microbial functions, is essential (Mao et al. 1992, Perry 1994, Haselwandter and Bowen 1996, Perry and Amaranthus 1997, Amaranthus 1998).

Recovery Potential

To determine if a site is at high risk for losing its ability to recover after a severe fire, we analyzed the effects of severe fire on site nutrients. Nutrient capital on a site is partitioned among the above-ground vegetation; snags and downed wood; forest floor; root systems; mineral soil; soil organic matter and its associated fungi and bacteria; and the animal species affiliated with each ecosystem component (Waring and Running 1998). In grassland ecosystems, a larger amount of nutrient capital is found below ground. In xeric, forested ecosystems, a greater proportion of nutrient capital is found below ground compared to what is found in more mesic, forested sites (Keyes and Grier 1981), where the greatest amount of nutrient capital is found in the foliage and forest floor strata.

The nutrient capital stored above ground and in the forest floor is at a greater risk for loss from a severe fire–loss through volatilization, ash redistribution, or transport off site (Raison et al. 1985, DeBano et al. 1998). On nutrient-poor sites, loss of the above ground nutrient capital produces a greater loss proportionately than when above-ground nutrient capital is consumed on a nutrient-rich site. The amounts of nutrients stored above ground in woodland and forest settings are greater than in grass and shrub communities (Waring and Running 1998).

In absolute amounts, the nutrients in the tree and forest floor biomass of spruce-fir forests exceed that of all other types listed: spruce-fir > ponderosa pine > lodgepole pine=aspen > pinyon-juniper > oak steppe > sagebrush/grass.[3] Spruce-fir forests have evolved under a stand-replacement fire regime with a greater proportion of site nutrient capital in the soil than in above ground biomass; this may indicate a lower vulnerability to the effects of a severe fire for this forest type than for others. In other vegetation types, including ponderosa pine and pinyon-juniper invasion sites, or oak steppe, fire-caused losses may be higher because of the larger amounts of nutrients stored above ground. Nutrient content in forest floor material can be substantial in these forest vegetation types (Table 2).

The tree species predominating in an area contribute significantly to the burning characteristics of the forest floor material that influence forest floor consumption, and consequently, nutrient release. These burning characteristics include: flame height, flame time, ember time, total burn time, percent combusted, and mean rate of weight loss (Fonda et al. 1998). Of the conifer species evaluated that grow in Colorado, ponderosa pine has the most flammable forest floor needle material. Its needles ranked in the top half of all of the burn characteristics. Ponderosa pine is predominant in communities that had a fire-return interval of two to four decades. This short fire-return interval presented repeated opportunities for nutrient loss from the frequent fires (Fonda et al. 1998). Nutrient losses from fire may be greater than subsequent

TABLE 2. Forest floor biomass and nutrient capital in western forest ecosystem forest floors.

Forest Type	Location	Biomass	N	P	K	S	Ca	References
		(tons ac[-1])	<----------- (lbs ac[-1]) ----------->					
Ponderosa pine	Arizona	14.1	412	56	133	–	496	Covington & Sackett 1984
Ponderosa pine	New Mexico	11.2	170	13	71	33	380	Wollum 1973
Ponderosa pine	New Mexico	14.4	279	21	251	34	195	Wollum 1973
Ponderosa pine	New Mexico	12.3	223	13	126	28	124	Wollum 1973
Ponderosa pine	Colorado	12.0[1]	271[1]	26[1]	145[1]	32[1]	298[1]	This paper
Lodgepole pine	Colorado	9.8-15.6	295	23	54	–	116	Moir & Grier 1969
Lodgepole pine	Wyoming	7.7	192	13	19	–	159	Fahey 1983
Lodgepole pine	Wyoming	15.1	414	25	33	–	269	Fahey 1983
White fir	New Mexico	36.0	788	76	209	154	2380	Wollum 1973
Subalpine fir/	Colorado	30.3	683	48	70	–	456	Arthur & Fahey 1990
Engelman spruce	Alberta	49.1	764	45	–	–	–	Prescott et al. 1992
Douglas-fir	Washington	16.6	596	55	245	–	–	Woolridge 1970
Douglas-fir	New Mexico	17.0	239	25	301	52	714	Wollum 1973
Douglas-fir	New Mexico	19.8	254	24	207	48	718	Wollum 1973
Utah juniper	Arizona	6.4[2]	146[2]	39[2]	58[2]	–	893[2]	DeBano et al. 1987
Pinyon/Juniper	New Mexico	4.2	71	11	19	16	193	Wollum 1973
Singleleaf pinyon	Nevada	26.8	910	39	–	50	–	Everett & Thran 1992
Aspen	Alaska	12.6	299	34	47	–	271	van Cleve & Noonan 1971
Aspen	Alaska	33.9	1007	79	118	–	827	van Cleve & Noonan 1971
Aspen	Minnesota	12.1	596	53	70	–	962	Perala & Alban 1982
Oak/Shrub	Utah	16.6	582	43	128	41	1042	Tiedemann & Clary 1996
Oak/Shrub	Arizona	15.1	228	31	–	–	–	DeBano 1990
Sagebrush/Grass	Idaho	8.0[3]	116[3]	–	–	–	–	Tiedemann 1987a

[1]Nutrient data for Buffalo Creek Fire area, CO, are means of nutrient data from Covington and Sackett (1984) and Wollum (1973). Forest floor biomass data are from Table 7 for Buffalo Creek.
[2]Data are for area around base of sampled trees and within crown area.
[3]Data are for total aboveground biomass and N.

leaching losses, particularly for N in western forest ecosystems (Belillas and Feller 1998, Johnson et al. 1998).

The other conifers evaluated and found commonly in Colorado–lodgepole pine, Douglas-fir, and subalpine fir–seldom rank in the upper half of any burn category (Fonda et al. 1998). The fire-return intervals for the communities in which these species grow are commonly two or more centuries, indicating the potential for greater amounts of nutrient-rich material to accumulate in soil organic matter and the forest floor and then be lost from the site in a stand-replacing fire. Forest community types with fire-return intervals in the decades, or those with intervals in the centuries, are both at risk for serious nutrient loss. The short-interval communities are at risk because nutrient-containing fuels have accumulated since the beginning of effective fire suppression, while the long-interval communities are at risk due to the historic absence of fire and the stand-replacement fire regimes natural to these communities.

Ponderosa pine, with its high flammability rating, will be vulnerable to fire wherever ignition is a possibility. With the increase of ladder fuel in ponderosa pine stands because of the absence of decadal fire, the possibility of a crown fire, and its concomitant loss of nutrients and structure, in this historic surface-fire type is now greater.

Nutrient demand of post-fire plant communities may be reflected in the rapidity with which new stands are developed. Aspen (*Populus tremuloides*) stands and oak shrub communities regenerate rapidly from suckers, with the aspen stands generating the greater biomass (Bartos and Johnston 1978, Schier et al. 1985, Tiedemann et al. 1987, DeBano et al. 1998). Lodgepole pine can have exceedingly high numbers of seedlings that are known to stagnate because of nutrient deficiency (Baumgartner et al. 1985, Smith et al. 1997). Post-burn communities dominated by grasses and shrubs, or by young pinyon-juniper woodlands, produce less biomass and have an opportunity to recover site nutrients prior to maximum nutrient demand by tree species (Tiedemann 1987a, Klopatek 1987).

A severe fire is characterized by combustion of large amounts of organic material and substantial nutrient losses (Grier 1975, Raison 1979). Temperatures are high, and dwell times may be long, or both. Organic material is combusted from above ground vegetation components, forest floor, and soil organic material. Heat flux is sufficient to combust below ground organic material consisting of soil organic matter, fine and coarse roots, and below ground stump components (Agee 1993). Soil hydrophobicity may result from such fires (Everett et al. 1995, DeBano et al. 1998).

On the other hand, a light burn is characterized by combustion of grasses, herbs and small amounts of shrubs and small diameter live or dead-and-down woody fuels, although flare-ups into the overstory will occur occasionally (Landsberg 1992). Fire temperatures are generally low to moderate, with a minimum heat flux into the soil. In a light burn, with only a portion of the above ground organic material combusted, the potential for nutrient loss both immediately, and subsequently, because of erosion, is reduced.

High-fire severity implies greater biomass and nutrient mass volatilized from the site. Nitrogen begins to volatilize at > 200°C, S at > 375°C, P and K at > 774°C (DeBano et al. 1998). Prescribed fires in ponderosa pine second-growth stands produced soil temperatures sometimes reaching 200°C (Shea 1993) but not always (Landsberg 1992). Soil temperatures >700°C are reached in stump holes, under piles of built-up fuel, and under downed boles (DeBano et al. 1998). Such high temperatures not only combust organic matter and nutrients, but also destroy soil aggregate structure (Giovannini et al. 1988, DeBano et al. 1998).

Because of their low volatilization temperatures, N and S, particularly, can be lost as the result of wildfire (Raison 1979, Raison et al. 1985). In a central Oregon ponderosa pine stand, prescribed underburns removed from 232 lb (Landsberg 1992) to 364 lb total N ac^{-1} (Shea 1993) from the forest floor and the top 2.4 in. of mineral soil. Grier (1975) reported losses of 760 lb N ac^{-1} from an intense wildfire in a mixed conifer stand in eastern Washington; an estimated 71 lb N ac^{-1} came from surface mineral soil. Some effects of prescribed underburning on litter N mineralization and soil N availability in ponderosa pine stands endured for 12 years (Monleon and Cromack 1996, Monleon et al. 1997). These stands would probably have lost even more nutrient capital if they had been consumed by a wildfire. In contrast, post fire leaching losses, particularly for N, are generally modest in comparison to nutrient losses by volatilization (Belillas and Feller 1998, Johnson et al. 1998). Nutrients such as P are more vulnerable to losses during intense fires (Giardina and Rhoades 2000), while K is prone to leaching losses from some soils (Jordan 1985).

Everett and Thran (1992) estimated that if more than one-third of the forest floor is lost in burning post-harvest slash in singleleaf pinyon (*Pinus monophylla*) woodland sites, a negative site N balance could occur, assuming a 100 yr rotation, and then current estimates of N replacement rates. Prescribed fires impact tree growth in ponderosa pine (Landsberg et al. 1984, Sutherland et al. 1991, Landsberg 1994); N losses from fire may contribute to some losses in ponderosa pine tree growth (Monleon et al. 1997). Virtually complete consumption of the forest floor, in addition to consumption of nutrient-rich foliage, occurred in the intense wildfire studied by Grier (1975). It seems clear that wildfires can potentially degrade sites. Interestingly, even residual fire products, such as charcoal, influence forest succession and soil ecosystem processes (Moore 1996, Zackrisson et al. 1996).

Maintenance of soil organic matter enhances colonization of soil micro-habitats by symbiotic ectomycorrhizal fungi in western coniferous forest ecosystems (Harvey et al. 1981, 1986, 1987). Furthermore, maintenance of soil organic nutrient reserves, such as N and S, is critical to those western forest ecosystems having N-based limits to forest productivity (Harvey et al. 1987).

Severe wildfires can also affect residual nutrient pools through convection of ash particles and through localized redistribution of ash layers, especially on steeper slopes (Grier 1975, Raison 1979, McNabb and Cromack 1990). If hydrophobic soil surface layers are formed following wildfires, there is an increased risk of soil-surface erosion (Giovannini and Lucchesi 1983, Giovannini et al. 1988, McNabb and Swanson 1990, Rab 1996, DeBano et al. 1998), especially during intense rainstorms on the unprotected soil prior to adequate revegetation.[4] The common practice of reseeding wildfire sites with nonnative grasses, such as the annual ryegrass species (*Lolium multiflorum*), may not provide adequate plant cover in sufficient time to reduce soil erosion during fall and winter rains (Amaranthus et al. 1993). Annual ryegrass also may retard native plant regeneration and compete for water with newly established tree seedlings (Amaranthus et al. 1993). Additionally, when forested sites are seeded with ryegrass, interspecific competition reduces the regeneration of N-fixing shrubs such as *Ceanothus integerrimus*. This could lead to the failure of a site to regain some of the N which was removed, such as when wildfires occurred in Oregon and California during the summer and fall of 1987 (Amaranthus et al. 1993).

Soil moisture can influence nodulation of N-fixing *Ceanothus* species (Pratt et al. 1997); thus the competition by exotic species for soil moisture (Amaranthus et al. 1993) may retard the normal ecosystem functions of shrubs such as *Ceanothus*. Early successional plants which can fix N and are stimulated to germinate by fire, such as *Ceanothus velutinus*, can add substantial quantities of both N and soil organic C to forest ecosystems (Youngberg and Wollum 1976, Binkley et al. 1982, Johnson 1995, Johnson et al. 1998), potentially replacing N losses due to fire. New experimental field research using N_2 fixing plants in agricultural systems has shown that C and N retention is greater with legumes used to supply soil N, than with conventional cropping using mineral N fertilizer (Drinkwater et al. 1998). Long-term comparisons of experimental plots in ponderosa pine forest ecosystems have shown that N-fixing understory species such as *Purshia tridentata* and *C. velutinus* can increase soil fertility, including soil C and N, and enhance tree growth (Busse et al. 1996). Experimental research with prescribed underburning suggests that both *C. velutinus* and *P. tridentata* can be retained in the understory, through resprouting, or from the existing seed-bank (Ruha et al. 1996). A legume, *Robinia neomexicana*, adds N and C to soil in ponderosa pine forests in Arizona (Klemmedson 1994). In addition, *P. tridentata*, *Ceanothus fendleri*, *C. velutinus*, and *Cercocarpus montanus* are important for wildlife habitat and food resources (Cromack et al. 1979, Conard et al. 1985, Paschke 1997).

The combined effects of severe wildfire and soil erosion may retard reco-lonization of a site with native vegetation and inhibit succession. In addition to increased losses of nutrients, there are potential impacts on seed-banks for plants (Morgan and Neuenschwander 1988) and on mycorrhizal spore diver-sity (Read 1998, van der Heijden et al. 1998) due to soil heating. Just as a more complete complement of native plant species in ecosystems may influ-ence above ground productivity (Tilman et al. 1996), a larger number of mycorrhizal species may be important for normal rates of secondary succes-sion to occur, and for maintenance of ecosystem productivity (van der Heij-den et al. 1998). Ecosystem stability may be enhanced by retention of ecosys-tem species diversity (McCann et al. 1998, Polis 1998). The interlinking of below ground species diversity and ecosystem functions affecting ecosystem productivity (Amaranthus and Perry 1994, Perry and Amaranthus 1997, Read 1998), together with field evidence to demonstrate how increasingly severe wildfires could retard recovery of sites (Amaranthus et al. 1993), is essential if we are to understand how global changes in climate might affect future forest productivity and carbon sequestration by forests (Schimel 1998, Cao and Woodward 1998). How severe fire affects soil nutrients and residual soil organic matter, and long-term replacement of soil organic matter will also be important to consider in future fire research (Almendros et al. 1990, Knicker et al. 1996, Sollins et al. 1996, DeBano et al. 1998).

EFFECTS OF SEVERE FIRE AND PRECIPITATION EVENTS ON THE RIO GRANDE PORTION OF THE SAN JUAN–RIO GRANDE NATIONAL FOREST

One objective of this paper is to explore the use of a GIS to evaluate risk and identify potential effects of a severe fire on a chosen area. The use of GIS in forest analysis has increased greatly in the last few years. In 1994, GIS was employed in developing recovery plans after the 1994 Tyee and Hatchery Complex fires on the Wenatchee NF.[5] Analysis using GIS has been used in several very large ecosystem projects, including the Eastside Assessment (Everett 1993, Everett et al. 1993) and the Interior Columbia Basin Ecosys-tem Management Project. Ecosystem studies in other areas include an evalua-tion of potential soil erosion in a 2815 ac watershed of the Coweeta Experi-mental Forest in North Carolina (McNulty and Swank 1996). A soils map was developed using GIS data for the Libby Creek drainage of the Snowy Mountains in the Medicine Bow NF of Wyoming by Rahman et al. (1997). The map was validated using field transect data. The GIS-created map was shown to be as or more representative of the large area it covered than maps generated using standard processes.

Case-Study Area

In this paper, we apply GIS data before a severe fire event to evaluate the potential outcome in a case-study area–the former Rio Grande NF portion of the San Juan–Rio Grande NF in south central Colorado. This area is bounded on the west by the Continental Divide (Figure 8.1). We refer to this area as the case-study area or the former Rio Grande NF, interchangeably. We evaluate the available GIS data layers to determine the relative risk of fire, the resource values at risk, and to produce information of potential value to land managers, especially fuels and fire managers. Land and fire management planning involves determining the objectives for a given area, defining the resource values at risk, and assessing the risk. GIS data provide information on the risk and on resources at risk which managers can use to set objectives and priorities. A fundamental concept of ecosystem studies is the interrelation of ecosystem components and processes. This is the case here–an outcome affecting one component has implications for many other components and processes of the system. Disturbance processes rarely act alone (Rogers 1996).

The GIS layers for western and central Colorado–a 44,460,000 acre area–were provided by Colorado State University and for the case-study area by the San Juan–Rio Grande NF. The data for western and central Colorado were derived from AVHRR satellite imagery (Loveland et al. 1991) and have a resolution of 1 km^2. Data for the former Rio Grande NF were from the Rocky Mountain Resource Information Survey and were developed from 1:24,000 aerial photos.

The georeferenced data for western and central Colorado and the former Rio Grande NF are shown at the fourth-field, or eight-digit, subbasin level. As indicated from the hydrologic unit codes (HUCs) from the U.S. Geological Survey (Seaber et al. 1987), the subbasins of the former Rio Grande NF are very similar (Table 3). The case-study area contains nearly all of one subbasin, the Headwaters of the Rio Grande River (Table 3), and portions of five additional subbasins (Figure 8.2). Four of the five additional subbasins contain considerable land that is flat and outside the National Forest boundaries (Figure 8.1). The main forest vegetation types in the case-study area are Engelmann spruce-subalpine fir, aspen, and mixed conifer dominated by Douglas-fir (Table 4).

Risk of Ignition

The risk of ignition by subbasin in western and central Colorado (Figure 3.4, Neuenschwander et al. 2000) was used as the starting point for evaluating the

TABLE 3. The case-study area–the former Rio Grande NF portion of the San Juan-Rio Grande NF–includes almost all of the HUC, Headwaters of the Rio Grande River; and parts of the other HUCs. Ignition frequencies are for the entire HUC, not only the part which is within the case-study area.

Hydrologic Unit Codes (HUCs)	Subbasin Names	Ignition frequencies for the HUC, ignitions 10 K ac^{-1} in 10 years, 1986-1995	Hydrophobicity composite risk index	Erosion composite risk index
13010001	Headwaters of the Rio Grande River, referred to as Headwaters	0.9	125.1	15.0
13010002	Lower Rio Grande River within Colorado, referred to as Lower Rio Grande	2.4	93.7	3.4
13010003	San Luis Creek and (lower part) Closed Basin, referred to as San Luis Creek	1.7	80.8	2.1
13010004	Saguache Creek	1.1	65.6	2.9
13010005	Conejos River	1.2	125.4	12.4
13020102	Chama River	0.8	206.5	11.2

TABLE 4. Timber harvesting since the 1980s has created a young vegetation structure on 27 percent of the respective forest types on the former Rio Grande NF.

Forest type	Forest area (ac)	Forest area (%)	Area harvested (ac)	Area harvested (%)
Englemann spruce-subalpine fir	564,071	30.5	84,071	~15
Aspen	265,115	14.1	unknown	–
Mixed conifer dominated by Douglas-fir	199,112	10.7	24,692	~12

comparative risk within the subbasins of the case-study area. The ignition frequency for each subbasin is an average over the entire subbasin and reflects all fire activity in that area. The data by subbasin showed ignition frequencies ranging from 0.8 to 2.4 ignitions per 10 K ac^{-1} in the ten years of record (Table 3). These were categorized by Neuenschwander et al. (2000) as "very low (<1)" to "low (1 to 5)," and are similar to the ignition frequencies for most of western and central Colorado (Figure 3.4, Neuenschwander et al. 2000). The highest ignition frequency for a subbasin with some land in

FIGURE 8.1. Satellite view of Colorado showing subbasin boundaries in black and the case study area outlined in white.

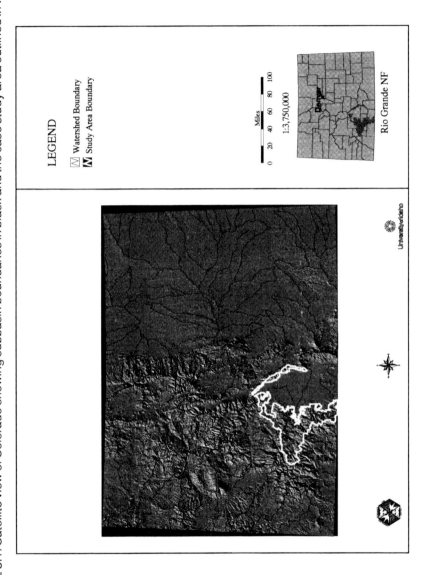

LEGEND

Watershed Boundary
Study Area Boundary

Miles

0   20   40   60   80   100

1:3,750,000

Rio Grande NF

FIGURE 8.2. The former Rio Grande National Forest portion of the San Juan-Rio Grande NF, outlined in black, with subbasins and hydrologic unit code numbers indicated.

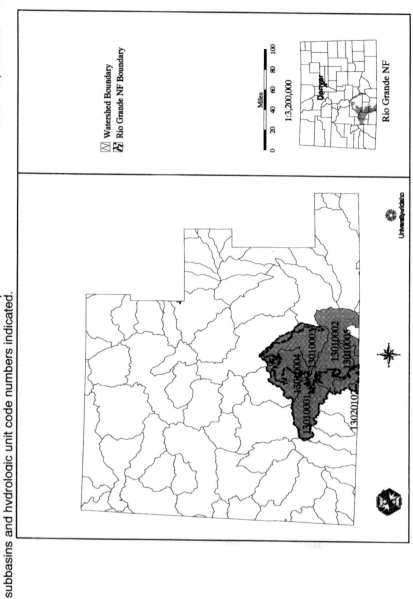

the case-study area was in the Lower-Rio Grande-within-Colorado subbasin, hereafter called the Lower Rio Grande, which had 2.4 ignitions per 10 K ac^{-1} in ten years. This subbasin contains mountainous terrain including Bennett (13,200 ft.) and Silver (12,420 ft.) Mountains, although much of the subbasin, which is outside the case-study area, is flat (Figure 8.3). The San Luis Creek subbasin, on the west-facing slope of the Sangre de Cristo Mountains, has the next highest ignition frequency, 1.7 ignitions per 10K ac^{-1} in ten years. When comparing the ignition frequencies with the corresponding ignition point locations in the same subbasin (Figure 8.3), it was apparent the majority of fire starts were in grasslands and outside the case-study area, but still within the subbasins that comprise the case-study area.

The ignition location points in the former Rio Grande NF (Figure 8.3) are scattered and do not appear to follow ridge lines. With the scattered locations of the ignition point data it is not possible to identify species at risk in the case-study area. Further, from the data, it appears that the former Rio Grande NF is not a high ignition frequency area. Data obtained from the Rio Grande NF validate this conclusion; they indicate the case-study area had only 19 fire starts per year from 1980 to 1990 with an average fire size of 0.5 to 1.0 ac.[6]

In Utah, on the Wasatch–Cache NF, Stratton (1998) used a similar process to evaluate the risk of fire occurrence. The GIS layers for the Logan Ranger District and adjacent state, county, and private lands included topography: slope, aspect, and elevation; infrastructure; vegetation; climate: precipitation, solar radiation, and lightning density; "sensitive" natural values; and fire history: source of ignition (human or lightning); and fires >10 acres. Using multiple logistic regression, the best predictor of fires for all causes combined was the proximity to trails and roads, indicating a large proportion of the fires in the area examined were human caused (Stratton 1999).

Resources at Risk

Soil resources–If a severe fire were to occur on the former Rio Grande NF, many resources would be at risk, including soil resources and the flora, fauna and habitats that are supported by these resources. After a severe fire which, by definition, removes most of the soils' protective mantle of forest floor material, the exposed, nutrient-rich soil horizons are vulnerable to transport off-site by water, gravity, and wind (McNabb and Swanson 1990). Soil resources lost because of a severe fire will include the above- and below-ground organic material consumed in the fire and the nutrients it contains. If a severe precipitation event follows a severe fire, off-site transport is increased. In addition to fluvial processes, on steep slopes dry raveling can occur, removing additional resources from the site (McNabb and Swanson 1990). On sites with shallow soils, a greater percentage of site nutrients can be lost than when soil is removed from a nutrient-rich site.

Erosion risk–The risk of soil erosion is based on the assumption that the surface cover of vegetation or leaf litter has been disturbed or destroyed, exposing bare surface soils to the elements of erosion (USDA 1996a, 1996b). In the case-study area, the risk of soil erosion after a disturbance event is classified as moderate to high on 94 percent of the land area (Table 5). This is risk beyond the natural, or on-going background erosion on the Rio Grande NF. The erosion risk is greater on steeper slopes and on soils with inherent natural erodibility (Figure 8.4). The increases in erosion and runoff are attributed to several processes (MacDonald et al. 1999). With nearly a third of the Rio Grande NF in the high erosion risk category, a severe fire could produce costly and irreparable soil erosion with the upper soil layers transported off site and mixed with less nutrient-rich horizons. The only large, contiguous area with low erosion risk is in the rock formations of the Sangre de Cristo Mountains at the eastern edge of the forest, which is, interestingly, adjacent to areas with steep slopes and very high erosion risk.

Serious erosion losses after fire have been quantified at other locations, including the Buffalo Creek Fire in Colorado. This 1996 fire burned about 11,800 ac on the Pike NF about 30 miles south of Denver in HUC 10190002 (Agnew et al. 1997). Of this area, about 8,000 ac burned as a high-severity crown fire and the remainder as an underburn (Table 6). Basal area of the stands at Buffalo Creek suggest that the forest in the crown fire area would have had a crown cover approaching 50%, based on Mitchell and Popovich's (1997) work, and assuming that the Douglas-fir component had a substantial effect on crown cover (Table 6). This is the type of wildfire predicted to occur in a ponderosa pine forest having a substantial understory of small trees and fuel accumulation due to fire exclusion (Wright and Bailey 1982). Less than two months after the fire, a severe thunderstorm dumped 2.5 in. of rain on the burned area, causing severe erosion losses. Those areas burned by the crown fire had much greater erosion losses than the underburned areas. In the crown-fire area, erosion losses averaged 1.4 in. of topsoil (Table 7), with substantial estimated nutrient and organic matter losses (Table 8). Organic matter and nutrient losses are further compounded by potential losses from tree crowns, understory vegetation and from soil erosion (Table 8); losses in other forest types would depend upon nutrient accumulation in the forest floor (Table 2) and other components, such as vegetation, coarse woody debris, and mineral soils, if erosion were to occur. In the underburned area, only one of the seven plots showed significant erosion, with a mean loss for all plots, of 0.04 in. of topsoil. A layer of ash and litter from 0.02 to 0.8 in. remained on the underburned area after the fire and thunderstorm. The soils in this area are characterized by Sphinx series, sandy-skeletal, mixed, frigid, shallow Typic Ustorthents. Slopes are 15 to 80 percent with approximately 20 percent of the area as exposed bedrock. "The lithology, bedrock physical

FIGURE 8.3. Fire ignitions in grasslands outside the case study area contributed substantially to the fire-ignition frequency it experienced.

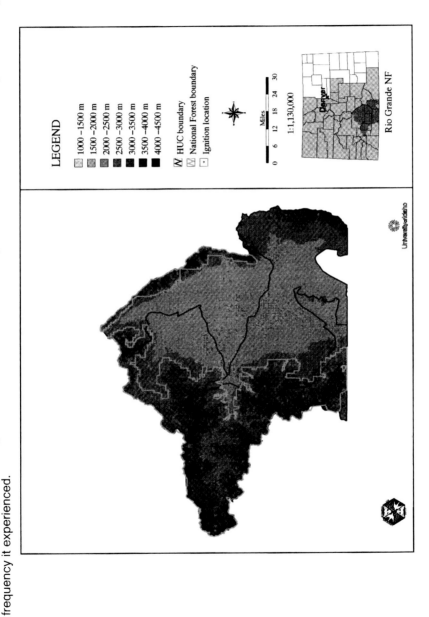

197

FIGURE 8.4. Risk of soil erosion is medium to high on 94 percent of the case study area.

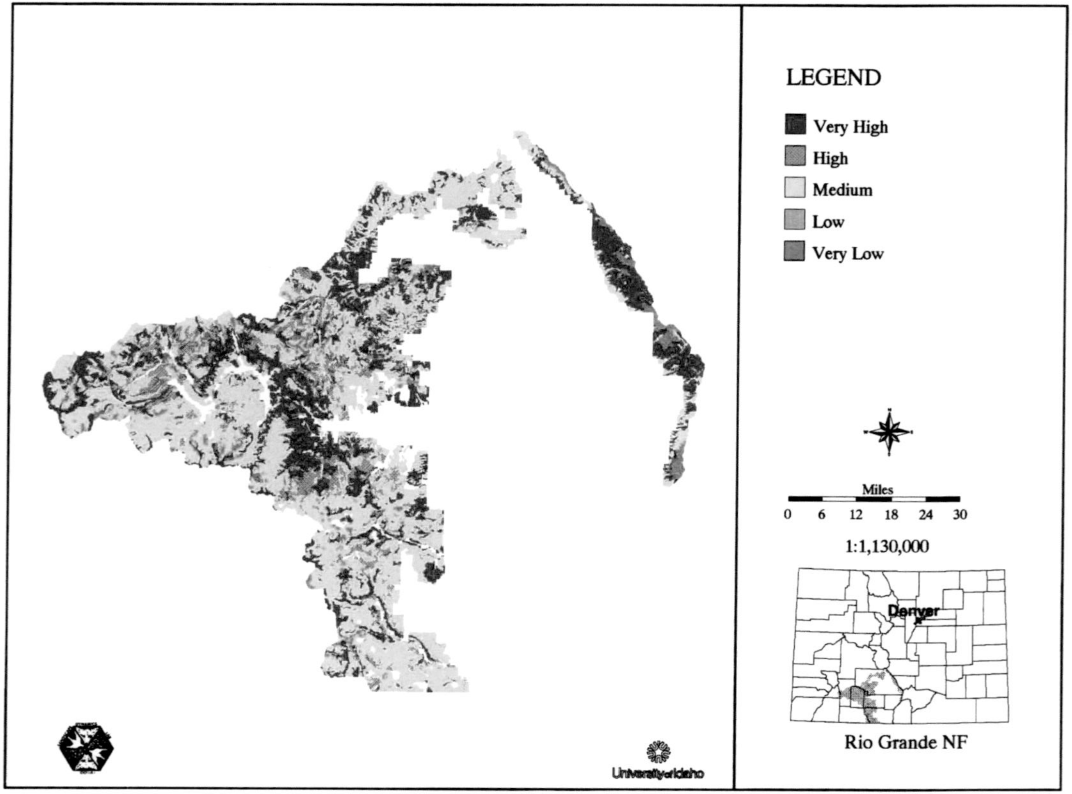

LEGEND

- Very High
- High
- Medium
- Low
- Very Low

Miles

0   6   12   18   24   30

1:1,130,000

Denver

Rio Grande NF

University of Idaho

TABLE 5. Moderate to high erosion risk exists on 94 percent of the former Rio Grande National Forest (NF) portion of the San Juan-Rio Grande NF.

Erosion risk category, definition from USDA (1996b)	Area (ac)	Area (%)
High: Moderate to high inherent erodibility; usually on moderate to steep slopes. Soil particles readily moved by overland flow after disturbance. Soils may require considerable expense to control erosion and sedimentation.	598,797	32
Moderately high	225,014	12
Moderate: Moderate inherent erodibility and are generally on moderate to steep slopes. Soils are more easily dispersed by raindrop impact and may require more expense in control of erosion and sedimentation	919,131	50
Low moderate	20,441	1
Low: Soils have a mixture of sand, silt, and clay and have relatively high organic matter content, creating strong structure. Generally on moderate slopes and do not usually require costly erosion control measures	90,498	5
Total	1,853,881	100

features, and past management history seem to be the controlling factors in the magnitude of the erosion and runoff."[7]

The severe fire at Buffalo Creek, coupled with an intense precipitation event, caused losses both on and off site. Downstream impacts were severe: "Hundreds of thousands of tons of sediment were dumped into a downstream reservoir–a 15-year sediment load" (Agnew et al. 1997). After the fire, three floods "deposited tons of ash and silt in the beds of [Buffalo] Creek and the North Fork of the South Platte River, smothering the food chain, filling the protective holes in which the fish once hid" (Able 1997a). In addition, the financial burden was high: "The cost of the three-phase reseeding and soil reclamation program will approach $1 million to $1.5 million" (Lovato[8] cited in Able 1997b).

A similar sequence of events–a wildfire followed by an intense storm–occurred in southern Oregon in 1987. After the Longwood Complex Fire, an intense December storm caused an estimated loss of 0.8 to 1.6 in. of surface soil (Amaranthus and Trappe 1993). Soil erosional losses were estimated as 42 ton ac^{-1} with substantial erosional losses of soil organic matter and N occurring in addition to those lost directly from the wildfire. Both the Buffalo Creek and the Longwood Complex wildfires show that erosion losses after fire can be severe, have extensive impacts, and be costly. In Yellowstone National Park, substantial erosion has occurred following the 1988 wildfire and subsequent rain events, together with a long record of earlier fires and erosion sequences dated using ^{14}C charcoal (Meyer et al. 1992, 1995). Other

TABLE 6. Buffalo Creek Fire. Mean tree density and size comparisons from unburned, surface fire, and crown fire areas.

Plot means	Unburned area	Fire area	
		Surface	Crown
Mean no. of trees (ac^{-1})			
Douglas-fir	13	5	76
Ponderosa pine	<u>67</u>	<u>186</u>	<u>128</u>
Total	80	191	204
Mean tree diam. (in.)			
Douglas-fir	15.3	9.3	9.1
Ponderosa pine	11.7	8.6	8.8
Combined	12.2	8.8	8.9
Mean tree height (ft)			
Douglas-fir	49.5	25.9	29.2
Ponderosa pine	33.1	25.6	30.8
Combined	35.8	25.6	30.1
Mean basal area (ft^2 ac^{-1})			
Douglas-fir	16.7	3.0	34.0
Ponderosa pine	50.0	79.0	54.0
Combined	66.7	82.0	88.0
Small trees <10 ft (ac^{-1})			
Douglas-fir	225	86	[1]
Ponderosa pine	35	17	[1]
Combined	260	103	155
Shrubs	Unburned	Mostly burned	Totally consumed

[1]Denotes that only small tree stobs were present in the crown fire area, thus limiting species identification.

recent work has documented watershed catchment effects of fire and soil erosion (Scott 1993).

To more accurately assess the erosion risk of an area, a new erosion risk equation (HYRISK) was developed (MacDonald et al. 1999). HYRISK incorporates the risk of hydrophobicity into the revised universal soil loss equation. By combining HYRISK with slope, soil erodibility, and a factor representing the likely increase in soil wetness due to removal of vegetation, MacDonald et al. (1999) found that two subbasins on the Rio Grande NF, the Headwaters of the Rio Grande (HUC 13010001) and the Conejos (HUC 13010005), have the second and fourth highest composite indices for post-

TABLE 7. Buffalo Creek Fire. Forest floor, woody fuel mass, and surface soil erosion from unburned, surface fire and crown fire areas.

Plot means	Unburned area	Fire area	
		Surface	Crown
Forest floor depth (in)	0.8	0.23	0.0
Mass (ton ac^{-1})	12.0	3.5	0.0
Woody fuels mass (ton ac^{-1})			
Wood diam. (in.)			
0-0.25	0.1	0.0	0.0
0.25-1.0	0.3	0.6	0.0
1.0-3.0	0.5	0.2	0.1
> 3.0 sound	1.3	0.0	0.2
> 3.0 rotten	<u>1.2</u>	<u>0.0</u>	<u>0.5</u>
Total woody fuels (ton ac^{-1})	3.4	0.3	0.8
Surface soil erosion–depth (in)	0.0	0.04	1.4

TABLE 8. Potential losses of organic matter and nutrients from a severe wildfire with subsequent erosion in a Colorado ponderosa pine forest.

Organic matter and nutrients	Losses from litter layer	Losses from tree canopy and understory[1]	Losses from erosion[2]
	< - (lbs ac^{-1}) - >		
Organic matter[3]	24000	7500	5800
N[3]	270	90	150
P[3,4]	25	10	70
K[3,4]	145	45	595
S[3]	30	10	40
Ca[3,4]	300	80	285

[1]Forest vegetation nutrient data are representative of typical forest ecosystems (Waring and Running 1998).
[2]Assumes a mineral soil loss of one inch, a bulk density of 1.0, and that the main losses of soil organic matter and nutrients are from erosion of the fine soil size fraction (< 0.08 in.), comprising about 60% of the soil mass (Cromack et al. 1999). Soil organic matter and total N concentrations are from Amaranthus and Trappe (1993); data for total K, Ca, and P are from a granitic soil (Attiwill and Leeper 1987). Soil total S concentrations are based on Cromack et al. (1999).
[3]Based on means for forest floor biomass and nutrients in Table 2 for ponderosa pine. Additional losses could occur from coarse woody debris and from buried decayed wood (Harvey et al. 1986, 1987).
[4]Assumes that these elements will be mainly deposited as ash, from total consumption of tree crowns understory vegetation, and the forest floor, with major losses during intense storm event(s), or rapid snowmelt, with severe surface erosion of one inch or more.

fire erosion risk of the 72 subbasins in central and western Colorado (Table 1, and Figure 4.6, MacDonald et al. 1999). These two subbasins on the former Rio Grande NF, although overall a low ignition-risk area, could rightly anticipate substantial erosional impacts in the steeper sloped areas if a severe precipitation event came after a wildfire.

For a watershed on the Coweeta Hydrological Experiment Station in North Carolina, GIS was used to predict soil erosion under a variety of management practices, and for a range of seasonal storm intensities. McNulty and Swank (1996) ran their GIS model, based on the universal soil loss equation, under different watershed management strategies. They suggest that a GIS model for a specific location be run iteratively using different scenarios until a management strategy is found that minimizes soil erosion.

Mass movement risk–While 94 percent of the Rio Grande NF has a moderate to severe erosion risk, only 24 percent is in medium to high mass-movement risk classes (Table 9). Mass movement generally occurs on slopes steeper than 60 percent (USDA 1996b). The largest areas of high mass-movement risk occur along the divides between subbasins, including along the Continental Divide and the eastern face of the Sangre de Cristo Mountains (Figure 8.5). Soil moisture, soil physical properties, bedrock type, and the rooting characteristics of understory vegetation influence the risk of mass movement and its severity (Pritchett and Fisher 1987, USDA 1996b). Loss of root strength after a severe fire can contribute to mass movement, especially if a severe precipitation event saturates the soils. After a severe fire and precipitation event, mass movement, such as debris flows, ". . . can severely scour existing channels and deliver large amounts of material to debris fans or downstream reaches, often with severe consequences to life, property, and aquatic ecosystems" (MacDonald et al. 1999).

Risk of nutrient loss–Nutrients in the vegetation or soil, or both, can be lost during and after a wildland fire by direct combustion of the organic matter and nutrient volatilization, or by erosion or mass movement, which physically moves the nutrients off site. The nutrients most vulnerable to direct loss by

TABLE 9. Mass movement risk is moderate to high on 24 percent of the former Rio Grande National Forest (NF) portion of the San Juan-Rio Grande NF.

Mass movement risk category	Probability of mass movement (%)	Area (ac)	Area (%)
High	75 to 100	209,945	11
Moderate	50 to 75	233,062	13
Low	25 to 50	705,172	38
Very low	0 to 25	705,743	38

volatilization include N and S, which have volatilization temperatures < 375°C, with P and K volatilizing at < 774°C (DeBano et al. 1998). In pinyon-pine woodlands in the Great Basin, Everett and Thran (1992) found that the forest floor contains 77 percent of the above ground N capital, equivalent to 910 lbs. N ac^{-1}. Loss of N in a fire is generally proportionate to the mass of fuel combusted. In the case-study area, *Ceanothus fendleri*, a species that may fix N (Conard et al. 1985), has been found to colonize the recent Buffalo Creek fire site.[9] More information about N-fixing species in Colorado forests would be valuable. For example, we observed *C. velutinus* growing in the understory of the lodgepole pine wildfire area of the Hourglass Fire at Pingree Park, Colorado, during the fall, 1996 workshop field trip. Nitrogen fixing species such as *P. tridentata* occur in southwestern US pinyon-juniper woodlands (Monsen 1987), and also in Colorado.[10] Other N-fixing species such as *Cercocarpus montanus* Raf. and *Shepherdia canadensis* Nutt. occur widely in Colorado and other areas of the intermountain west (Paschke 1997).

Sulfur (S) capital in a site is found predominately in the soil and litter, and will volatilize at a temperature as low as 375°C (Tiedemann 1987b). The exothermic peak of fire is reached at about 320°C (DeBano et al. 1998), and flaming combustion can boost temperatures to 800 to 1500°C (Philpot 1965). In a severe burn, soil temperatures would likely be sufficient to volatilize S over much of the burned area. Flaming combustion would consume much of the above ground live and dead fuels resulting in loss of above ground stores of S.

Although K and P volatilize at higher temperatures (> 774°C), their losses may be accelerated in severe wildfires, including nutrient losses by convection. White et al. (1973) suggested that K can volatilize when ponderosa pine burns. A crown fire would risk losses of about an additional 90 lb N ac^{-1}, 10 lb P ac^{-1}, and 45 lb K ac^{-1}, for example, based on typical foliar biomass and nutrient concentrations in coniferous forests (Waring and Running 1998). Increases in available soil K concentration have been reported (Isaac and Hopkins 1937, Austin and Baisinger 1955) with degree of soil heating and consumption of forest floor likely affecting the results obtained. Surface deposited K in ash could then be transported into the soil profile or off site with surface erosion. This also has been observed to occur with the soluble fractions of soil P (Gachene et al. 1997).

Loss of organic carbon (C)–The C reserves above ground and in the forest floor are an essential source of energy for the nutrient cycling process. Loss of these C reserves because of fire can alter site quality. The organisms which carry out nutrient cycling processes–the shredders, decomposers, and microorganisms–rely on site C for energy. The loss of site C, as a fuel for driving ecosystem processes, can slow recovery from disturbance. Losses of large masses of forest floor material from wildfire and from prescribed fire are common. In a ponderosa pine stand in Oregon, forest floor organic matter mass was 13.4 ton ac^{-1} (Landsberg 1992) before burning; after a high-fuel-consumption prescribed fire, only 2.2 ton ac^{-1} forest floor organic matter remained. This high-fuel consumption prescribed fire removed much of the forest floor material and subordinate vegetation but did not enter the canopy except in small, isolated areas; because it did not enter the crown in any significant amount, this fire was classified as moderate.

Forest floor C comprises roughly half the forest floor organic matter (Nelson and Sommers 1996); therefore, forest floor C loss at the aforementioned Oregon site was about 6.7 tons C ac^{-1}. Busse (1994) evaluated the role of decomposing downed boles in central Oregon lodgepole pine stands. He found that the downed boles comprised 23.5 percent of detrital C. The microbial biomass C and the ratio of microbial C to total organic C were significantly greater in the surface 0 to 1.6 in. of soil beneath rapidly decaying boles than in soil without a bole-wood component. The ratio remained elevated throughout the advanced stages of wood decomposition, suggesting a long-term change in the efficiency of C use by microbial communities associated with decaying wood (Busse 1994). Thus, it appears that in lodgepole pine stands, and probably in other types as well, decaying boles are an important C source for microorganisms (Harvey et al. 1978, 1987). In a severe fire, the organic C in vegetation, forest floor, soil, and downed boles will be volatilized, including increased losses of rotten wood buried in the soil (Harvey et al. 1987). Though the high C:N ratio of less decayed wood may initially decrease soil N availability (Giardina et al. 2000), the value of more decayed wood as a soil biotic habitat needs to be considered (Harvey et al. 1987).

Risk to vegetation–On the former Rio Grande NF, 63 percent of the land area is in tree species (Figure 8.6). The principal stand types, in decreasing area, are Engelmann spruce–subalpine fir (S–F) (*Picea engelmannii–Abies lasiocarpa*), aspen, and mixed conifer with Douglas-fir (*Pseudotsuga menziesii*) predominating. Portions of each stand type, totaling 27 percent of the forested area on the former Rio Grande NF, were harvested in the 1980s (Table 4). Because the treed portion of the case-study area is at relatively higher elevations and therefore cooler and drier, decomposition rates are lower, and fuels will remain in these harvested areas for many years, setting the stage for greater heat release and more severe effects from any subsequent fire. Of the forested area that was harvested, young vegetation will predominate and crown heights will be close to the ground. An ignition could quickly move into these crowns, producing flame lengths longer than those from a surface fire. Therefore, a severe fire event may result. The resulting mix of forest age

FIGURE 8.5. Risk of mass movement is medium to high on 24 percent of the case study area.

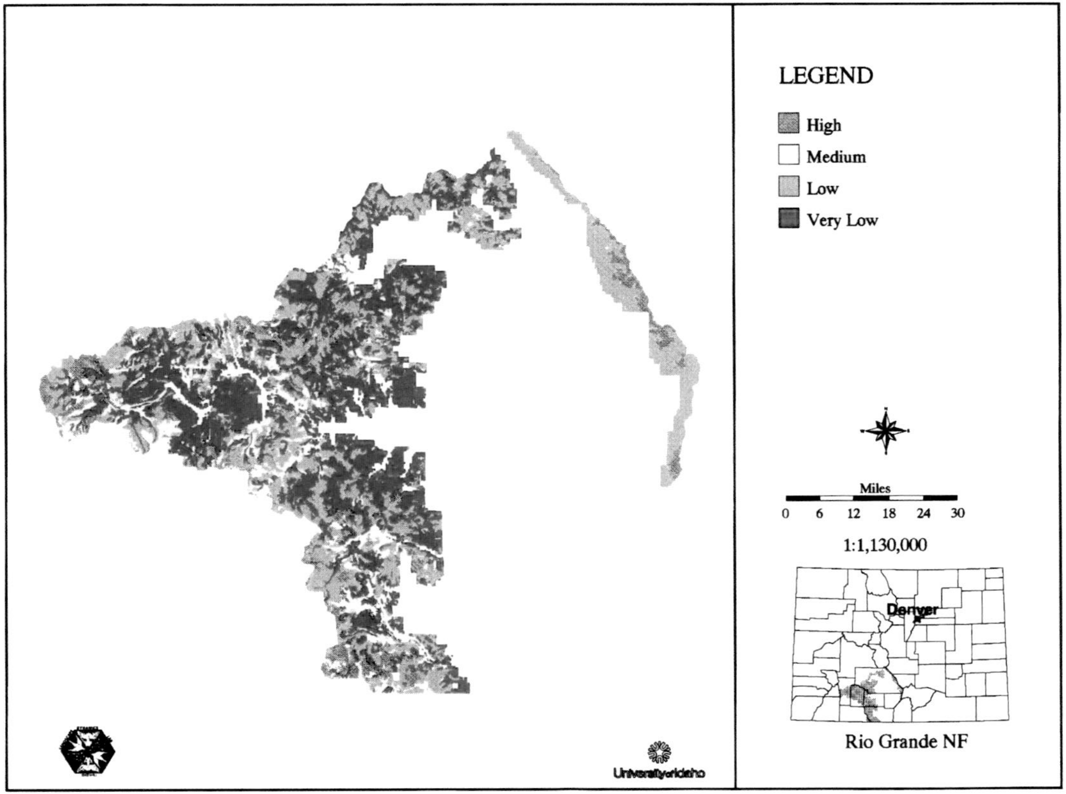

LEGEND

- High
- Medium
- Low
- Very Low

Miles

0   6   12   18   24   30

1:1,130,000

Denver

Rio Grande NF

FIGURE 8.6. The case study area supports 31 vegetation types, of which 10 are shown.

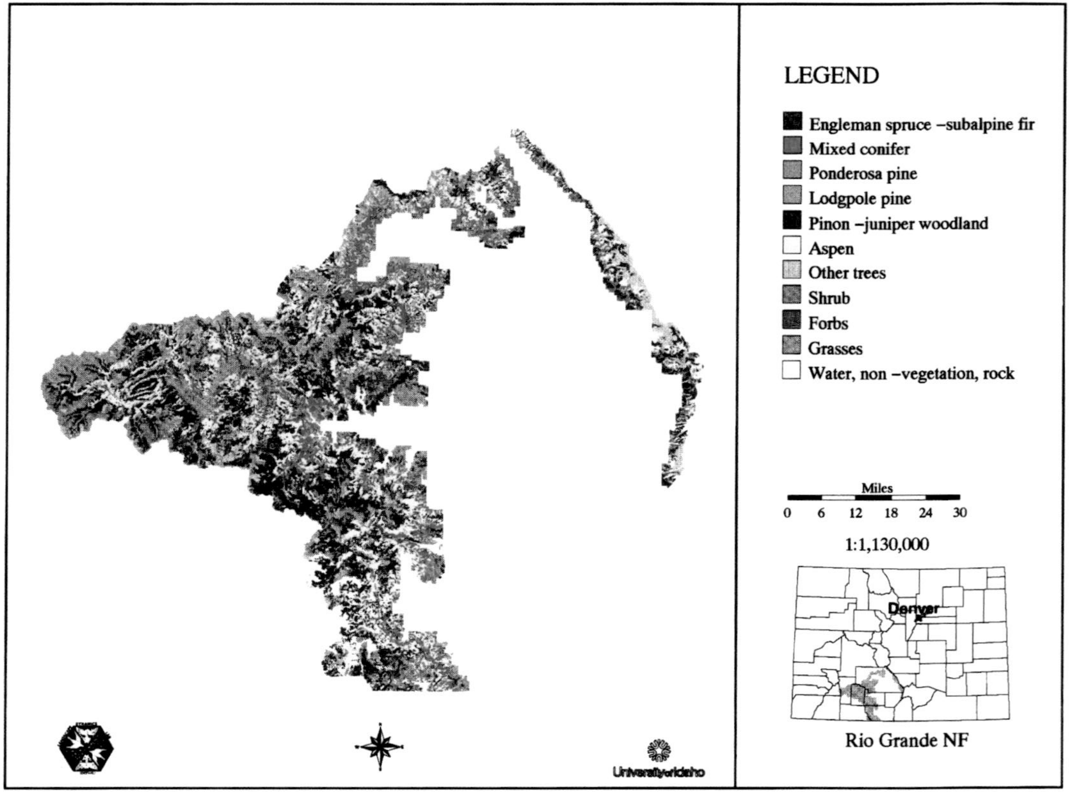
LEGEND
Engleman spruce –subalpine fir
Mixed conifer
Ponderosa pine
Lodgpole pine
Pinon –juniper woodland
Aspen
Other trees
Shrub
Forbs
Grasses
Water, non –vegetation, rock
Miles
0 6 12 18 24 30
1:1,130,000
Denver
Rio Grande NF
UniversityofIdaho

classes will afford opportunities for innovative forest management (Graham 1994, Smith et al. 1997).

Aspen, which is the second most common forest type in the Rio Grande NF, may be at an additional risk–a risk due to the absence of fire (Kay 1993, 1997, Bartos 1998). In Colorado, aspen has declined from 2,188,003 ac in 1900 to 1,077,239 ac in the 1990s, a loss of 49 percent. This loss, and future potential losses, of aspen can be attributed primarily to the successional processes that occur with the reduction (or elimination) of fire and by excessive browsing. Aspen clones require some disturbance or die-back in order to alter the hormonal balance so roots are stimulated to form suckers (Schier et al. 1985, Bancroft 1989). Historically, Native Americans set the aspen clones afire as they were leaving the high country in the fall when the aspen leaves were on the ground and dry enough to burn. This practice produced suckering in the spring, drawing ungulates into these areas, thereby making hunting more efficient. Now, the combination of fire absence, resulting in reduced numbers of new suckers, and increased browsing by wildlife and cattle (Kay 1990) of the few new suckers produced, is lowering the numbers and viability of aspen clones in some locations in the western United States (Bartos 1998). Local area losses of aspen could result in decreases in water, forage, and biodiversity, as well as other benefits, including aesthetic values (Jones and DeByle 1985). In the Colorado Front Range, however, there has been substantial aspen regeneration during the last 30 years, except where impacted by elk overgrazing (Suzuki et al. 1999).

POST-FIRE RECOVERY POTENTIAL AFTER A SEVERE FIRE IN WESTERN AND CENTRAL COLORADO

Risk of Ignition

Of the 72 subbasins in western and central Colorado, six have ignition frequencies in the "very high (>10 ignitions 10 K ac^{-1} in 10 years)" or "high (7 to 10 ignitions 10 K ac^{-1} in 10 years)" classifications (Figure 3.4, Neuenschwander et al. 1999). The subbasin with the very highest ignition frequency (17.4 ignitions 10 K ac^{-1} in 10 years) incorporates Mesa Verde National Park in the Mancos River watershed, and the next highest (11.3 ignitions 10 K ac^{-1} in 10 years) covers the Boulder area (Table 10). Those in the "high" category include the Denver area (7.4 ignitions 10 K ac^{-1} in 10 years) and two subbasins in northwestern Colorado (8.6 and 8.3 ignitions 10 K ac^{-1} in 10 years) (Table 10).

A comparison of the ignition points (Figure 3.1, Neuenschwander et al. 2000) and the ignition frequencies (ignitions 10 K ac^{-1} in 10 years) by subbasin (Figure 3.4, Neuenschwander et al. 2000) shows a very high density

TABLE 10. Western and central Colorado subbasins having both high or very high ignition frequencies or hydrophobicity composite risk indices (first two subbasin). Additional subbasins given have a high or very high risk in one or more risk category combinations and are identified by footnotes preceding the first HUC number in each table section.

HUC number and, for subbasins with highest risk or in case-study area, name or location		Fire ignitions per 10 K ac^{-1}, 1986-1995	Hydrophobicity composite risk index	Erosion composite risk index	Vegetation type in subbasins
[1]10190005	Includes Boulder, CO	11.3[2]	130.2[3]	0.9	Ponderosa pine, lodgepole pine, mixed conifer, spruce-fir
10190002	contains Buffalo Creek	7.4[2]	137.7[3]	2.0	Ponderosa pine, spruce-fir, pinyon-juniper
[4]14080107	SE of Cortez, CO	17.4[2]	98.2	3.7	Sage, pinyon-juniper, chaparral, ponderosa pine, mixed conifer, aspen
14040106	Yellow and Black Sulfur Rivers draining into White River	8.6[2]	29.3	0.1	Perennial grass, annual grass, sage, pinyon-juniper
14050006	NW corner of CO	8.3[2]	16.3	0.1	Perrenial grass, annual grass, pinyon-juniper
10190006	south of Fort Collins, CO	7.7[2]	103.6	0.8	Chaparral, lodgepole pine, ponderosa pine, spruce-fir
[5]13020102	Chama River, RGNF[6]	0.8	206.6[3]	11.2[7]	Lodgepole pine
11080001		1.6	186.0[3]	7.3[7]	Mixed conifer with Douglas-fir predominating
10180002		1.1	150.6[3]	0.4	Spruce-fir, lodgepole pine, aspen
14010004		1.3	149.7[3]	19.1[7]	Lodgepole pine, spruce-fir
14010003		4.4	139.8[3]	7.0[7]	Lodgepole pine, spruce-fir
10180001		1.0	134.5[3]	0.6	Lodgepole pine
14010002		2.7	129.4[3]	6.9[7]	Spruce-fir, lodgepole pine
14010001		1.5	126.7[3]	0.9	Lodgepole pine, spruce-fir
13010005	Conejos River, RGNF	1.2	125.4[3]	12.4[7]	Lodgepole pine
13010001	Headwaters of the Rio Grande River, RGNF	0.9	125.1[3]	15.0[7]	Spruce-fir, lodgepole pine
10190004		1.7	120.9[3]	2.5	Spruce-fir, ponderosa pine

HUC number and, for subbasins with highest risk or in case-study area, name or location	Fire ignitions per 10 K ac^{-1}, 1986-1995	Hydrophobicity composite risk index	Erosion composite risk index	Vegetation type in subbasins
[8]14010006	1.8	113.2	12.6[7]	Ponderosa pine, annual grass, perennial grass
14080104	4.2	103.1	8.5[7]	Mixed conifer with Douglas-fir predominating, spruce-fir, ponderosa pine, aspen
14080101	3.5	102.6	7.1[7]	Mixed conifer with Douglas-fir predominating, spruce-fir, lodgepole pine, aspen
14080102	4.3	81.7	5.2[7]	Mixed conifer with Douglas-fir predominating, spruce-fir, ponderosa pine, aspen
14020002	1.0	87.4	4.8[7]	Lodgepole pine, mixed conifer with Douglas-fir predominating
14020003	0.8	94.4	4.2[7]	Spruce-fir
14020006	2.4	81.4	4.2[7]	Lodgepole and ponderosa pine, mixed conifer with Douglas-fir predominating, pinyon-juniper

[1]Subbasins in this section have both high (7 to 10 ignitions 10 K ac^{-1}) or very high (>10 ignitions 10 K ac^{-1}) past fire ignitions *and* high (120 to 150) or very high (>150) hydrophobicity composite risk indices.
[2]Subbasins with high or very high past fire ignitions.
[3]Subbasins with high or very high hydrophobicity composite risk indices.
[4]Subbasins in this section have high or very high past fire ignitions only. Listed in decreasing order by number of fire ignitions.
[5]Subbasins in this section have high or very high hydrophobicity composite risk indices and may have high (4 to 8) or very high (8 to 16) erosion composite risk, but not high or very high past fire ignitions.
[6]RGNF = Former Rio Grande National Forest portion of the San Juan-Rio Grande, NF.
[7]Subbasins with high or very high erosion composite risk indices.
[8]Subbasins in this section have a high or very high erosion composite risk only

of ignition points in the subbasins cited above, but again, as in the Rio Grande NF case-study area, the ignition location points are clustered and are not distributed uniformly within the subbasins.

Resources at Risk

Soil resources–The soil resources at risk will be those located where there is a likelihood of wildfire; at even greater risk will be those located where there is likelihood of both severe fire and erosion.

Risk of hydrophobicity–Where hydrophobic layers form in a soil because of fire, those soil layers are more vulnerable to sheet erosion. Hydrophobic soils increase run-off, providing the opportunity for downslope and down-stream soil movement when a precipitation event comes after a fire. The hydrophobicity composite risk index for western and central Colorado is mapped by pixel (Figure 4.2, MacDonald et al. 2000). From Table 1 of MacDonald et al. (1999), the subbasins with the highest hydrophobicity composite risk index were selected (Table 10). Two of the subbasins with very high or high ignition risk frequency categories also had high hydrophobicity composite risk indices; both of these subbasins include metropolitan population centers–Boulder and Denver.

Risk of erosion–Thirteen subbasins in western and central Colorado are classified as very high or high using the erosion composite risk index (Figure 4.6, MacDonald et al. 2000). The very high category of erosion composite risk includes three subbasins affiliated with the Rio Grande NF case-study area. In decreasing order of risk they are: Headwaters of the Rio Grande River, Conejos River, and the Chama River. In a comparison of subbasins with the highest ignition frequencies (Figure 3.4, Neuenschwander et al. 2000) and those with the highest erosion composite risk index (Figure 4.6, MacDonald et al. 2000), it became apparent that there was no overlap between the subbasins with very high and high ignition frequencies and those with very high and high erosion composite risk indices. There are seven subbasins with both high hydrophobicity composite risk indices and high erosion composite risk indices. This is expected because the hydrophobicity composite risk index is a factor in calculating the erosion composite risk index. Surprisingly, seven subbasins have a high erosion composite risk without showing a high hydrophobicity composite risk index.

Severe surface soil movement risk–The highest risk of severe surface soil movement occurred in only a few locations in western and central Colorado (Figure 8.7). These locations are concentrated in southern Colorado, along the Front Range, and along the eastern edge of the former San Isabel NF. The areas at greatest risk are most often found along divides between subbasins where the slopes are potentially the greatest.

Nutrient risk–The risks to site nutrients for western and central Colorado will be similar to those previously discussed in the case-study area.

The Buffalo Creek Fire area, south of Denver, is in a subbasin (HUC 1019002) with a "high" ignition frequency (7.4 ignitions 10 K ac^{-1} in 10 years). This subbasin is one of two in all western and central Colorado with a very high or high ignition frequency and a high hydrophobicity composite risk index as well. At the Buffalo Creek Fire site, the area that burned in a crown fire averaged a loss of 1.4 in. of topsoil, while the area that burned in an underburn had very limited erosional losses (Table 7). The losses of forest floor and topsoil measured at Buffalo Creek attest to the serious outcome when a severe fire is followed by a severe thunderstorm. The results from Buffalo Creek suggest that subbasins with very high or high ignition frequencies and hydrophobicity indices need to be carefully evaluated.

The two subbasins with both high or very high fire ignition frequencies and a high hydrophobicity index either encompass the Boulder area or are adjacent to Denver, as mentioned above. These two subbasins need to be evaluated at a finer scale to see if the risks from future fire and precipitation events would adversely affect infrastructure or areas of human habitation.

Two more subbasins with high ignition frequencies and hydrophobicity composite risk indices that are nearly sufficient to place them in the above category are the subbasin with the very highest ignition frequency, SE of Cortez, Colorado, and the subbasin south of Fort Collins, Colorado (Table 10). These two subbasins also need to be evaluated at a scale finer than subbasin level so the specific areas at risk can be identified. This finer scale evaluation then needs to be compared with both the hydrophobicity composite risk index and the soil erosion composite risk index. Such an analysis will likely reveal areas of ignition density at less than the subbasin scale that also have a propensity for the formation of hydrophobic soils or for soil erosion, or both. These areas would then be targeted for remedial action.

Risk to vegetation–The disturbance type considered here that is most likely to affect vegetation is fire, rather than erosion or mass movement. The vegetation in areas with ignition risks in the very high and high categories will be most vulnerable. Those subbasins have been identified and their vegetation classified from AVHRR satellite imagery. The vegetation of western and central Colorado has been composited into 12 groups (Figure 2.3, Sampson and Neuenschwander, 2000). No single type of vegetation is at the highest risk. The vegetation, rather, is strongly related to the area in which it grows; and the risk to that vegetation is strongly related to the ignition frequency of that area. Thus, ponderosa pine growing in one subbasin may be very vulnerable because of a high ignition frequency there, while ponderosa pine in another subbasin may be at a much lower risk. The vegetation in the subbasins with very high or high ignition frequencies (Table 10) spans the spectrum from grasses to shrubs to woodlands to stands of aspen and softwoods. An examination of the ignition frequencies at a scale finer than subbasin will identify areas, and their species, smaller than subbasins that have significant ignition frequencies. Once ignition occurs, ponderosa pine stands (among the softwoods) have the most flammable needles on the forest floor compared with lodgepole pine, Douglas-fir, and subalpine fir (Fonda et al. 1998). This high flammability found in ponderosa pine litter may contribute to higher vulnerability of ponderosa pine stands.

Assessment of Overall Risk

The portions of Colorado which may show the most serious effects from a severe fire and a subsequent intense precipitation event are shown in Figure 8.8. This map was developed using MacDonald et al.'s (2000) SOILEROD index.

The index includes a factor, s, for slope, and a factor, k, for inherent erodibility; calculation of the values, s and k, are given in MacDonald et al. (2000). The value of k is based on information from the state soils database (STATS-GO), which is limited by its 6 km^2 minimum map unit. A SOILEROD value of 1.5 was chosen,[11] which included the most highly erodible soils on the steepest slopes. Only those areas with a SOILEROD value ≥ 1.5, which included approximately 30% of western and central Colorado that is most likely to show serious effects, are shown in Figure 8.8. A value >1.5 for SOILEROD would have produced an area with a higher average risk, and similarly, a lower value would have produced a larger area with a lower average risk.

All the terrain shown in Figure 8.8 is at risk to erosion because of its steep slopes and inherently erodible soils. The conifer species dominating this vulnerable terrain include Engelmann spruce-subalpine fir, mixed conifer with Douglas-fir predominating, ponderosa and lodgepole pine, and pinyon-juniper woodlands. Along the Front Range, from near Fort Collins south toward Colorado Springs, ponderosa pine is the predominant species. Engelmann spruce-subalpine fir are found extensively along the backbone of the Continental Divide at upper elevations. Lodgepole pine grow generally along the lower slopes which grade to either mixed conifer or Engelmann spruce-subalpine fir forests. Pinyon-juniper woodlands occur in scattered locations throughout western and central Colorado and also form a periphery along Colorado's western border.

Superimposed on the distribution of conifer species are the locations of fire ignitions that occurred during 10 years on these highly vulnerable areas (Figure 8.8). Again, the Front Range area stands out with its high density of ignition points in areas highly vulnerable to soil erosion. This area is adjacent to the densely populated centers of Fort Collins, Boulder, and Denver–a cause for concern. The Buffalo Creek Fire area, located toward the southern end of the Front Range, burned in terrain highly vulnerable to soil erosion (Figure 8.8). After the Buffalo Creek Fire, an intense thunderstorm pounded the area with 2.5 in. of rain and produced average erosion losses of 1.4 in. of topsoil. This erosion and resulting sediment has forced the city of Denver to utilize alternative water sources.

In the southwestern corner of Colorado, the area near Mesa Verde National Park and south of the city of Cortez, also showed a high incidence of fire ignitions on terrain vulnerable to high and very high erosion risk; the vegetation in this area included pinyon-juniper.

We determined those areas where potential serious effects resulting from fire and subsequent erosion after fire might occur (Figure 8.8). Neuenschwander et al. (2000), in using a different approach, determined all areas where high-probability, large wildfires might occur. Their results (Figure 3.5,

FIGURE 8.7. Subbasins with risk of severe soil movement in western and central Colorado.

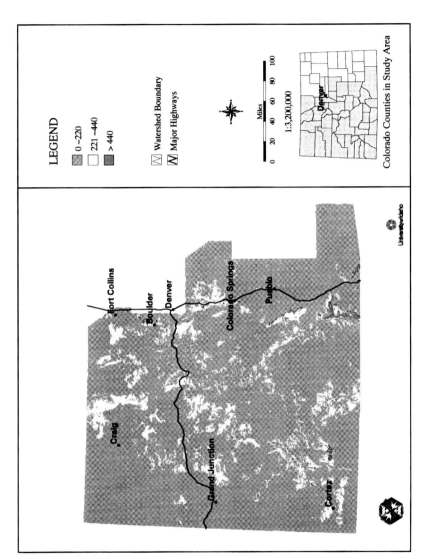

FIGURE 8.8. Areas of high or very high vulnerability to erosion, by forest type, with recent ignition locations.

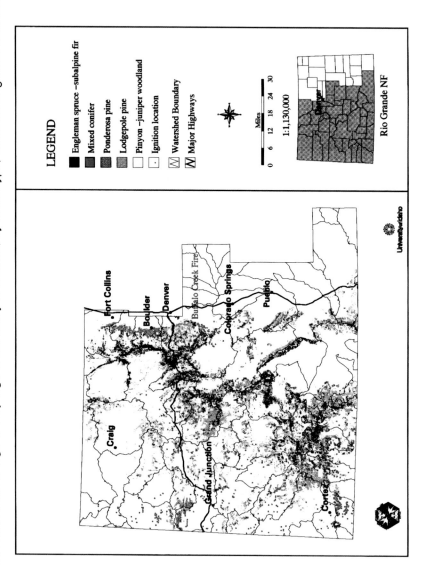

Neuenschwander et al. 2000) include a zone in northwestern Colorado. This zone does not appear in Figure 8.8 because very little terrain in northwestern Colorado has a SOILEROD value ≥ 1.5 because of lesser slopes or less inherently erodible soils.

The incidence of ignition points gives indications of past fire history. The zones where ignition points are densely packed can anticipate future fires based on the availability of ignition sources and sufficient fuels. The conifer species in these areas can be located in Figure 8.8. These are the areas where erosion can be substantial if an intense precipitation event occurs after a fire. Other terrain, too, not indicated either by propensity to erosion or to fire ignition, may experience similar effects. A fire in an area that does not normally burn, plus an unusual, intense thunderstorm, can combine to produce erosion almost anywhere.

The Next Steps

We have demonstrated the usefulness of GIS in determining the vulnerabilities of a case-study area, and of western and central Colorado. For the case-study area, input data had been taken from aerial photos. The input data for western and central Colorado had been taken from AVHRR satellite imagery. The finer scale data for the case-study area permitted the identification of areas of potential risk on a finer-than-subbasin basis. For the much larger region encompassing western and central Colorado, incorporating input based on aerial photos, and in the case of soils input at a scale finer than that available in STATSGO (6 km^2 basis), would increase accuracy and identify areas smaller than subbasins for targeted remedial work.

Geographic information systems are powerful tools and will be more closely integrated into land management planning as GIS data layers have greater resolution and are available for more characteristics. The accuracy of GIS input data needs to be systematically verified by ground truthing–sampling of the area–or other means, before GIS data layers can be considered fully functional and the output more reliable (Bolstad et al. 1998, Waring and Running 1998).

Figure 8.8 shows the portions of western and central Colorado most susceptible to fire and erosional losses after a sudden thunderstorm based on inherent soil erodibility and slope. This analysis can be improved when more precise soil maps and other data layers are available. The addition of meteorological data identifying locations with numerous 2-hour precipitation events to the ignition frequency data will identify the areas likely to get both severe fire and severe precipitation events–events that set them up for severe erosional disturbances. In addition, GIS databases and modeling simulations incorporating information about topography, meteorology, and fire dynamics will be important in improving future assessments of risks from fire (Kushla and Ripple 1997, Keane and Long 1998, Waring and Running 1998). GIS is also proving to be a valuable tool in assessing fire damage on burned watersheds (Lachowski et al. 1997). Geographic information systems analyses of these suggested input layers may indicate that not all of the above factors are necessary for predicting the locations with the greatest vulnerabilities, or that additional input layers are needed.

Analyses at the landscape level are needed to identify large areas such as these, which present specific fire problems so action can be taken (Keane and Long 1998). Officials in one area of Colorado with a known high ignition risk have begun evaluating the conditions in their locality. Near Boulder, in the "Red Zone," which covers 1,528 square miles along the Front Range in Colorado, the Forest Service has already begun measures to reduce the risk of fire. These measures include inventorying the area, identifying unhealthy stands of trees, and thinning trees (Rocky Mountain News 1998). Ultimately, progress in forest ecosystem management should help to enhance benefits of fire in forest types where it is an important component of ecosystem disturbance (Attiwill 1994, Kimmins 1996, Landsberg 1997). This includes integrating management effects of insects, diseases, and fire (Anderson et al. 1987, Baker and Veblen 1990, Filip and Yang-Erve 1997, Waring and Running 1998).

Summary

Changes in climate and changes brought about by people of European origin beginning in the 16th and 17th centuries may make problematic the potential return of some of the ecosystems in western and central Colorado to their current condition, or to their pre-European-settlement condition after a severe fire and precipitation disturbance. The combined effects of grazing, introduced in the 1700s, and the climate changes since the end of the Little Ice Age in the 1870s, may have altered the trajectory of some ecosystems to the extent that the systems currently on the sites, but initiated under different prior circumstances, are no longer attainable. These factors, the introduction of grazing, and the inability to reestablish some species in some locations, need to be considered when the ability of an ecosystem to recover after a severe fire and intense precipitation is evaluated.

Severe wildfires followed by intense precipitation events can have serious ecological consequences in western and central Colorado. The 1996 Buffalo Creek Fire near Denver serves as an example: In the crown-consuming portion of the Buffalo Creek Fire, nutrient-rich foliage was lost to direct volatilization or to ash and particulate production. The sudden thunderstorm after the fire produced additional nutrient losses through off-site transport of the nutrient-rich upper layers of the soil and forest floor material. Nutrients that are vulnerable to loss by both mechanisms, volatilization and off-site transport, are those that have significant capital in or near the forest floor and low volatilization temperatures, especially organic C, N, P, S, and K. Thus, the site capital of several essential nutrients can be seriously compromised by a severe fire and precipitation disturbance. Losses of important nutrients can negatively affect the ability of a site to recover following serious disturbance.

Fire-free intervals have increased in many forest types of the intermountain west since the beginning of fire suppression. Without frequent fire, fuel loads have increased, which could result in hotter, more severe fires. Severe wildfires are currently increasing in acreage and numbers, setting the stage for significant losses of nutrients and soil quality. These losses can be anticipated to occur in the future, potentially limiting the long-term sustainability of the ecosystems affected.

We found two zones in western and central Colorado to be the most susceptible to losses from fire and intense precipitation. These are located in the Front Range and near Mesa Verde National Park, south of the city of Cortez.

AUTHORS NOTE

The authors thank John Dobson, hydrologist; John Rawinski, soil scientist; Bob Tribble, operations research analyst; Katie Stuart, zone timber management specialist; and Todd Pechota, assistant fire management officer for their generous cooperation and contribution of information for the case-study area. All are Forest Service employees located on the former Rio Grande National Forest portion of the San Juan–Rio Grande NF. They also thank Ron Gosnell, Colorado State Forest Service, and Dr. Kenneth Poppe, and his students, Ted Brash, Bobby Davey, Summer Davey and Kathleen Gosnell from Lyons High School, Lyons, Colorado, and Henry Goehle from the South Platte Ranger District, for their invaluable assistance in obtaining the plot surveys for the Buffalo Creek Fire. Further, they thank Lee MacDonald for his very necessary assistance in selecting the value for SOILEROD used in preparing Figure 8.8. William O. Russell, III made valuable suggestions concerning erosion effects. Angeline S. Cromack and Kevin T. Cromack provided assistance with editing and manuscript preparation. Some support for the senior author was from NSF Ecosystem Studies. Any errors or misinterpretations remain the sole responsibility of the authors.

NOTES

1. Personal communication. 1997. Marvin Hoover, Retired Hydrologist, Rocky Mountain Forest and Range Experiment Station, Fort Collins, CO, now living in Wenatchee, WA.

2. Landsberg, personal observation.

3. Professional judgement of the authors.

4. Personal communication. 1996. R. Averill, Program Manager, Forest Health Management, Lakewood, CO.

5. Everett, R.L., June 23, 1995. Preliminary information on historical, pre-fire, and post-fire conditions for the 1994 Tyee and Hatchery Creek burn areas. Portland, OR. U.S. Department of Agriculture, Forest Service, Pacific Northwest Research Station. 21 pp. Report on file with: Forestry Sciences Laboratory, Pacific Northwest Research Station, 1133 N. Western Avenue, Wenatchee, WA.

6. Personal Communication. 1997. Todd Pechota, Assistant Fire Management Officer, Saguache Ranger District, San Juan–Rio Grande National Forest, Saguache, CO.

7. Personal communication. 1998. Jeffrey Bruggink, Soil Scientist, Pike National Forest, CO.

8. Dan Lovato, Deputy Ranger, South Platte Ranger District, Pike National Forest, Morrison, CO.

9. Personal communication. 1998. M.W. Paschke, Rangeland Ecologist, Department of Rangeland Ecosystem Science, Colorado State University, Ft. Collins, CO.

10. Personal communication. 1998. M.W. Paschke, Rangeland Ecologist, Department of Rangeland Ecosystem Science, Colorado State University, Ft. Collins, CO.

11. Personal communication. 1998. L.H. MacDonald, Associate Professor–Watershed Science Program, Department of Earth Resources, Colorado State University, Fort Collins, CO.

REFERENCES

Able, C. 1997a. Floods, fire destroyed habitat. *Rocky Mountain News*. Denver, CO. May 11.

Able, C. 1997b. Officials pinning hopes on current soil-saving effort. *Rocky Mountain News*. Denver, CO. May 11.

Agee, J.K. 1993. Fire Ecology of Pacific Northwest Forests. Inland Press, Washington, DC. 493 p.

Agnew, W., Lahn, R.E. and M.V. Harding. 1997. Buffalo Creek, Colorado, fire and flood of 1996. *Land and Water* 4: 27-29.

Almendros, G., F.J. Gonzalez-Vila, and F. Martin. 1990. Fire-induced transformation of soil organic matter from an oak forest: an experimental approach to the effects of fire on humic substances. *Soil Science* 149: 158-168.

Amaranthus, M.P. 1998. The importance and conservation of ectomycorrhizal fungal diversity in forest ecosystems: lessons from Europe and the Pacific Northwest. USDA, Forest Service, Pacific Northwest Research Station, Portland, OR. Gen. Tech. Rep. PNW-GTR-431. 15 p.

Amaranthus, M.P. and D.A. Perry. 1994. The functioning of ectomycorrhizal fungi in the field: linkages in space and time. *Plant and Soil* 159: 133-140.

Amaranthus, M.P. and J.M. Trappe. 1993. Effects of ecto- and VA-mycorrhizal inoculum potential of soil following forest fire in southwest Oregon. *Plant and Soil* 150: 41-49.

Amaranthus, M.P., J.M. Trappe and D.A. Perry. 1993. Soil moisture, native revegetation, *Pinus lambertiana* seedling growth, and mycorrhiza formation following wildfire and grass seeding. *Restoration Ecology* 9: 188-195.

Anderson, L., C.E. Carlson and R.H. Wakimoto. 1987. Forest fire frequency and western spruce budworm outbreaks in western Montana. *Forest Ecology and Management* 22: 251-260.

Arno, S.F. and T.D. Peterson. 1983. Variation in estimates of fire intervals: A closer look at fire history on the Bitterroot National Forest. USDA, Forest Service, Intermountain Research Station, Ogden, UT. Res. Pap. INT-301. 8 p.

Arno, S.F., J.H. Scott and M.G. Hartwell. 1995. Age-class structure of old growth ponderosa pine/Douglas-fir stands and its relationship to fire history. USDA, Forest Service, Intermountain Research Station, Ogden, UT. Research Paper INT-RP-481. 25 p.

Arno, S.F., H.Y. Smith and M.A. Krebs. 1997. Old growth ponderosa pine and western larch stand stuctures: Influences of pre-1990 fires and fire exclusion. USDA, Forest Service, Intermountain Research Station, Ogden, UT. Research Paper INT-RP-495. 20 p.

Arthur, M.A. and T.J. Fahey. 1990. Mass and nutrient content of decaying boles in an Engelmann spruce–subalpine fir forest, Rocky Mountain National Park, Colorado. Canadian Journal of Forest Research 20: 730-737.

Attiwill, P.M. 1994. The disturbance of forest ecosystems: the ecological basis for conservative management. *Forest Ecology and Management* 63: 247-300.

Attiwill, P.M. and G.W. Leeper. 1987. Forest Soils and Nutrient Cycles. Melbourne University Press, Carlton, Victoria, Australia. 202 p.

Austin, R.C. and D.H. Baisinger. 1955. Some effects of burning on forest soils of western Oregon and Washington. *Journal of Forestry* 53: 275-280.

Baker, W.L. and T.T. Veblen. 1990. Spruce beetles and fires in the nineteenth century subalpine forests of western Colorado, U.S.A. *Arctic and Alpine Research* 22: 65-80.

Bancroft, B. 1989. Response of aspen suckering to pre-harvest stem treatments: A literature review. Forestry Canada and B.C. Ministry of Forests, Research Branch, Victoria, B.C., 55 p.

Bartos, D. 1998. Aspen, fire, and wildlife. In: pp. 44-48. Fire and Wildlife in the Pacific Northwest– Research, Policy, and Management. April 6-8, 1998. The Wildlife Society, Spokane, WA.

Bartos, D.L. and R.S. Johnston. 1978. Biomass and nutrient content of quaking aspen at two sites in the western United States. *Forest Science* 24: 273-280.

Barrett, J.W. 1994. Regional Silviculture of the United States, 3rd ed. John Wiley, NY, 643 p.

Barrett, S.W., S.F. Arno and C.H. Key. 1991. Fire regimes of western larch-lodgepole pine forests in Glacier National Park, Montana. Canadian Journal of Forest Research 21: 1711-1720.

Baumgartner, D.M., R.G. Krebill, J.T. Arnott and G.F. Weetman (Eds.). 1985. Lodgepole Pine: The Species and Its Management, Symposium Proceedings May 8-10, 1984, Spokane, WA and repeated May 14-16, 1984, Vancouver, B.C. Published by: Office of Conferences and Institutes, Cooperative Extension, Washington State University, Pullman, WA.

Belillas, C.M. and M.C. Feller. 1998. Relationships between fire severity and atmospheric and leaching nutrient losses in British Columbia's coastal western hemlock zone forests. *International Journal of Wildland Fire* 8: 87-101.

Binkley, D., K. Cromack, Jr. and R.L. Fredriksen, R.L. 1982. Nitrogen accretion and availability in some snowbrush ecosystems. *Forest Science* 28: 720-724.

Bolstad, P.V., W. Swank and J. Vose. 1998. Predicting southern Appalachian overstory vegetation with digital terrain data. *Landscape Ecology* 13: 271-283.

Brown, J.K. and N.V. DeByle. 1987. Fire damage, mortality, and suckering in aspen. *Canadian Journal of Forest Research* 17: 1100-1109.

Busse, M.D. 1994. Downed bole-wood decomposition in lodgepole pine forests of central Oregon. *Soil Science Society of America Journal* 58: 221-227.

Busse, M.D., P.H. Cochran and J.W. Barrett. 1996. Changes in ponderosa pine site productivity following removal of understory vegetation. *Soil Science Society of America Journal* 60: 1614-1621.

Cao, M. and F.I. Woodward. 1998. Dynamic responses of terrestrial ecosystem carbon cycling to global climate change. *Nature* 393: 249-252.

Conard, S.G., A.E. Jaramillo, K. Cromack, Jr. and S. Rose (Compilers). 1985. The Role of the Genus *Ceanothus* in Western Forest Ecosystems. USDA, Forest Service, PNW Forest and Range Experiment Station, Portland, OR. GTR PNW-182. 72 p.

Cooper, C.F. 1960. Changes in vegetation, structure, and growth of southwestern pine forests since white settlement. *Ecological Monographs* 30: 129-164.

Covington, W.W. and M.M. Moore. 1994. Southwestern ponderosa forest structure–changes since Euro-American settlement. *Journal of Forestry* 92: 39-47.

Covington, W.W. and S.S. Sackett. 1984. The effect of a prescribed burn in southwestern ponderosa pine on organic matter and nutrients in woody debris and forest floor. *Forest Science* 30: 183-192.

Covington, W.W., R.L. Everett, R. Steele, L.L. Irwin, T.A. Daer and A.N.D. Auclair. 1994. Historical and anticipated changes in forest ecosystems of the inland west of the United States. In: pp. 13-63. R.N. Sampson, D.L. Adams and M.J. Enzer (Eds.), Assessing Forest Ecosystem Health in the Inland West. Food Products Press, Binghamton, NY.

Cromack, K. Jr., C.C. Delwiche and D.H. McNabb. 1979. Prospects and problems of nitrogen management using symbiotic nitrogen fixers. In: pp. 210-223. J.C. Gordon, C.T. Wheeler, D.A. Perry (Eds.). Symbiotic Nitrogen Fixation in the Management of Temperate Forests. Forest Research Lab., Oregon State Univ., Corvallis, OR.

Cromack, K. Jr., R.E. Miller, O.T. Helgerson, R.B. Smith and H.W. Anderson. 1999. Soil carbon and nutrients in a coastal Oregon Douglas-fir plantation with red alder. *Soil Science Society of America Journal* 63: 232-239.

DeBano, L.F. 1981. Water repellent soils: a state-of-the-art. USDA, Forest Service, Pacific Southwest Forest and Range Experiment Station, Berkeley, CA. Gen. Tech. Rep. PSW-46. 21 p.

DeBano, L.F. 1990. Effects of fire on the soil resource in Arizona chaparral. In: pp. 65-77. J.S. Krammes (Tech. Coord.). Effects of Fire Management of Southwestern Natural Resources. USDA, Forest Service, Gen. Tech. Rep. RM-191.

DeBano, L.F., D.G. Neary and P.F. Ffolliott. 1998. Fire: Its Effects on Soil and Other Ecosystem Resources. John Wiley and Sons, NY, 612 p.

DeBano, L.F., H.M. Perry and S.T. Overby. 1987. Effects of fuelwood and slash burning on biomass and nutrient relationships in a pinyon-juniper stand. In: pp. 382-386. Proceedings: Pinyon-Juniper Conference. Jan. 13-16, 1986. Reno, NV. USDA, Forest Service, Intermountain Research Station. Ogden, UT. Gen. Tech. Rept. INT-215.

Dieterich, J.M. 1980. Chimney Spring Forest Fire History. USDA, Forest Service, Rocky Mountain Research Station, Fort Collins, CO. Research paper RM-220. 8 p.

Drinkwater, L.E., P. Wagoner and M. Sarrantonio. 1998. Legume-based cropping systems have reduced carbon and nitrogen losses. *Nature* 396: 262-265.

Everett, R.L. (compiler). 1993. Eastside Forest Ecosystem Health Assessment. Vol. IV. Restoration of stressed sites and processes. USDA, Forest Service, Pacific Northwest Research Station, Portland, OR. PNW-GTR-330.

Everett, R.L. and D.F. Thran. 1992. Above-ground biomass and N, P, and S capital in singleleaf pinyon woodlands. *Journal of Environmental Management* 34: 137-147.

Everett, R.L., P.F. Hessberg, M.E. Jensen and B.T. Bormann. 1993. Eastside Forest Ecosystem Health Assessment. Vol. I. Executive summary. USDA, Forest Service, Pacific Northwest Research Station, Portland, OR. 57 p.

Everett, R.L., B.J. Java-Sharpe, G.R. Sherer, F.M. Wilt and R.D. Ottmar. 1995. Co-occurrence of hydrophobicity and allelopathy in sand pits under burned slash. *Soil Science Society of America Journal* 59: 1176-1183.

Fahey, T. J. 1983. Nutrient dynamics of aboveground detritus in lodgepole pine (*Pinus contorta* ssp. *latifolia*) ecosystems, southeastern Wyoming. Ecological Monographs 53: 51-72.

Filip, G.M. and L.Yang-Erve. 1997. Effects of prescribed burning on the viability of *Armillaria ostoyae* in mixed-conifer forest soils in the Blue Mountains of Oregon. *Northwest Science* 71: 137-144.

Fonda, R.W., L.A. Belanger and L.L. Burley. 1998. Burning characteristics of western conifer needles. *Northwest Science* 72: 1-9.

Frank, D.A., S.J. McNaughton and B.F. Tracy. 1998. The ecology of the earth's grazing ecosystems. *Bioscience* 48: 513-521.

Gachene, C.K.K., N.J. Jarvis, H. Linner and J.P. Mbuvi. 1997. Soil erosion effects on soil properties in a highland area of central Kenya. *Soil Science Society of America Journal* 61: 559-564.

Giardina, C.P. and C.C. Rhoades. 2000. Clear-cutting and burning on soil nitrogen supply and seedling growth in a lodgepole pine forest of southeastern Wyoming. *Canadian Journal of Forest Research* (in press).

Giardina, C.P., R.L. Sanford, Jr. and I.C. Dockersmith. 2000. Soil phosphorus, carbon, and nitrogen changes during slash-and-burn conversion of a dry tropical forest. *Soil Science Society of America Journal* 64: 399-405.

Giovannini, G. and S. Lucchesi. 1983. Effect of hydrophobic and cementing substances of soil aggregates. *Soil Science* 136: 231-236.

Giovannini, G., S. Lucchesi and M. Giachett. 1988. Effect of heating on some physical and chemical parameters related to soil aggregation and erodibility. *Soil Science* 146: 255-261.

Goldblum, D. and T.T. Veblen. 1992. Fire history of a ponderosa pine/Douglas-fir forest in the Colorado Front Range. Physical Geography 13: 133-148.

Graham, R.T. 1994. Silviculture, fire and ecosystem management. In: pp. 339-351. R.N. Sampson and D.L. Adams (Eds.). Assessing Forest Ecosystem Health in the Inland West. The Haworth Press Inc., NY.

Grier, C.C. 1975. Wildfire effects on nutrient distribution and leaching in a coniferous ecosystem. *Canadian Journal of Forest Research* 5: 599-607.

Harvey, A.E., M.F. Jurgensen and M.J. Larsen. 1978. Role of residue in, and impacts of its management on forest soil biology. In: UN Food and Agriculture Organization, 8th World Forestry Congress, Special Paper. 11 p.

Harvey, A.E., M.F. Jurgensen and M.J. Larsen. 1981. Organic reserves: importance to ectomycorrhizae in forest soils of western Montana. *Forest Science* 27: 442-445.

Harvey, A.E., M.F. Jurgensen, M.J. Larsen and J.A. Schlieter. 1986. Distribution of active ectomycorrhizal short roots in forest soils of the inland northwest: effects of site and disturbance. USDA, Forest Service, Intermountain Research Station, Ogden, UT. Gen. Tech. Rep. INT-374, 8 p.

Harvey, A.E., M.F. Jurgensen, M.J. Larsen and R.T. Graham. 1987. Decaying organic materials and soil quality in the inland northwest: a management opportunity. USDA, Forest Service, Intermountain Research Station. Ogden, UT. Gen Tech. Rep. INT-225, 16 p.

Haselwandter, K. and G.D. Bowen. 1996. Mycorrhizal relations in trees for agroforestry and land rehabilitation. *Forest Ecology and Management* 81: 1-17.

Isaac, L.A. and G. Hopkins. 1937. The forest soil of the Douglas-fir (sic) region and changes wrought upon it by logging and slash burning. *Ecology* 18: 264-279.

Jakubos, B. and W.H. Romme. 1993. Invasion of subalpine meadows by lodgepole pine in Yellowstone National Park, WY, USA. *Arctic and Alpine Research* 25: 382-390.

Johnson, D.W. 1995. Soil properties beneath *Ceanothus* and pine stands in the eastern Sierra Nevada. *Soil Science Society of America Journal* 59: 918-924.

Johnson , D.W., R.B. Susfalk, R.A. Dahlgren and J.M. Klopatek. 1998. Fire is more important than water for nitrogen fluxes in semi-arid forests. *Environmental Science and Policy* 1: 79-86.

Jones, J.R. and N.V. DeByle. 1985. Soils, In: pp. 65-70. N.V. DeByle and R.P. Winokur (Eds.). Aspen Ecology and Management in the Western United States. USDA, Forest Service, Rocky Mountain Research Station, Ft. Collins, CO. Research Paper RM-119.

Jordan, C.F. 1985. Nutrient Cycling in Tropical Forest Ecosystems. John Wiley and Sons, NY, 190 p.

Kay, C.E. 1990. Yellowstone's northern elk herd: A critical evaluation of the "natural regulation" paradigm. Ph.D. Dissertation. Utah State University, Logan, UT. 490 p.

Kay, C.E. 1993. Aspen seedlings in recently burned areas in Grand Teton and Yellowstone National Parks. *Northwest Science* 67: 94-104.

Kay, C.E. 1997. Is aspen doomed? *Journal of Forestry* 95: 4-11.

Keane, R.E. and D.G. Long. 1998. A comparison of coarse scale fire effects simulation strategies. *Northwest Science* 72: 76-90.

Keyes, M.R. and C.C. Grier. 1981. Above-and below-ground net production in 40-year-old Douglas-fir stands on low and high productivity sites. *Canadian Journal of Forest Research* 11: 599-605.

Kimmins, J.P. 1996. Importance of soil and role of ecosystem disturbance for sustained productivity of cool temperate and boreal forests. *Soil Science Society of America Journal* 60: 1643-1654.

Klemmedson, J.O. 1994. New Mexican locust and parent material: influence on forest floor and soil macronutrients. *Soil Science Society of America Journal* 58: 974-980.

Klopatek, J.M. 1987. Nutrient patterns and succession in pinyon-juniper ecosystems of northern Arizona. In: pp. 391-396. R.L. Everett (Comp.). Proceedings–Pinyon Juniper Conference, Reno, NV, January 13-16, 1986. USDA, Forest Service, Intermountain Research Station, Ogden, UT. Gen Tech. Rep. INT-215.

Knicker, H., G. Almendros, F.J. Gonzalez-Vila, F. Martin and H.-D. Ludemann. 1996. ^{13}C and ^{15}N- NMR spectroscopic examination of the transformation of organic nitrogen in plant biomass during thermal treatment. *Soil Biology and Biochemistry* 28: 1053-1060.

Kushla, J.D. and W.J. Ripple. 1997. The role of terrain in a fire mosaic of a temperate coniferous forest. *Forest Ecology and Management* 95: 97-107.

Lachowski, H., P. Hardwick, R. Griffith, A. Parsons and R. Warbington. 1997. Faster, better data for burned watersheds needing emergency rehab. *Journal of Forestry* 95: 4-9.

Landsberg, J.D. 1992. Response of ponderosa pine forests in central Oregon to prescribed underburning. Oregon State University, Corvallis. Dissertation Abstracts. 282 p.

Landsberg, J.D. 1994. A review of prescribed fire and tree growth response in the genus *Pinus*. In: pp. 326-346. J. Cohen (Tech. Coord.). Proceedings of the Twelfth Conference on Fire and Forest Meteorology. (Ed.) J. Cohen. Oct. 26-28, 1993. Jekyll Is., GA, Soc. of American Foresters, Baltimore, MD.

Landsberg, J.D. 1997. Fire and forests: fire–a good servant or a bad master. In: pp. 209-213. Forest and Tree Resources. October 13-22, 1997. Proceedings of the XI World Forestry Congress. Antalya, Turkey.

Landsberg, J.D., P.H. Cochran, M.M. Finck and R.E. Martin. 1984. Foliar nitrogen content and tree growth after prescribed fire in ponderosa pine. USDA, Forest Service, Pacific Northwest Forest and Range Experiment Station Research Note, PNW-412, 15 p.

Laven, R.D., P.N. Omi, J.G. Wyatt and A.S. Pinkerton. 1980. Interpretation of the fire scar data from a ponderosa pine ecosystem in the central Rocky Mountains, Colorado. In: pp. 46-49. Proceedings of the Fire History Workshop, Oct. 19-24, 1980. Tucson, AZ. USDA, Forest Service, Rocky Mountain Forest and Range Experiment Station, Ft. Collins, CO. GTR-RM-81.

Loveland, T.R., J.M. Merchant, D.O. Ohlen and J.F. Brown. 1991. Development of a land-cover characteristics database for the conterminous U.S. *Photogrammetric Engineering and Remote Sensing* 57: 1453-1463.

MacDonald, L. H., R. Sampson, D. Brady, L. Juarros and D. Martin. 2000. Predicting Erosion and Sedimentation Risk from Wildfires: A Case Study from Western

Colorado. In Sampson, R.N., R.D. Atkinson, and J.W. Lewis (Eds.), Mapping Wildfire Hazards and Risks. Papers from the American Forests Scientific Workshop, September 29-October 5, 1996, Pingree Park, CO. The Haworth Press, Inc., New York.

Mao, D.M., Y.W. Min, L.L. Yu, R. Martens and H. Insam. 1992. Effect of afforestation on microbial biomass and activity in soils of tropical China. *Soil Biology and Biochemistry* 24: 865-877.

McCann, K., A. Hastings and G.R. Huxel. 1998. Weak trophic interactions and the balance of nature. *Nature* 395: 794-798.

McNabb, D.H. and K. Cromack, Jr. 1990. Effects of prescribed fire on nutrients and soil productivity. In: pp. 125-142. J.D. Walstad, S.R. Radosevich, and D.V. Sandberg (Eds.), Natural and Prescribed Fire in Pacific Northwest Forests. Oregon State University Press, Corvallis, OR.

McNabb, D.H. and F.J. Swanson. 1990. Effects of fire on soil erosion. In: pp. 159-176. J.D. Walstad, S.R. Radosevich, and D.V. Sandberg (Eds.), Natural and Prescribed Fire in Pacific Northwest Forests. Oregon State University Press, Corvallis, OR.

McNulty, S.G. and W.T. Swank. 1996. Analysis of ecosystems using GIS. In: pp. 167-172. Proceedings of Eco-Informa '96 Global Network for Environmental Information. November 4-7, 1996. Lake Buena Vista, FL, Volume 10, Ann Arbor, MI, Environmental Research Institute of Michigan.

Meyer, G.A., S.G. Wells and A.J.T. Jull. 1995. Fire and alluvial chronology in Yellowstone National Park: climatic and intrinsic controls on Holocene geomorphic processes. *Geological Society of America Bulletin* 107: 1211-1230.

Meyer, G.A., S.G. Wells, R.C. Balling, Jr. and A.J.T. Jull. 1992. Response of alluvial systems to fire and climate change in Yellowstone National Park. *Nature* 357: 147-149.

Millspaugh, S.H. and C. Whitlock. 1995. A 750-year fire history based on lake sediment records in central Yellowstone National Park, USA. *The Holocene* 5: 283-292.

Mitchell, J.E. and S.J. Popovich. 1997. Effectiveness of basal area for estimating canopy cover of ponderosa pine. *Forest Ecology and Management* 95: 45-51.

Moir, W.H. and H. Grier. 1969. Weight and nitrogen, phosphorus, potassium and calcium content of forest floor humus of lodgepole pine stands in Colorado. Soil Science Society of America Proceedings 33: 137-140.

Monleon, V.J. and K. Cromack, Jr. 1996. Long-term effects of prescribed underburning on litter decomposition and nutrient release in ponderosa pine stands in central Oregon. *Forest Ecology and Management* 81: 143-152.

Monleon, V.J., K. Cromack, Jr. and J.D. Landsberg. 1997. Short- and long-term effects of prescribed underburning on nitrogen availability in ponderosa pine stands in central Oregon. *Canadian Journal of Forest Research* 27: 369-378.

Monsen, S.B. 1987. Shrub selections for pinyon-juniper plantings. In: pp. 316-329. R.L. Everett, (Comp.), Proceedings–Pinyon Juniper Conference, Reno, NV, January 13-16,1986. USDA, Forest Service, Intermountain Research Station, Ogden, UT. Gen. Tech. Rep. INT-215.

Moore, P.D. 1996. Fire damage soils our forests. *Nature* 384: 312-313.

Morgan, P. and L.F. Neuenschwander. 1988. Seed-bank contributions to regeneration of shrub species after clear-cutting and burning. *Canadian Journal of Botany* 66: 169-172.

Mutch, R.W. 1994. A return to ecosystem health. *Journal of Forestry* 92: 31-33.

Nelson, D.W. and L.E. Sommers. 1996. Total carbon, organic carbon, and organic matter. In: pp. 961-1010. Methods of Soil Analysis, Part 3–Chemical Methods. Soil Science Society of America, Inc. and American Society of Agronomy. Madison, WI.

Neuenschwander, L.F., J.P. Menakis, M. Miller, R.N. Sampson, C. Hardy, R. Averill and R. Mask. 2000. Indexing Colorado Watersheds to Risk of Wildfire. In Sampson, R.N., R.D. Atkinson, and J.W. Lewis (Eds.), Mapping Wildfire Hazards and Risks. Papers from the American Forests Scientific Workshop, September 29-October 5, 1996, Pingree Park, CO. The Haworth Press, Inc., New York.

Ottmar, R.D., E. Alvarado, P.F. Hessburg [and others]. 2000. Historical and current forest and range landscapes in the Interior Columbia River Basin and portions of the Klamath and Great Basins. Part II: Linking vegetation patterns and potential smoke production and fire behavior. USDA, Forest Service, Pacific Northwest Research Station, Portland, OR. Gen. Tech. Rep. PNW-GTR-000. xx pp. (Thomas M. Quigley, Science Team Leader, Interior Columbia River Basin Ecosystem Management Project: Scientific Assessment; Paul Hessburg, Landscape Ecology Staff Co-Leader and Technical Editor (in press).

Paschke, M.W. 1997. Actinorhizal plants in rangelands of the western United States. *Journal of Range Management* 50: 62-72.

Perala, D.A. and D.H. Alban. 1982. Biomass, nutrient distribution and litterfall in Populus, Pinus and Picea stands on two different soils in Minnesota. Plant and Soil 64: 177-192.

Perry, D.A. 1994. Forest Ecosystems. Johns Hopkins Univ. Press, Baltimore.

Perry, D.A. and M..P. Amaranthus. 1997. Disturbance, recovery and stability. In: pp. 31-56. K.A. Kohm and J.F. Franklin (Eds.). Creating a Forestry for the 21st Century. Island Press, Washington, DC.

Philpot, C.W. 1965. Temperatures in a large natural fire. USDA, Forest Service, Pacific Southwest Forest and Range Experiment Station, Berkeley, CA. Research Note PSW-90. 15 p.

Polis, G.A. 1998. Stability is woven by complex webs. *Nature* 395: 744-745.

Pratt, S.D., A.S. Konopka, M.A. Murry, F.W. Ewers and S.D. Davis. 1997. Influence of soil moisture on the nodulation of post fire seedlings of *Ceanothus* spp. growing in the Santa Monica Mountains of southern California. *Physiologia Plantarum* 99: 673-679.

Prescott, C.E., J.P. Corbin and D. Parkison. 1992. Availability of carbon, nitrogen and phosphorus in the forest floor of Rocky Mountain coniferous forests. Canadian Journal of Forest Research 22: 593-600.

Pritchett, W.L. and R.F. Fisher. 1987. Properties and Management of Forest Soils, 2nd Ed. John Wiley & Sons, New York. 494 p.

Rab, M.A. 1996. Soil physical and hydrological properties following logging and slash burning in the Eucalyptus regans forest of southeastern Australia. Forest Ecology and Management 84: 159-176.

Rahman, S., L.C. Munn, G.F. Vance and C. Arneson. 1997. Wyoming Rocky Mountain forest soils: Mapping using an ARC/INFO geographic information system. *Soil Science Society of America Journal* 61: 1730-1737.

Raison, R.J. 1979. Modification of the soil environment by vegetation fires, with particular reference to nitrogen transformations–a review. *Plant and Soil* 51: 73-108.

Raison, R.J., P.K. Khanna and P.V. Woods. 1985. Mechanisms of element transfer to the atmosphere during vegetation fires. *Canadian Journal of Forest Research* 15: 132-140.

Rocky Mountain News. 5 April 1998. Denver, CO.

Read, D. 1998. Plants on the web. *Nature* 396: 22-23.

Rogers, P. 1996. Disturbance ecology and forest management: A review of the literature. USDA, Forest Service, Intermountain Research Station, Fort Collins, CO. Gen. Tech. Rep. INT-GTR-336. 16 p.

Ruha, T.L.A., J.D. Landsberg and R.E. Martin. 1996. Influence of fire on understory shrub vegetation in ponderosa pine stands. In: pp. 108-113. J.R. Barrow, E.D. McArthur, R.E. Sosebee and R.J. Tausch (Compilers), Proceedings: Shrubland Ecosystem Dynamics in a Changing Environment, May 23-25, 1995, Las Cruces, NM. Gen Tech. Rep. INT-GTR-338. Ogden, UT. USDA, Forest Service, Intermountain Research Station.

Ryan, K.C. and E.D. Reinhardt. 1988. Predicting postfire mortality of seven western conifers. *Canadian Journal of Forest Research* 18: 1291-1297.

Sampson, R.N. and L.F. Neuenschwander. 2000. Characteristics of the study area, data utilized, and modeling approach. In Sampson, R.N., R.D. Atkinson, and J.W. Lewis (Eds.), Mapping Wildfire Hazards and Risks. Papers from the American Forests Scientific Workshop, September 29-October 5, 1996, Pingree Park, CO. The Haworth Press, Inc., New York.

Savage, M. 1991. Structural dynamics of a southwestern pine forest under chronic human influence. *Annals of the Association of American Geographers* 81: 271-289.

Savage, M., P.M. Brown and J. Feddema. 1996. The role of climate in a pine forest regeneration pulse in the southwestern United States. *Ecoscience* 3: 310-318.

Schier, G.A., J.R. Joanes and R.P. Winokur. 1985. Vegetative regeneration. In: pp. 29-35. N.F. DeByle and R.P. Winokur (Eds). Aspen: Ecology and Management in the Western United States. USDA, Forest Service, Rocky Mountain Research Station, Ogden, UT. Gen. Tech. Rep. RM-119.

Schimel, D.S. 1998. The carbon equation. *Nature* 393: 208-209.

Scott, D.F. 1993. The hydrological effects of fire in South African mountain catchments. *Journal of Hydrology* 150: 409-432.

Seaber, P.R., F.P. Kapinos and G.L. Knapp. 1987. Hydrologic Unit Maps. U.S. Geological Survey Water-Supply Paper 2294. Denver, CO: USGS, Books and Open-file Reports Section.

Shea, R.W. 1993. Effects of prescribed fire and silvicultural activities on fuel mass and nitrogen redistribution in *Pinus ponderosa* ecosystems of central Oregon. Master's Thesis, Oregon State University, Corvallis. OR. 163 p.

Skinner, T.V. and R.D. Laven. 1983. A fire history of the Longs Peak region of Rocky Mountain National Park. In: pp. 71-74. Seventh Conference on Fire and Forest

Meteorology. April 25-28, 1983. Ft. Collins, CO. Boston: American Meteorological Society.

Smith, D.M., B.C. Larson, M.J. Kelty and P.M.S. Ashton. 1997. The Practice of Silviculture, 9th Ed. John Wiley, NY. 537 p.

Sollins, P., P. Homann and B. A. Caldwell. 1996. Stabilization and destabilization of soil organic matter: mechanisms and controls. *Geoderma* 74: 65-105.

Stratton, R.D. 1999. A GIS assessment method for fire management: Identifying fire danger areas. In: Proceedings of the Fire Effects on Rare and Endangered Species and Habitats–1998 conference. April 5-8, 1998. Coeur d'Alene, ID. The International Association of Wildland Fire. Fairfield, WA (in press).

Sutherland, E.K., W.W. Covington and S. Andariese. 1991. A model of ponderosa pine growth response to prescribed burning. *Forest Ecology and Management* 44: 161-173.

Suzuki, K., H. Suzuki, D. Binkley and T.J. Stohlgren. 1999. Aspen regeneration in the Colorado Front Range: differences at local and landscape scales. *Landscape Ecology* 14: 231-237.

Swetnam, T.W. and C.H. Baisan. 1996. Fire histories of montane forests in the Madrean Borderlands. In: pp. 15-36. P.F. Folliott et al. (Tech. Coords.). Effects of Fire on Madrean Province Ecosystems A Symposium Proceedings. March 11-15, 1996. Tucson, AZ. USDA, Forest Service, Rocky Mountain Forest and Range Experiment Station, Ft. Collins, CO. Gen. Tech. Rep. RM-GTR-289. 277 p.

Swetnam, T.W. 1990. Fire history and climate in the southwestern United States. In: Effects of fire management of southwestern natural resources. USDA, Forest Service, Intermountain Research Station, Ft. Collins, CO. Gen Tech. Rep. RM-191. 293 p.

Swetnam, T.W. and J. H. Dieterich. 1985. Fire history of ponderosa pine forests in the Gila Wilderness, New Mexico. In: pp. 390-397. J.E. Lotan et al. (Tech. Coords.) Proceedings–Symposium and Workshop on Wilderness Fire. USDA, Forest Service, Intermountain Research Station, Ft. Collins, CO. Gen. Tech. Rep. INT-182.

Tiedemann, A.R. 1987a. Nutrient accumulations in pinyon-juniper ecosystems– managing for future site productivity. In: pp. 352-359. R.L. Everett, compiler. Proceedings–Pinyon Juniper Conference, Reno, NV, January 13-16, 1986. USDA, Forest Service, Intermountain Research Station, Ogden, UT. Gen. Tech. Rep. INT-215.

Tiedemann, A.R. 1987b. Combustion losses of sulfur from forest foliage and litter. *Forest Science* 33: 216- 233.

Tiedemann, A.R. and W.P. Clary. 1996. Nutrient distribution in *Quercus gambelii* stands in central Utah. *Great Basin Naturalist* 56: 119-128.

Tiedemann, A.R., W.P. Clary and R.J. Barbour. 1987. Underground systems of Gambel oak (*Quercus gambelii*) in central Utah. *American Journal of Botany* 47: 1065-1071.

Tilman, D., D. Wedin and J. Knops. 1996. Productivity and sustainability influenced by biodiversity in grassland ecosystems. *Nature* 379: 718-720.

U.S. Department of Agriculture, Forest Service, Rocky Mountain Region, Rio Grande National Forest. 1996a. Final Environmental Impact Statement. Fort Collins, CO.

U.S. Department of Agriculture, Forest Service, Rio Grande National Forest portion of the San Juan and Rio Grande National Forests. 1996b. Draft Soil Resource and Ecological Inventory of Rio Grande National Forest–West Part, CO. 350 p.

van Cleve, K. and L.L. Noonan. 1971. Physical and chemical properties of the forest floor in birch and aspen stands in interior Alaska. Soil Science Society of America Proceedings 35: 356-360.

Valette, J.-C., V. Gomendy, J. Marechal, C. Houssard and D. Gillon. 1994. Heat transfer in the soil during very low-intensity experimental fires: the role of duff and soil moisture content. *International Journal of Wildland Fire* 4: 225-237.

van der Heijden, M.G.A., J.N. Klironomos, M. Ursic, P. Moutoglia, R. Streitwolf-Engel, T. Boller, A. Wiemken and I.R. Sanders. 1998. Mycorrhizal fungal diversity determines plant biodiversity, ecosystem variability and productivity. *Nature* 396: 69-72.

Waring, R.H. and S.W. Running. 1998. Forest Ecosystems: Analysis at Multiple Scales. Academic Press, New York. 370 p.

Weaver, H. 1951. Fire as an ecological factor in the southwestern ponderosa pine forests. *Journal of Forestry* 49: 93-98.

White, E.M., W.W. Thompson and F.R. Gartner. 1973. Heat effects on nutrient release from soils under ponderosa pine. *Journal of Range Management* 26: 22-24.

Whitlock, C. 1998. An examination of Holocene fire history in the northwestern United States. In: pp. 7-9. Fire and Wildlife in the Pacific Northwest: Research, Policy, and Management. April 6-8, 1998. The Wildlife Society, Northwest Section, Oregon and Washington Chapters. Spokane, WA. Oregon State University Press. 141 p.

Whitlock, C. and S.H. Millspaugh. 1996. Testing the assumptions of fire-history studies: an examination of modern charcoal accumulation in Yellowstone National Park, USA. *The Holocene* 6: 7-15.

Wooldridge, D.D. 1970. Chemical and physical properties of forest litter layers in central Washington. In: pp. 327-337. C.T. Youngberg and C.B. Davey (Eds.) Tree Growth and Forest Soils. Proceedings of the Third North American Forest Soils Conference. Aug., 1988. North Carolina State University, Raleigh, NC.

Wollum, A.G., II. 1973. Characterization of the forest floor in stands along a moisture gradient in southern New Mexico. *Soil Science Society of America Proceedings* 33: 637-640.

Wright, H.A. and A.W. Bailey. 1982. Fire Ecology: United States and Southern Canada. John Wiley & Sons, NY, 501 p.

Youngberg, C.T. and A.G. Wollum, II. 1976. Nitrogen accretion in developing *Ceanothus velutinus* stands. *Soil Science Society of America Journal* 40: 109-112.

Zackrisson, O., M.C. Nilsson and D.A. Wardle. 1996. Key ecological function of charcoal from wildfire in the Boreal forest. *Oikos* 77: 10-19.

SECTION III

Chapter 9

A Database for Spatial Assessments of Fire Characteristics, Fuel Profiles, and PM$_{10}$ Emissions

Colin C. Hardy
Robert E. Burgan
Roger D. Ottmar

SUMMARY. This paper describes the procedures and data used to develop a database of 28 fire, fuels, and smoke attributes for the broad-scale scientific assessment of the Interior Columbia River Basin. These attributes relate to three general areas: (1) fire weather, fuel moisture, and fire characteristics; (2) fuel loading and fuel consumption; and

Colin C. Hardy and Robert E. Burgan are affiliated with USDA Forest Service, Intermountain Research Station, Intermountain Fire Sciences Laboratory, Missoula, MT 59807. Roger D. Ottmar is affiliated with USDA Forest Service, Pacific Northwest Research Station, Seattle, WA 98105.

[Haworth co-indexing entry note]: "Chapter 9. A Database for Spatial Assessments of Fire Characteristics, Fuel Profiles, and PM$_{10}$ Emissions." Hardy, Colin C., Robert E. Burgan, and Roger D. Ottmar. Co-published simultaneously in *Journal of Sustainable Forestry* (Food Products Press, an imprint of The Haworth Press, Inc.) Vol. 11, No. 1/2, 2000, pp. 229-244; and: *Mapping Wildfire Hazards and Risks* (ed: R. Neil Sampson, R. Dwight Atkinson, and Joe W. Lewis) Food Products Press, an imprint of The Haworth Press, Inc., 2000, pp. 229-244. Single or multiple copies of this article are available for a fee from The Haworth Document Delivery Service [1-800-342-9678, 9:00 a.m. - 5:00 p.m. (EST). E-mail address: getinfo@ haworthpressinc.com].

(3) PM_{10} smoke emissions. The process flow and development protocols for creation of the database are fully described and illustrated, with examples provided where appropriate. This database was developed for application to a certain geographic area with parameters specific to both the biophysical environment and the management issues of that area. However, the methods and protocols used to develop this comprehensive suite of fire-related data are applicable to any ecosystem for which predictions are needed for wildfire hazard, fire potential, biomass consumption, and smoke emissions. *[Article copies available for a fee from The Haworth Document Delivery Service: 1-800-342-9678. E-mail address: <getinfo@haworthpressinc.com> Website: <http://www.haworthpressinc.com>]*

KEYWORDS. Wildfire, smoke, particulates

INTRODUCTION

Consideration of the role and relative impact of fire is essential in landscape-scale assessments of ecosystem conditions and trends and in the development and assessment of alternative future management strategies. The need to recognize the role of fire as a critical agent of change was recognized nearly a century ago by Gifford Pinchot, who called fire "one of the great factors which govern the distribution and character of forest growth" (Pinchot 1899).

Indeed, the role of fire under historic, current, and future management scenarios has recently become a significant consideration in several regional and multi-regional ecosystem assessments. For example, potential fire behavior and related smoke production from historical and current landscapes in eastern Oregon and Washington were recently studied in conjunction with the Eastside Forest Ecosystem Health Assessment (Huff et al. 1995). For forests of the Sierra Nevada, the ecological functions of fire and their associated management implications were recently addressed in the Sierra Nevada Ecosystem Project technical assessments (University of California, Davis 1996). On an even larger scale, the Grand Canyon Visibility Transport Commission has completed a draft assessment of smoke emissions from various wildfire and prescribed fire treatment scenarios for a ten-state transport region across the western United States (Grand Canyon Visibility Transport Commission 1996). Most recently, an integrated scientific assessment has been completed for ecosystems in the Interior Columbia River Basin (ICRB). This multiagency assessment was chartered to support the evaluation of alternatives for management of all federal lands within the ICRB (USDA Forest Service and USDI Bureau of Land Management 1994).

The Fire-Fuels-Emissions Database described in this paper was developed for the broadscale scientific assessment of the ICRB. The database developed for the ICRB provides 28 fire- and fuels-related attributes which relate to three general areas: (1) fire weather, fuel moisture, and fire characteristics; (2) fuel loading and fuel consumption; and (3) PM_{10} smoke emissions. While this particular project relates to the ICRB effort, the methods and protocols used to develop this comprehensive suite of fire-related data are applicable to any ecosystem for which predictions are needed of wildfire hazard, fire potential, biomass consumption, and smoke emissions.

For the broadscale ICRB assessment, this database was initially linked to data from two other map layers: (1) Version 1.0 of the ICRB Current Cover-type Map, which has 46 vegetation cover types–43 wildland vegetation types and three other types (water, barren ground, and agriculture); and (2) The Structural Development Stage for each respective Current Covertype–there are seven possible structural development stages for each of the 46 cover-types. For purposes of successional modeling, these linkages were subsequently extended to other historic and current potential vegetation data through a series of crosswalks. Neither these secondary linkages nor their application to the successional modeling will be discussed in this paper.

The database provides estimates of fire, fuels, and emissions characteristics for any classification category, whether pixel or polygon, within which a fire (wildfire or prescribed fire) may occur. The database is designed to provide estimates for any of three unique fire-weather scenarios: wet, normal, and dry. The occurrence of a prescribed or wildfire is invoked through probabilities set within each of several management scenarios developed independently of this database. The effects from a specific fire type (prescribed versus wildfire) are determined by triggering one of two subsets of this database: wildfires trigger the *"Dry Scenario"* data subset, and prescribed fire triggers the *"Normal Scenario"* data subset. These weather-dependent subsets are explained later in this description.

The individual activities comprising the development process for this database are diagrammed in Figure 9.1. Combining the covertype-structural development stage couplets with each of the three weather scenarios results in a 966-cell matrix. Each cell in this matrix could thereby be populated with a unique set of fire, fuels, and emissions characteristics.

METHODS

There are many interdepencies within the database. These dependencies are generally illustrated by the heavy lines in the diagram of activities shown in Figure 9.1. The fire weather and fire characteristics (behavior) were determined first (right-side column of Figure 9.1). These data were then used in the fuel consumption calculations (middle column of Figure 9.1), which

FIGURE 9.1. The general flow of development activities for the database.

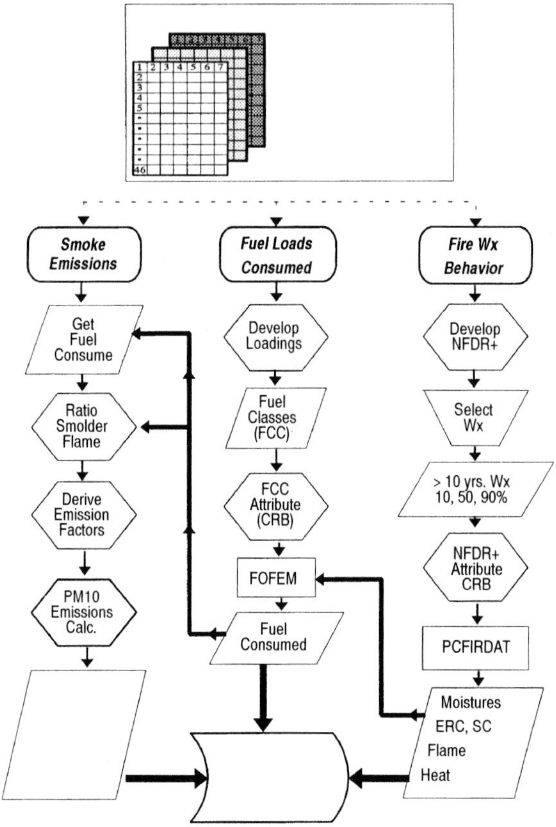

ultimately were used to derive estimates of smoke emissions production (left-side column of Figure 9.1).

Each of the procedures used in the development of the Fire-Fuels-Emissions Database for the broadscale ICRB assessment are discussed in the following sections. Throughout the discussion, reference will be made to individual elements of the detailed process-flow diagram shown in Figure 9.2, where numbers for the elements relating to emissions or fuels are prefaced by the letters 'E' or 'F,' respectively.

Fuel Moisture and Fire Characteristics

Two distinct sets of fuels attributes were developed for the database: (1) stylized fuel models for determining fire characteristics such as wildfire

FIGURE 9.2. The specific database development process and elements.

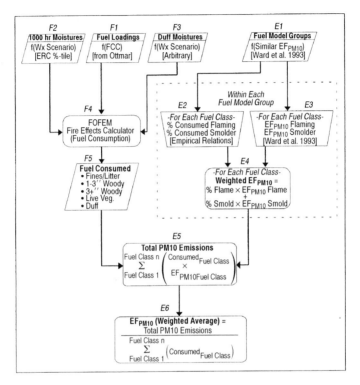

hazard; and (2) fuels attributes for calculating fuel consumption and smoke emissions. The stylized fuel models represent only biomass in the upper duff, surface litter, and vertically oriented vegetation within about six feet of the ground. They are not appropriate for estimating smoke production that results from long term fuel consumption behind the fire front. Therefore, the following discussion is limited to the first set of fuels attributes (the stylized fuel models). Fuels attributes relating to fire effects such as fuel consumption and smoke emissions will be discussed under the section heading "Fuel Attributes for Fuel Consumption."

Stylized fuel models–Fuel models are a set of numbers that describe vegetation characteristics in terms that are required by mathematical fire models for computing fire potential. There are two fuel model sets that have been in use for many years–those used in the National Fire Danger Rating System, called the "1978 NFDRS" models (Deeming et al. 1977), and those used in the fire behavior system (Albini 1976). Each fuel model set is specifically designed for the system in which it is used. Thus, there was an initial choice to be made between using fire behavior fuel models and fire danger fuel models. The NFDRS fuel model set was selected because it has two more fuel components than the fire behavior fuel model set–one more dead fuel component and one more live fuel component. These extra fuel components improve the capability to portray seasonal fire potential variation. The NFDRS fuel model parameters are: (1) dead fuel loads by size class, (2) live herbaceous and shrub loads, (3) fuel bed depth, (4) fuel heat content, and other parameters not directly related to biomass. Ten of the twenty standard 1978 NFDRS fuel models were used to represent fire characteristics in the database (Table 1).

These ten base NFDRS fuel models did not provide enough resolution to represent all the required vegetation covertype/structural stage combinations for the ICRB assessment. Therefore, we used an expanded set of the 1978 NFDRS fuel models produced by both increasing and decreasing the loading (mass) assigned to each live and dead fuel class by one-third. By adding new fuel models with one third greater and one third less fuel (by mass) we tripled the resolution of the fuels information and decreased the corresponding intervals between adjacent models. Fuel bed depth was adjusted to maintain a constant packing ratio (pounds of fuel per cubic foot of fuel bed) and a characteristic surface area-to-volume ratio to mitigate the potential of producing an aberrant fuel model. These fuel models were labeled to indicate low, medium and high loadings of the original NFDRS fuel models. The two-letter model designations are of the form *model-load*; that is, the model letter precedes the loading level, as in "CL," "CM," and "CH" where the letter "C" defines the NFDRS fuel model, and "L," "M," and "H" represent 1/3 less than the original load (low), the original load (medium), and 1/3 more than original load (high), respectively. Four of the ten NFDRS fuel

TABLE 1. Ten National Fire Danger Rating System (NFDRS) fuel models used to represent fire characteristics in the database.

NFDRS fuel model	Fuel model description
A	Western annual grasses and forbs
C	Open pine stands with litter and branchwood fuels
F	Mature, closed chamise and oakbrush
G	Dense conifer stands with heavy litter and woody debris
H	Short-needled, healthy conifers
L	Perennial western grasslands
R	Hardwood areas after leafout
S	Alpine tundra; lichens and mosses
T	Sagebrush-grass types
U	Western long-needled pines; closed canopy

models were assigned fewer than three scaled levels. The resulting 25 scaled-models derived for the ICRB are described in Table 2. These 25 scaled models were then linked to each vegetation covertype/structural stage combination on the basis of the fuel model descriptions and expert knowledge.

Fire-weather scenarios–It is assumed that climatology and site conditions over the last several hundred years have defined the geographic location of the various vegetation types within the ICRB and that individual weather stations can represent the vegetation types within which they occur. Weather data from National Fire Danger Rating System (NFDRS) weather stations were used because these data provide measures of all the inputs required by the NFDRS processor (Deeming et al. 1977). Weather data sets from 17 NFDRS fire weather stations identified as being among the best agency-operated weather stations in California, Oregon, Idaho, and Washington provided the base data for the ICRB analysis area (Figure 9.3).

The period July 1 through August 31 for the years 1978 through 1992

TABLE 2. Scaled NFDRS fuel models for the ICRB assessment.

ICRB fuel model	Fuel mass (loading) by NFDRS fuel component										Fuel depth	Packing ratio	Relative packing ratio	Surface area/ volume
	1 hr	10 hr	100 hr	1000 hr	Wood	Herbs	1 hr + herbs	Total live	Total 0-3''	Grand total				
	------ tons of fuel per acre ------										(feet)			(ft²/ft³)
AL	0.13	0.00	0.00	0.00	0.00	0.20	0.33	0.20	0.13	0.33	0.53	0.0007	0.151	3000
AM	0.20	0.00	0.00	0.00	0.00	0.30	0.50	0.30	0.20	0.50	0.80	0.0007	0.151	3000
AH	0.27	0.00	0.00	0.00	0.00	0.40	0.67	0.40	0.27	0.67	1.07	0.0007	0.151	3000
CL	0.27	0.67	0.00	0.00	0.33	0.53	0.80	0.87	0.93	1.80	0.50	0.0034	0.544	2114
CM	0.40	1.00	0.00	0.00	0.50	0.80	1.20	1.30	1.40	2.70	0.75	0.0034	0.544	2114
CH	0.53	1.33	0.00	0.00	0.67	1.07	1.60	1.73	1.87	3.60	1.00	0.0034	0.544	2114
FL	1.67	1.33	1.00	0.00	6.00	0.00	1.67	6.00	4.00	10.00	3.00	0.0027	0.261	1155
FM	2.50	2.00	1.50	0.00	9.00	0.00	2.50	9.00	6.00	15.00	4.50	0.0027	0.261	1155
FH	3.33	2.67	2.00	0.00	12.00	0.00	3.33	12.00	8.00	20.00	6.00	0.0027	0.261	1155
GM	2.50	2.00	5.00	12.00	0.50	0.50	3.00	1.00	9.50	22.50	1.00	0.0172	2.434	1848
GH	3.33	2.67	6.67	16.00	0.67	0.67	4.00	1.33	12.67	30.00	1.33	0.0172	2.434	1848
HL	1.00	0.67	1.33	1.33	0.33	0.33	1.33	0.67	3.00	5.00	0.20	0.0287	4.074	1858
HM	1.50	1.00	2.00	2.00	0.50	0.50	2.00	1.00	4.50	7.50	0.30	0.0287	4.074	1858
HH	2.00	1.33	2.67	2.67	0.67	0.67	2.67	1.33	6.00	10.00	0.40	0.0287	4.074	1858
LL	0.17	0.00	0.00	0.00	0.00	0.33	0.50	0.33	0.17	0.50	0.67	0.0007	0.108	2000
LM	0.25	0.00	0.00	0.00	0.00	0.50	0.75	0.50	0.25	0.75	1.00	0.0007	0.108	2000
LH	0.33	0.00	0.00	0.00	0.00	0.67	1.00	0.67	0.33	1.00	1.33	0.0007	0.108	2000
RM	0.50	0.50	0.50	0.00	0.50	0.50	1.00	1.00	1.50	2.50	0.25	0.0015	1.484	1657
SL	0.33	0.33	0.33	0.33	0.33	0.33	0.67	0.67	1.00	2.00	0.27	0.0072	0.795	1372
SM	0.50	0.50	0.50	0.50	0.50	0.50	1.00	1.00	1.50	3.00	0.40	0.0072	0.795	1372
TL	0.67	0.33	0.00	0.00	1.67	0.33	1.00	2.00	1.00	3.00	0.83	0.0029	0.415	1900
TM	1.00	0.50	0.00	0.00	2.50	0.50	1.50	3.00	1.50	4.50	1.25	0.0029	0.415	1900
TH	1.33	0.67	0.00	0.00	3.33	0.67	2.00	4.00	2.00	6.00	1.67	0.0029	0.415	1900
UL	1.00	1.00	0.67	0.00	0.33	0.33	1.33	0.67	2.67	3.33	0.33	0.0158	2.077	1694
UM	1.50	1.50	1.00	0.00	0.50	0.50	2.00	1.00	4.00	5.00	0.50	0.0158	2.077	1694

FIGURE 9.3. Location of the 17 NFDRS fire-weather stations used in the analysis.

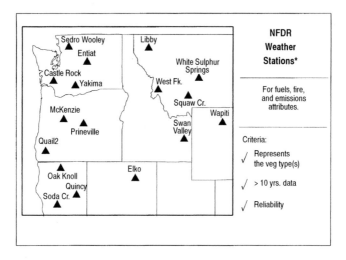

defined the temporal scope of the analysis. The weather data were analyzed with respect to fire potential using a predicted NFDRS index called the energy release component (ERC) which provides a measure of energy release–BTUs released per unit area burned within the fire front. The analysis was based on ERC because it is largely a function of fuel load and moisture, and is not influenced by wind speed. The database was designed to represent the broadest possible range of realistic fire-weather scenarios for the analysis area, so it was determined that fuel moistures for the wet, normal, and dry fire-weather scenarios would be derived from median weather data for which the percentile ERCs were 10, 50, and 90, respectively. For those models that have a small ERC range (AL, AM, AH), another NFDRS index called the burning index, or BI (which does consider wind speed) was used to provide a larger range from which to select fuel moistures.

Processing of the weather data to derive fire- and fuels-related attributes was done using the programs *pcFIRDAT* and *pcSEASON* (California Department of Forestry and Fire Protection 1994), which are adaptations of the original *FIREFAMILY* set of historical fire danger analysis programs (Main et al. 1990). The program *pcFIRDAT* was used to calculate fuel moistures and NFDRS indices and components (Deeming et al. 1977). The resulting output file was processed using *pcSEASON* to delineate the median values of the lowest 20 percent, middle 20 percent, and highest 20 percent daily ERC. These median values were also the 10th, 50th, and 90th ERC percentiles.

National fire danger rating weather stations are always located in the open because they are meant to monitor "near worst case" conditions. It was therefore necessary to adjust the fuel moistures for the effect of shading by forest canopies. No shading adjustment was made for shrub or grass vegetation types. A fixed increase of 4.0 percent was made for all the dead fuel moisture classes to adjust the moistures for shaded conditions. For those fuel models requiring a BI calculation, the 20 foot wind speeds were adjusted to account for differences between ICRB cover types and the NFDRS fuel model-dependent wind reduction factors. The adjusted fuel moistures and associated NFDRS indices (heat per unit area, fireline intensity, flame lengths) occurring at the 10th, 50th, and 90th percentile ERC or BI values for each ICRB vegetation cover type/structural stage combination were then entered into the database for the wet, normal, and dry fire-weather scenarios (Figure 9.2, elements F2, F3). The fuel moistures were subsequently used to calculate fuel consumption and immediate fire effects for each weather scenario, and the NFDRS indices were used to display tradeoffs in fire potential among the various weather and management scenarios.

Fuel Attributes for Fuel Consumption

Fuel loading–Seven attributes were used to describe the fuel loading characteristics of each vegetation covertype/structural stage combination. The loading attribute data were parsed from an existing matrix of predetermined fuel condition classes (FCC) developed for a Fire Emissions Tradeoff Model used in another study (Schaaf 1996). For that study, a team of fuel specialists, fire managers, and research personnel developed the fuel profiles assigned to each FCC. The resulting 188 FCCs represent nine general vegetation types, four age classes, three levels of fuel loading, and nine harvest and/or fuel management activities. Each FCC has nine fuel loading components–two live, six dead and downed woody, and duff. Since these included all of the attributes needed for the ICRB assessment, it was thereby possible to link appropriate FCC classes to each broadscale ICRB vegetation covertype/structural stage combination. These loading estimates were only used to predict first-order fire effects such as fuel consumption and emissions production, and the values are not necessarily comparable to those described in the stylized NFDRS fuel models.

Fuel consumption–The fuel loading attributes and respective fuel moisture values for each weather scenario (Figure 9.2, elements F1, F2, F3) were used with the First Order Fire Effects Model (FOFEM) (Reinhardt et al. 1996) to calculate fuel consumption, by fuel component, for each unique fuels/weather combination (Figure 9.2, element F4). The consumption values for the eight FOFEM fuels components were subsequently consolidated into five components for the ICRB assessment (Figure 9.2, element F5); the results of this consolidation are shown in Table 3. For each of the three weather scenar-

TABLE 3. Fuel loading components for NFDRS, FOFEM, and ICRB assessment.

Fuel profile component	Model application		
	NFDRS	FOFEM	CRB
Litter		Litter	Fine
1-hr (0-1/4″)	X	Fine	Fine
10-hr (1/4-1″)	X	Fine	Fine
100-hr (1-3″)	X	Small	Small
1000-hr (3-9″)	X	Large	Large
Woody (> 9″)	X	Large	Large
Duff		Duff	Duff
Herbs	X	Herbs	Live
Shrubs		Shrubs	Live
Regen.		Regen.	Live

ios, each vegetation covertype/structural stage combination in the database was then attributed with the respective set of fuel consumption values.

PM_{10} Emissions

Aggregating the data by emissions group–Many of the vegetation covertype/structural stage combinations shared common emissions characteristics. The NFDRS fuel model assignments were used to key the database into three groupings of the original 10 base NFDRS fuel models from Table 1 (Figure 9.2, element E1). The three groupings are referenced by their dominant vegetation: conifers–NFDRS models G, H, R, U; shrubs–NFDRS models F, T; and grasses–NFDRS models A, C, L, S.

Emission factors–Emission factors for smoke from wildland fires are strongly related to the fire conditions associated with the combustion of a given fuel component. Combustion efficiency α is a term used to describe the fire condition relative to its emission source strength. Combustion efficiency is the proportion of carbonaceous emissions from combustion that are converted into CO_2. For example, perfect combustion would produce only CO_2 and water, and $\alpha = 1.00$. Anything less than perfect combustion creates other products, such as CO, CH_4, and particulate matter. Ward and Hardy (1991) have synthesized various emissions data into linear functions used to predict emission factors from α. While a function using α to predict PM_{10} (particulate matter smaller then 10 micrometers in mean mass-diameter) has not been directly derived from observations, one can be estimated from known size-class distributions of particulate matter (Ward et al. 1993). A linear function derived for PM_{10} is shown in Figure 9.4. Rather than estimating an average α for each fuel class, a separate α was estimated for each of two phases of

FIGURE 9.4. The linear function for predicting an emission factor for PM_{10} as derived from combustion efficiency.

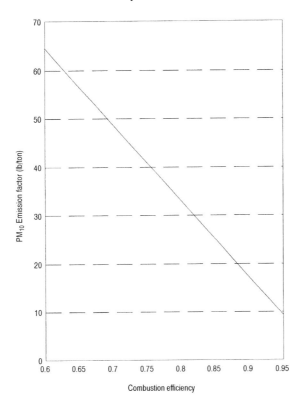

combustion (flaming and smoldering) for each fuel class, within each of the three emissions groups (Table 4). A weighted-average α was then computed for each fuel class using the proportions of consumption expected to occur in each of the combustion phases. Finally, the weighted-average α for each of the five fuel classes was used in the linear function (Figure 9.4) to derive specific PM_{10} emission factors for the wet, normal, and dry fire-weather scenarios (Table 4) (Figure 9.2, elements E2, E3, E4).

Total PM_{10} Emissions per Unit Area

The total mass of PM_{10} produced per unit area from each fuel component is the product of the mass of the fuel component consumed and the weighted-average emission factor for the respective fuel component. The total mass of PM_{10} produced per unit area from all fuel within the vegetation type/structural stage combination is the sum of the PM_{10} emissions from the five fuel components. The example shown in Table 5 lists the weighted-average emission factors for each fuel component (PM_{10} EF-bar) and their "grand weighted-average" (24.0 lb/ton) for the conifer emissions group. The grand weighted-average value is the mean of the individual PM_{10} emission factors for each fuel component, weighted by the total fuel consumed within each fuel component. Also given in the example shown in Table 5 are the total PM_{10} emissions for each fuel component (Total PM_{10}) and their sum (419.7 lb/acre).

TABLE 4. Combustion efficiencies, flaming/smoldering (smold) proportions, and PM_{10} emission factors for each fire-weather scenario.

Emission group	CRP fuels	Combustion efficiency Flame	Combustion efficiency Smold	Wet % consumed Flame	Wet % consumed Smold	Wet PM_{10}	Normal % consumed Flame	Normal % consumed Smold	Normal PM_{10}	Dry % consumed Flame	Dry % consumed Smold	Dry PM_{10}
Grass	Fines	0.95	0.76	1.0	0.0	09.3	1.0	0.0	09.3	1.0	0.0	09.3
	Small	0.92	0.76	1.0	0.0	14.0	1.0	0.0	14.0	1.0	0.0	14.0
	Large	0.92	0.76	0.5	0.5	26.6	0.7	0.3	21.6	0.8	0.2	19.1
	Live	0.85	0.76	1.0	0.0	25.1	1.0	0.0	25.1	1.0	0.0	25.1
	Duff	0.90	0.76	0.5	0.5	28.2	0.4	0.6	30.4	0.4	0.6	30.4
Shrubs	Fines	0.95	0.76	1.0	0.0	09.3	1.0	0.0	09.3	1.0	0.0	09.3
	Small	0.92	0.76	1.0	0.0	14.0	1.0	0.0	14.0	1.0	0.0	14.0
	Large	0.92	0.76	0.5	0.5	26.6	0.7	0.3	21.6	0.8	0.2	19.1
	Live	0.91	0.76	1.0	0.0	15.6	1.0	0.0	15.6	1.0	0.0	15.6
	Duff	0.90	0.76	0.5	0.5	28.2	0.4	0.6	30.4	0.4	0.6	30.4
Conifers	Fines	0.95	0.76	1.0	0.0	09.3	1.0	0.0	09.3	1.0	0.0	09.3
	Small	0.92	0.76	0.9	0.1	16.6	1.0	0.0	14.0	1.0	0.0	14.0
	Large	0.92	0.76	0.5	0.5	26.6	0.7	0.3	21.6	0.8	0.2	19.1
	Live	0.85	0.76	1.0	0.0	25.1	1.0	0.0	25.1	1.0	0.0	25.1
	Duff	0.90	0.76	0.5	0.5	28.2	0.4	0.6	30.4	0.4	0.6	30.4

TABLE 5. Example calculations of weighted-average PM_{10} emission factors and total PM_{10} emissions for the conifers emissions group.

Fuelbed component	Total fuel consumed	Fraction consumed Flame	Fraction consumed Smolder	Consumed Flame	Consumed Smolder	Combustion Effcy. Flame	Combustion Effcy. Smolder	PM_{10} -bar	Total PM_{10}
				- - (tons/acre) - -				(lb/ton)	(lb/acre)
Fine Woody	1.90	1.0	0.0	1.90	0.00	0.95	0.76	9.3	17.7
1-3'' Woody	0.70	1.0	0.0	0.70	0.00	0.92	0.76	14.0	9.8
3''+ Woody	6.50	0.7	0.3	4.55	1.95	0.92	0.76	21.6	140.4
Live Veg	0.70	1.0	0.0	0.70	0.00	0.85	0.76	25.1	17.5
Duff/Litter	7.70	0.4	0.6	3.08	4.62	0.90	0.76	30.4	234.3
Grand-average								24.0	419.7

RESULTS AND DISCUSSION

The data dictionary for the Fire-Fuels-Emissions Database is presented in Table 6. The database contains the complete set of attributes relating to fire characteristics, fuel components, and PM_{10} emissions for each of the three weather scenarios (dry, normal, and wet). The assumption used for the ICRB assessment was that the data for the dry scenario represented wildfire conditions and the data for the normal scenario were appropriate for prescribed fire conditions. The data from the wet scenario would not apply to most fire events, since the fuel moisture values are typically too wet to sustain combustion.

Spatial Analysis of Emissions Production

The total PM_{10} values in the Fire-Fuels-Emissions Database represent the total mass of PM_{10} emissions produced per unit area, expressed in units of

TABLE 6. Data definition table for the ICRB Fire-Fuels-Emissions Database.

Data Name	Columns	Description
Data ID#	1-5	Unique identifier.
CRB ID#	9-10	Current VegType Map Value.
Str. Stg	15	Structural Development Stage.
Wx % tile	19-21	Weather scenario (Dry, Normal, Wet).
CRB Name	24-29	SAF, SRM or CRB (custom) Classification.
Weather Station	32-39	Representative fire weather station.
Fuel Row	42-44	Cross-reference to "Ottmar's FCC" Table.
Loading:		
0-1	46-49	Oven-dry mass of 0-1″ biomass (tons/acre).
1-3	51-54	Oven-dry mass of 1-3″ biomass (tons/acre).
3+	56-59	Oven-dry mass of biomass > than 3″ diameter (tons/acre).
Live	61-64	Oven-dry mass of live vegetation (tons/acre).
Duff	66-69	Oven-dry mass of duff (tons/acre).
Totl	71-74	Total oven-dry biomass (tons/acre).
Fuel Model	78-79	"Scaled" NFDR Fuel Model.
Moist 3+	84-85	Moisture content of 3+ material (percent).
PCFIRDAT:		
ERC	88-90	Energy Release Component (dimensionless).
SC	93-94	Spread Component (dimensionless).
Heat	97-99	Heat perUnit Area (BTU/foot^2).
I-B	101-103	Byrom's Fireline Intensity (BTU/fireline foot/sec).
FL	107-109	Flame Length (feet).
Consumed:		
0-1	110-113	Consumption of 0-1″ biomass (tons/acre).
1-3	115-118	Consumption of 1-3″ biomass (tons/acre).
3+	120-123	Consumption of biomass > than 3″ diameter (tons/acre).
Duff	125-128	Consumption of duff (tons/acre).
Live	130-133	Consumption of live vegetation (tons/acre).
Totl	136-139	Total consumption of biomass (tons/acre).
PM_{10}:		
Total	142-146	Total mass of PM_{10} emissions (pounds/acre).
EFbar	149-152	Weighted-average emission factor for PM_{10}.

pounds of PM_{10} per acre burned (lb/acre). In a spatial analysis, the area burned (in this example expressed in acres) within each respective vegetation covertype/structural stage combination is multiplied by the total PM_{10} values (lb/acre) to calculate total PM_{10} emissions from a fire within the vegetation covertype/structural stage combination. The sum of PM_{10} emissions produced from all the vegetation covertype/structural stage combinations burned is the total PM_{10} emission from the fire event(s).

An example of the application of these data for the spatial assessment of the ICRB is the comparison of both energy release component (ERC) and PM_{10} emissions between wildfire and prescribed fire under three different management scenarios. The three management scenarios in this example include: (1) Current Management; (2) Active Management; and (3) Passive Management. The long-term consequences of managing dry forests in the ICRB under each of these scenarios were predicted using the vegetation dynamics model CRBSUM (Keane et al. 1996), which is linked to the Fire-Fuels-Emissions Database through a matrix of cross-walks. Table 7 lists the predicted total burned area of dry forest having 80th percentile values for ERC and PM_{10}, expressed as hectares per decade. In this example, the Active Management scenario would be selected if the only objective were to minimize increases in ERC–note the dramatic shift in 80th percentile ERC from wildfire to prescribed fire when comparing Current and Active Management scenarios, respectively. While PM_{10} emissions are least under the Current Management scenario, both ERC and PM_{10} emissions are highest (by approximately three times) under the Passive Management scenario. The Fire-Fuels-Emission Database has supported dozens of similar analyses for the ICRB assessment and for the subsequent development of the draft environmental impact statement.

Other Applications for These Procedures

The 966-cell database developed for the ICRB scientific assessment is specific to the ICRB with respect to the vegetation covertypes and structural

TABLE 7. An example analysis product from the Fire-Fuels-Emission Database comparing the long-term consequences of three management scenarios on ERC and PM_{10} emissions from wildfire and prescribed (Rx) fire.

Data	Current Management			Active Management			Passive Management		
	Wild fire	Rx fire	Total	Wild fire	Rx fire	Total	Wildfire	Rx	Total fire
	----------------------------------- Hectares per decade -----------------------------								
ERC 80th percentile	332,000	61,000	393,000	143,000	209,000	352,000	1,154,000	13,000	1,167,000
PM_{10} 80th percentile	257,000	101,000	358,000	106,000	275,000	381,000	965,000	16,000	981,000

stages defined for the area. However, many (if not most) of the covertype/ structural stage combinations referenced in the database exist elsewhere across the western United States. While much of the current data prepared for the broadscale ICRB assessment are transportable to many other spatial landscape assessments, perhaps the most signficant product from this effort is the methodology and protocols used in the database development. Similar protocols were recently used in an analysis of emissions tradeoffs between prescribed fires and wildfires (Schaaf 1996). Work recently reported for the Grand Canyon Visibility Transport Commission utilized many components of these data as well as the procedures used to derive them (Grand Canyon Visibility Transport Commission 1996). The methodology used here closely follows the procedures used in regional and national emissions inventory efforts (Ward et al. 1993). Current model simulations of the effects of fire management alternatives on landscape processes and the concomitant implications for global change rely extensively on both the data and procedures presented here. The Western States Air Resource Council (WESTAR) is currently developing modeling capabilities for assessing regional transport and impacts of smoke from prescribed burning. The WESTAR effort will utilize some of these database components and procedures. The fuel condition classes, emission factors, and fuel consumption values are timeless and can be applied in a broad spectrum of landscape analysis activities.

REFERENCES

Albini, F.A. 1976. Estimating wildfire behavior and effects. Gen. Tech. Rep. INT-30. Ogden, UT: U.S. Department of Agriculture, Forest Service, Intermountain Research Station. 92 p.

California Department of Forestry and Fire Protection. 1994. pcFIRDAT/pcSEASON user's guide: FIREFAMILY for personal computers. Sacramento, CA. California Department of Forestry and Fire Protection. 45 p.

Deeming, J.E., R.E. Burgan, and J.D. Cohen. 1977. The national fire danger rating system–1978. Gen. Tech. Rep. INT-39. Ogden, UT: U.S. Department of Agriculture, Forest Service, Intermountain Research Station. 63 p.

Grand Canyon Visibility Transport Commission. 1996. Report of the Grand Canyon Visibility Transport Commission: recommendations for improving western vistas. Denver, CO: Western Governor's Association. 91 p.

Huff, M.H., R.D. Ottmar, E. Alvarado, R.E. Vihnanek, J.F. Lehmkuhl, P.F. Hessburg, and R.L. Everett. 1995. Historical and current forest landscape changes in eastern Oregon and Washington. Part II: Linking vegetation characteristics to potential fire behavior and related smoke production. Gen. Tech. Rep. PNW-GTR-355. Portland, OR: US Department of Agriculture, Forest Service, Pacific Northwest Research Station. 43 p.

Keane, R.E., J.P. Menakis, D.G. Long, W.J. Hann, and C. Bevins. 1996. Simulating coarse-scale vegetation dynamics with the Columbia River Basin succession model–CRBSUM. Gen. Tech. Rep. INT-GTR-340. Ogden, UT: U.S. Department of Agriculture, Forest Service, Intermountain Research Station. 50 p.

Main, W.A., D.M. Paananen, and R.E. Burgan. 1990. FIREFAMILY 1988. Gen. Tech. Rep. NC-138. St. Paul, MN: U.S. Department of Agriculture, Forest Service, North Central Forest Experimental Station. 35 p.

Pinchot, G. 1899. The relation of forests and forest fire. National Geographic. 10:393-403.

Reinhardt, E.D., R.E. Keane, and J.K. Brown. 1996. First order fire effects model–FOFEM 4.0–User's guide. Gen. Tech. Rep. INT-GTR-344. Ogden, UT: U.S. Department of Agriculture, Forest Service, Intermountain Research Station. 65 p.

Schaaf, M.D. 1996. Fire emissions tradeoff model (FETM) and application to the Grand Ronde River Basin, Oregon. Portland, OR. U.S. Department of Agriculture, Forest Service, Pacific Northwest Research Station; technical report; CH2MHill contract 53-82FT-03-2: [not paged].

University of California, Davis. 1996. Summary of the Sierra Nevada ecosystem project report. Davis, CA: University of California, Davis, Centers for water and wildland resources. 22 p.

USDA Forest Service; USDI Bureau of Land Management. 1994. Charter–Interior Columbia River Basin ecosystem management framework and scientific assessment and eastside Oregon and Washington environmental impact statement. Walla Walla, WA: US Department of Agriculture, Forest Service and US Department of Interior, Bureau of Land Management. 12 p.

Ward, D.E., and C.C. Hardy. 1991. Smoke emissions from wildland fires. Environment International. 17:117-134.

Ward, D.E., J.L. Peterson, and W. Hao. 1993. An inventory of particulate matter and air toxic emissions from prescribed fire in the USA for 1989. In: proceedings of the 86th annual meeting and exhibition of the Air and Waste Management Association; 1993 June 13-18; Denver, CO. Pittsburgh, PA. Air and Waste Management Association; paper #93-MP-6.04. [not paged].

Chapter 10

Mapping Ecological Attributes Using an Integrated Vegetation Classification System Approach

James P. Menakis
Robert E. Keane
Donald G. Long

SUMMARY. Land managers need vegetation maps to inventory, monitor, and manage ecological resources across multiple spatial and temporal scales. Current vegetation maps usually only describe one vegetation characteristic, such as cover types, across the landscape. Although these maps provide important information for land management, they often fall short of addressing key issues like forest health and ecosystem management. In this paper we present an integrated approach where three different vegetation classifications are used in concert to spatially characterize many ecological attributes such as snag densities, insect susceptibility, and fire behavior across the landscape. Two examples from the Pacific Northwest are used to illustrate how this approach can be used to describe fuel characteristics and resource hazard across multiple scales. *[Article copies available for a fee from The Haworth Document Delivery Service: 1-800-342-9678. E-mail address: <getinfo@haworthpressinc. com> Website: <http://www.haworthpressinc.com>]*

James P. Menakis is Forester, Robert E. Keane is Research Ecologist, and Donald G. Long is Forester, U.S. Department of Agriculture Forest Service, Intermountain Research Station, Intermountain Fire Sciences Laboratory, P.O. Box 8089, Missoula, MT 59807.

[Haworth co-indexing entry note]: "Chapter 10. Mapping Ecological Attributes Using an Integrated Vegetation Classification System Approach." Menakis, James P., Robert E. Keane, and Donald G. Long. Co-published simultaneously in *Journal of Sustainable Forestry* (Food Products Press, an imprint of The Haworth Press, Inc.) Vol. 11, No. 1/2, 2000, pp. 245-263; and: *Mapping Wildfire Hazards and Risks* (ed: R. Neil Sampson, R. Dwight Atkinson, and Joe W. Lewis) Food Products Press, an imprint of The Haworth Press, Inc., 2000, pp. 245-263. Single or multiple copies of this article are available for a fee from The Haworth Document Delivery Service [1-800-342-9678, 9:00 a.m. - 5:00 p.m. (EST). E-mail address: getinfo@haworthpressinc.com].

KEYWORDS. Vegetation condition, geographic information systems, GIS, fuel characteristics

INTRODUCTION

Land managers rely on maps of vegetation to inventory, monitor, and manage ecological resources. Most maps can be categorized into one of three meso- or coarse-scale classification systems–(1) potential vegetation types (Kuchler 1964), (2) existing vegetation or cover types (Eyre 1980; Shiflet 1994), and (3) ecological land types (combinations of climate, soils, geology, and landforms) (Bailey et al. 1996). Traditionally, maps using these classification systems have been created independently based on the specific criteria and objectives of the individuals developing the spatial classification (Pfister 1991). Moreover, these maps delineate only a single component or characteristic of the landscape, such as cover type or structural stage, and seldom provide insight in addressing complex land management issues like forest health and ecosystem management. For example, vegetation maps having attributes such as stand density, stand age, and species composition can be very useful for describing the spatial distribution of insect, disease, and wildfire hazard (Covington et al. 1994), but these attributes are not addressed in most existing vegetation type maps. In addition, existing vegetation maps (e.g., cover type map) rarely provide an assessment of trend or successional advancement. The successional sequence of a plant community such as a ponderosa pine cover type is mostly dictated by the type of site where it is found (Arno et al. 1985). Often, there is a great deal of variability of an ecological attribute (such as snag density) within a map category (such as ponderosa pine cover type) because of the wide variety of sites on which that category can occur. Ideally, a mapping system useful to many management applications would stratify the existing vegetation classification by some sort of site and structural classification. Some vegetation mapping schemes attempt to integrate these three attributes into one classification (e.g., Bailey et al. [1996] ecological land types), but usually only two of the three are implicitly represented by the resultant map categories.

Recently, researchers have found that the quality and predictive value of spatial maps can be improved by integrating several vegetation classifications together into one classification. Shao et al. (1996) used Potential Vegetation Types (PVT) to refine a satellite imagery classification of forest cover types for natural reserves in China. The Interior Columbia Basin Ecosystem Management Project (ICBEMP) integrated PVT, cover type, and structural stage classifications to map many ecological attributes describing hydrology, wildlife, fire, and fuels characteristics at a course scale (1:2,000,000 map scale) (Quigley et al. 1996). Keane et al. (1997) used the same procedure to map fuels for the Selway-Bitterroot Wilderness Complex (SBWC) at the mid-scale (1:50,000). This integration was also used to simulate vegetation succession across coarse scales (Keane et al. 1996). This spatial integration of several vegetation classifications has become feasible with the recent advancement and availability of the Geographic Information Systems (GIS).

The spatial distribution of ecological attributes across a landscape would provide land managers a tool for describing and understanding their ecosystems and a vehicle for solving key management problems. For example, mapping of ecological attributes such as fuels, stand density, and stand structure, could be used to delineate potential areas at risk to extreme wildfires. The purpose of this paper is to provide an overview of the methods used to derive and map ecological attributes across coarse to fine scale landscapes using an integrated vegetation classification approach. Two examples taken from the course scale ICBEMP and fine scale SBWC projects will illustrate how the integrated classifications of PVT, cover type, and structural stage are used to predict many other ecological characteristics across two spatial scales. In addition, we will show how an integrated vegetation classification mapping approach will increase the spatial accuracy of the original base layers.

BACKGROUND

This approach integrates the classifications of Potential Vegetation Type (PVT), cover type, and structural stage to predict numerous other ecological characteristics. It assumes that the combination of a site (PVT), dominant vegetation (cover type), and vertical structure (structural stage) will predict most ecosystem attributes important to management. These three classifications are intimately linked in time and space by a variety of ecosystem processes such as climate, fire and succession. The advantages of this approach are many. First, the concept can be used across spatial scales because each of the three classifications can be scaled to the appropriate level of application. For instance, a cover type category at a coarse scale may be "needleleaf conifer" whereas the same cover type at a mid or fine scale might be "ponderosa pine." Second, most land management agencies already use these classifications to some extent in their analyses. These classifications can be easily developed if they do not exist for some areas. Third, many resource professionals already use some form of these classifications to describe stands or watersheds. Lastly, there is a large body of research available on these classifications and their mapping.

The integration of diverse classification schemes has existed in vegetation science for decades (O'Hara et al. 1996), but it was rarely applied in the GIS environment. Past integrated classifications were designed from the start to be hierarchically nested in spatial and organizational scale. Classification categories (e.g., cover types) were designed to be finite, discrete types (e.g., lodgepole pine) that could be nested in scale and they comprehensively described the landscape to meet management objectives. A set of ecological descriptors computed from the analysis of sampled data could then be accurately assigned to each map category to widen the scope of the classification (Steele 1984; Arno et al. 1985).

The PVT is a site classification based on the habitat type concept of Daubenmire and Daubenmire (1968), where the PVT reflects the integration of vegetation and environmental factors (e.g., temperature, moisture, and soils) (Pfister et al. 1977). It is based upon potential or "projected" climax vegetation (Arno et al. 1985) that is self-regenerating in the absence of disturbances such as fire, grazing, cutting, etc. (Daubenmire and Daubenmire 1968). The foundation of habitat typing is based on (1) successional theory, (2) the role of disturbance in community organization and composition, and (3) the use of indicator plants (Cook 1996). Habitat type classifications are relatively stable barring disturbance (Daubenmire 1952; Steele and Geier-Hayes 1987, 1989, 1993). Habitat type classifications provide a logical framework for studying succession and occasionally infer successional relationships but offer only sporadic information on seral communities (Arno et al. 1985; Steele and Geier-Hayes 1987, 1989, 1993). Habitat types are useful site PVT classification categories for fine scale applications, whereas groups of habitat types (e.g., fire groups) are often used for mid-scale projects (Fischer and Bradley 1985). Coarse scale PVT categories usually describe biophysical properties of the site such as temperature, rainfall, and elevation (Quigley et al. 1996).

Cover types are based on the dominance of existing plant species (Eyre 1980; Shiflet 1994) and often describe successional stages within a PVT. Classifications of seral communities that occur within a habitat type have been developed by Arno et al. (1985) for Montana and Steele and Geier-Hayes (1987, 1989, 1993) for Idaho using the cone model described by Huschle and Hironaka (1980). Steele and Geier-Hayes (1987, 1989, 1993) further stratified seral communities into lifeform groups of trees, shrubs, and herbaceous. For each lifeform group the following management implications were summarized: tree damage from pocket gophers and snow damage; tree growth and yield capability, and age to breast height; competition between planted tree seedlings and shrubs; natural tree establishments; and, foraging preference for deer, elk, cattle, sheep, and black bear. Arno et al. (1985) developed seral community types based upon tree characteristic and undergrowth species. They further classified the seral community types into simple structural or developmental stages (e.g., seedling, sapling, and pole) such as those used by Thomas (1979). For each of the structural stages, which is the highest level of the classification hierarchy, Arno et al. (1985) summarized the stand characteristics of tree canopy coverage, average d.b.h. of dominant trees, basal area, and stand age.

The combination of habitat type, seral or climax plant communities, and structural phases has also been used by Kessell and Fischer (1981) in the development of conceptual models of plant succession. These models move vegetation forward in time using succession. Each model is based on a habitat type or fire group with changes in vegetation occurring between structural stages and eventually between cover types. Disturbances like fire were included in the model and showed several possible successional sequences or pathways for a plant community following a disturbance. The successional diagrams used to display the models have provided managers with an understanding of the role of disturbance and how it relates to basic trends in structural changes and trees species succession for a given habitat type or fire group.

The above classifications demonstrate the strong ecological relationship between site classifications, seral communities, and structural stages. Therefore, it would follow that many stand and landscape characteristics can be characterized from these three attributes. At the highest level of these complex hierarchal classifications (usually structural stages) researchers have summarized management implications (Steele and Geier-Hayes 1987, 1989, 1993) and tree characteristics (Arno et al. 1985). These summaries are essentially ecological characterizations or ecological attributes. By having GIS layer of PVT, cover type, and structural stages (the "vegetation triplet"), these ecological attributes then could be mapped using the GIS.

INTEGRATING THE VEGETATION TRIPLET

The combining of PVT, Cover Type, and Structural Stage layers into one integrated vegetation classification is efficiently accomplished using GIS. GIS tools allow the user to study the interactions between the layers by cross-referencing pixels in the same geographic location and to generate a list of all possible spatial combinations between three layers. This list is used to evaluate and correct inconsistencies between the layers that occur because of different mapping methodologies (e.g., satellite imagery, heuristic rule-based, and aerial photograph) and different vegetation classification used to create the individual layers. Inconsistencies can be isolated and corrected based on similar classifications found in the literature, expert opinion, and knowledge of the inherent problems associated with different map-making methodologies. Corrections are used in the GIS to generate new layers that agree spatially and become the integrated vegetation classification. We will refer to this spatial matching and correction of different layers as spatial agreement.

Ecological attributes can be mapped from an integrated vegetation classification of PVT, Cover Type, and Structural Stage layers was accomplished by assigning values to each combination of the three vegetation categories or vegetative triplet. For example, a stand height of 10 meters is assigned to all polygons that have the combination of a Douglas-fir PVT, Ponderosa pine cover type, and Pole tree structural stages. The primary source used to quantify these ecological attributes is georeferenced, ecological ground-truth data gathered for a particular project area. Ecological attribute measurements from this plot data can be summarized to PVT, cover type, and structural stage combinations, and these summaries can be entered into the GIS database for map creation. For example, Keane et al. (1997) averaged stand height across all plots in each vegetation triplet category and then entered these averages in the GIS database to predict stand height across the entire area. A second source used to quantify ecological attributes is existing databases. There are many existing databases that associate PVT, cover type, structural stage, and other fields to various ecosystem attributes. Ottmar et al. (1997) created a database that assigned fuel loading, smoke and fire behavior characteristics to cover type and structural stage categories. Lastly, a group of resource experts/specialists can assign many ecosystem characteristics to vegetation triplets based on past experience. This would only be used if no field data were available.

Terminology

In this paper, an integrated vegetation classification is defined as the combination of multiple spatial maps of ecological descriptors (or individual vegetation classification) mapped independently and then combined into one classification using the GIS. A hierarchial vegetation classification is defined as vegetation classification designed to be composed of two to three levels of ecological descriptors and traditionally were rarely mapped (described above).

A data layer or spatial map is defined as an electronic (or digital) map composed of a logical set of data used to delineate a piece of ground. A raster map is a digital map of pixels which are divided by a square grid. A raster layer is an electronic map (a spatial layer) divided into a grid with gridded divisions (square pieces) called pixels. A pixel, also referred to as a cell, is a square section of a raster map, the dimensions of which define the spatial resolution of the raster map.

EXAMPLES OF THE INTEGRATED APPROACH

Interior Columbia Basin Ecosystem Management Project

The objective of the ICBEMP scientific assessment was to develop a scientifically-sound, ecosystem-based strategy for management of federal lands in the Interior Columbia River Basin (Figure 10.1) (Quigley et al. 1996). This was accomplished by characterizing and assessing landscape, ecosystems, social, and economic processes and functions and describing probable outcomes of continued management practices and trends (Quigley et al. 1996). Most of these characterizations consisted of a group of raster GIS layers developed independently at a coarse scale (1:2,000,000). The raster layers developed for the ICBEMP were at one square kilometer pixel size.

The PVT spatial layer for the ICBEMP was created by assigning a PVT category for each unique combination of Biophysical Setting and Cover Type categories that occurred within each Ecoregion. Unique combinations of categories were determined by cross-referencing pixels between the spatial layers that occur within the same geographic location (called overlaying). Area ecologists and botanists made PVT category assignments during a series of workshops (Table 1) (Menakis et al. 1996; Keane et al. 1996). The Ecoregion layer is a map of broad classes of ecological climatic zones and vegetational macro-features (Figure 10.2), and represents one level of a hierarchal ecological land unit classification (Bailey et al. 1996; Bailey 1995).

The Biophysical Setting layer classifies the Interior Columbia River Basin into 48 categories composed of 16 temperature/moisture classes (Figure 10.3) defined for each of the physiognomic types: forests, shrubs, and herbaceous

FIGURE 10.1. The Interior Columbia Basin Ecosystem Management Project (ICBEMP) and the Selway-Bitterroot Wilderness Complex (SBWC) cover portions of 7 states in the western United States.

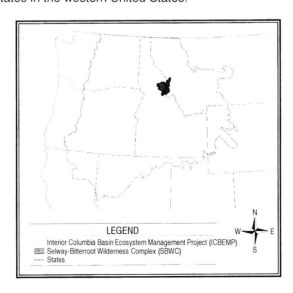

TABLE 1. Interior Columbia Basin Ecosystem Management Project (ICBEMP) potential vegetation types.

FOREST
Cedar/Hemlock Inland
Dry Douglas-Fir Without Ponderosa Pine
Dry Douglas-Fir with Ponderosa Pine
Dry Grand Fir/White Fir
Lodgepole Pine-Yellowstone
Lodgepole Pine-Oregon
Grand Fir/White Fir East Cascades
Grand Fir/White Fir Inland
Interior Ponderosa Pine
Moist Douglas-Fir
Mountain Hemlock East Cascades
Mountain Hemlock Inland
Mountain Hemlock/Shasta Red Fir
Pacific Ponderosa Pine/Sierra Mixed Conifer
Pacific Silver Fir
Spruce-Fir Dry with Aspen
Spruce-Fir Dry Without Aspen
Spruce-Fir Wet
Spruce-Fir(Whitebark Pine > Lodgepole Pine)
Spruce-Fir(Lodgepole Pine > Whitebark Pine)
Whitebark Pine/Alpine Larch North
Whitebark Pine/Alpine Larch South

OTHER
Alpine Shrub-Herbaceous
Barren
Dry Crop/Pasture Land
Irrigated Crop Land
Urban
Water

GRASS
Agropyron Steppe
Fescue Grassland
Fescue Grassland with Conifer

SHRUB
Antelope Bitterbrush
Big Sage-Cool
Big Sage-Warm
Big Sage Steppe
Low Sage-Mesic
Low Sage-Mesic with Juniper
Low Sage-Xeric
Mountain Big Sage-Mesic-East
Mountain Big Sage-Mesic-East w/Conifer
Mountain Big Sage-Mesic-West
Mountain Big Sage Mesic West w/Juniper
Mountain Shrub
Salt Desert Shrub
Three Tipp Sage

RIPARIAN
Aspen
Cottonwood Rivarian
Mountain Riparian Low Shrub
Salix/Carex
Saltbrush Riparian

WOODLAND
Juniper
Limber Pine
Mountain Mahogany
Mountain Mahogany with Sage
White Oak

(Reid et al. 1996). The layer was produced using a heuristic, rule-based approach developed for each Subsection (Figure 10.2). The Subsection layer maps geologic (e.g., lithology structure), physiographic (e.g., glaciated mountains), and statewide climatic zones, and represents a finer level of the same hierarchal ecological land unit classification used in mapping Ecoregions (Bailey et al. 1996; ECOMAP 1993). The heuristic rules were created from local expert ecologists and were based on all possible combinations of elevation, aspect, and slope classes within a Subsection (Reid et al. 1995).

The ICBEMP Cover Type layer was created by revising the Land Cover Characterization raster map developed at the EROS Data Center (Loveland et al. 1991). This map was developed from a classification of temporally stratified Advanced Very High Resolution Radiometer (AVHRR) satellite imagery

FIGURE 10.2. Ecoregions and subsections within the Interior Columbia Basin Ecosystem Management Project (ICBEMP).

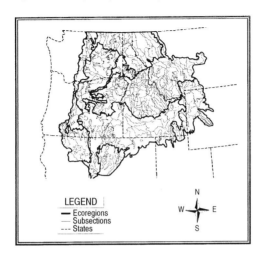

FIGURE 10.3. Temperature/Moisture Matrix used in the ICBEMP Biophysical Setting Layer.

TEMP:	MOISTURE:			
	WET (1)	MOIST (2)	DRY (3)	VERY DRY (4)
COLD (1)	COLD/WET	COLD/MOIST	COLD/DRY	COLD/VERY DRY
COOL (2)	COOL/WET	COOL/MOIST	COOL/DRY	COOL/VERY DRY
WARM (3)	WARM/WET	WARM/MOIST	WARM/DRY	WARM/VERY DRY
HOT (4)	HOT/WET	HOT/MOIST	HOT/DRY	HOT/VERY DRY

collected from March through October 1990 and other ancillary data (Loveland et al. 1991). Area ecologists revised and refined the 158 Land Cover Characterization classes into 48 cover types based on expert opinion and several additional ancillary layers (e.g., elevation, ownership, biophysical, and ecological mapping units) (Hardy et al. 1996; Menakis et al. 1996). The

ICBEMP Cover Type categories (Table 2) were based mostly on existing descriptions developed by the Society of American Foresters (Eyre 1980) and the Society of Range Management (Shiflet 1994).

The ICBEMP Structural Stage categories (Table 3) that describe the developmental changes in a coarse scale plant community's structure was based on Oliver's (1981) and Oliver and Larson's (1990) process-oriented, stand development classification. O'Hara et al. (1996) developed the ICBEMP classification for forest and woodland, and Willard and Villnow (1996) for rangeland. The ICBEMP Structural Stage categories were mapped using a discriminant analysis based on a subsample of watersheds that were mapped using aerial photography at scales mostly 1:24,000 but sometimes as high as 1:100,000 (Hessburg and Smith 1996). The independent variables used in the discriminant analysis were based on several continuous ancillary layers, including topographical, climatic, and vegetational indices (Menakis et al. 1996).

The initial ICBEMP layers of PVT, Cover Type, and Structural Stage were

TABLE 2. Interior Columbia Basin Ecosystem Management Project (ICBEMP) cover types.

CONIFEROUS FOREST	WOODLAND
Englemann Spruce/Subalpine Fir	Juniper Woodlands
Grand Fir/White Fir	Juniper/Sagebrush
Interior Douglas-Fir	Mixed Conifer Woodlands
Interior Ponderosa Pine	Oregon White Oak
Limber Pine	
Lodgepole Pine	**SHRUB**
Mt Hemlock	Antelope Bitterbrush/Bluebunch Wheatgrass
Pacific Ponderosa Pine	Big Sagebrush
Pacific Silver Fir/Mt Hemlock	Low Sage
Red Fir	Mountain Mahogany
Sierra Nevada Mixed Conifer	Mountain Big Sagebrush
Western Larch	Salt Desert Shrub
Western Redcedar/Western Hemlock	Seral Shrub-Regen
Western White Pine	
Whitebark Pine/Alpine Larch	**GRASSLAND**
Whitebark Pine	Agropyron Bunchgrass
	Exotic Forbs/Annual Grass
DECIDUOUS FOREST	Fescue-Bunchgrass
Aspen	Native Forb
Chokecherry/Serviceberry/Rose	
Cottonwood/Willow	**OTHER**
	Barren
HERBACEOUS	Cropland/Hay/Pasture
Alpine Tundra	Urban
Herbaceous Wetlands	Water
Shrub Wetlands	

TABLE 3. Interior Columbia Basin Ecosystem Management Project (ICBEMP) structural stages.

FOREST	SHRUB/HERBACEOUS
Stand Initiation Forest	Open Herbland
Stem Exclusion Open Canopy Forest	Closed Herbland
Stem Exclusion Closed Canopy Forest	Closed Low Shrub
Understory Reinitiation Forest	Open Low Shrub
Young Multi-Strata Forest	Open Mid Shrub
Old Multi-Strata Forest	Closed Mid Shrub
Old Single-Strata Forest	Open Tall Shrub
	Closed Tall Shrub
WOODLAND	
Stand Initiation Woodland	**OTHER**
Stem Exclusion Woodland	Agricultural
Understory Reinitiation Woodland	Urban
Young Multi-Strata Woodland	Water
Old Multi-Strata Woodland	Rock

combined to create a spatial layer of unique combinations of the vegetative triplets (PVT, cover type, and structural stage categories). The vegetative triplet categories were compared with the successional pathways developed for the Interior Columbia Basin SUccession Model (CRBSUM) (Keane et al. 1996). The successional pathways model changes in cover type and structural stage overtime and were developed for each PVT in the Interior Columbia Basin. Regional ecologists developed successional pathways based on existing classifications found in the literature and expert opinion (Figure 10.4). These successional pathways provided a complete list of all cover type and structural stage categories that occur in a PVT and were compared to the vegetative triplets. Inconsistencies found between the vegetative triplets and successional pathways (combinations of PVT, cover type, and structural stage categories found in the combined layers that did not occur in the successional pathways) were corrected based on the methods used in developing the layers (Menakis et al. 1996).

An example of the inconsistency problem was the occurrence of a Whitebark Pine Cover Type (a high elevation species) on a Dry Douglas-Fir PVT (a low elevation type). Based on the successional pathways, only a Ponderosa Pine or Douglas-Fir Cover Type can occur in Dry Douglas-Fir PVT. Since the PVT layer was based on the temperature/moisture gradients found in Biophysical Setting Layer (which did an extremely good job of delineating elevation breaks), it was unlikely the PVT layer was incorrectly mapped. The Cover Type layer was mapped with satellite imagery, which poorly discriminates between cover types with similar spectral signatures. Since Whitebark Pine and Ponderosa Pine Cover Type have similar spectral signatures (or

FIGURE 10.4. Hypothetical successional pathway for a potential vegetation type. Numbers in circles identify a cover type and structural stage combination.

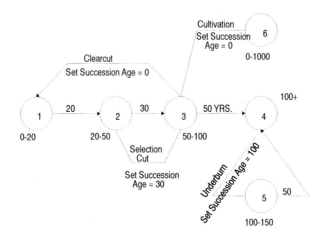

reflectance), the cover type was switched to Ponderosa Pine (which also agrees with the successional pathways).

The PVT layer was usually the layer corrected when problems occurred between lifeforms. For example, a Ponderosa Pine Cover Type was not allowed to occur on Bluebunch/Wheatgrass PVT, because only non-forest cover types can occur on this PVT based on the successional pathways. Since the Cover Type layer was mapped from satellite imagery, which accurately distinguishes between lifeforms, it was apparent the problem was associated with the PVT layer. The PVT layer was essentially mapped from an elevation, aspect, and slope classes. Since both the Ponderosa Pine and Bluebunch/ Wheatgrass PVT had overlapping topography, the PVT was switched to Ponderosa Pine, which matched the successional pathways.

Ecological attributes for ICBEMP were derived from existing intensive plot data–called ECODATA (Hann et al. 1988; Keane et al. 1990; Jensen et al. 1993)–collected throughout the assessment area. The ECODATA sampling methods provide a standard set of sampling procedures for the inventory and monitoring of vegetation that would be appropriate to all resource functions (Keane et al. 1990). This sampling method also creates a standard system for ecosystem characterization and analysis (Jensen et al. 1993). ECODATA plots for the ICBEMP was first classified into PVT, cover type, and structural stage combinations. This stratification was then used to summarize ecological attributes like snags per acre, density of live trees, fuel loadings, litter and duff depths, and percent bare ground, using several basic statistics (e.g., mean, standard deviation, minimum, and maximum). Each of these statistics could then be easily mapped for each of the ecological attributes on the basis of the vegetative triplet. In addition, overstory and understory species compositions were also mapped by summarizing the dominant and codominant species by each size classes (e.g., seedling, sapling, and pole). These summaries included trees per acre and basal area for forest types, and percent crown cover for all types.

Selway-Bitterroot Wilderness Complex Mapping Project

The objective of the SBWC Fuel Mapping Project (Figure 10.1) was to develop the spatial layers needed by the spatially explicit fire spread model FARSITE (Fire Area Simulator) which projects the growth and behavior of wildland fire across the landscape (Finney 1994). The model requires the GIS layers of elevation, aspect, slope, average canopy height, average crown base height, tree canopy cover, crown bulk density, and Anderson's Fire Behavior Models (Anderson 1982). The first three data layers were determined from the 30 meter resolution Digital Elevation Models (DEM) gathered for the SBWC. However, the last five layers did not exist for this study area. It seemed inefficient to create each map individually when all layers were related to vegetation, so these maps were created by assigning their attributes to the combination of PVT, cover type and structural stages categories that occurred in the SBWC.

Individual vegetation classifications of PVT, Cover Type, and Structural Stage were used to create the FARSITE vegetation layers, but only the PVT layer was developed specifically for this project. These original layers were combined and modified into one integrated vegetation classification where each pixel delineates a combination of PVT, cover type, and structural stage categories–the vegetative triplet. The methodologies used in developing these individual vegetation classifications and their integration will be described below. The PVT layer for the SBWC was created from a geographic and topographic classes using a heuristic, rule-based approach (Keane et al. 1997). First, all possible midscale PVT categories (Table 4) were compiled for the SBWC based on the opinions of local ecologists and a literature review (Habeck 1972). Then a set of rules was developed to predict PVT from slope, aspect, and elevation. The rules were based on a workshop of fire managers, ecologists, and scientists. The final PVT layer was then generated using a GIS and was based on elevation, aspect, and slope layers.

The SBWC Cover Type layer was derived from the Land Cover Database developed by the Wildlife Spatial Analysis Laboratory at the University of Montana from Landsat satellite imagery (Redmond and Prather 1996). Broad cover type (Table 5), structural stage, and canopy cover classes were assigned to polygons following the criteria based on ECODATA sampling methods (Hann et al. 1988; Keane et al. 1990; Jensen et al. 1993). The imagery data

TABLE 4. Selway-Bitterroot Wilderness Complex Project (SBWC) potential vegetation types.

FOREST	OTHER
Western Red Cedar	Persistent Herbland
Grand Fir	Rock/Alpine
Douglas-Fir	Developed Land
Lower Subalpine–Moist	Water
Lower Subalpine–Dry	
Upper Subalpine–Moist	
Upper Subalpine–Dry	

TABLE 5. Selway-Bitterroot Wilderness Complex Project (SBWC) cover types.

FOREST	SHRUB/HERBACEOUS
Douglas-Fir	Alpine Meadow
Douglas-Fir/Grand Fir	Disturbed Grasslands
Douglas-Fir/Lodgepole Pine	Cold Mesic Shrubland
Englemann Spruce	Grasslands
Grand Fir	Herbaceous Clearcuts
Lodgepole Pine	Subalpine Meadows and Montane Parklands
Ponderosa Pine	Warm Mesic Shrubland
Subalpine Fir	
Western Larch	**OTHER**
Western Larch/Douglas Fir	Agricultural
Western Larch/Lodgepole Pine	Exposed Rock
Western Red Cedar	Mixed Barren Land
Western Red Cedar/Grand Fir	Shoreline and Stream Gravel Bars
Broadleaf Forest	Snow or Ice
Mixed Alpine Forest	Urban or Developed Land
Mixed Mesic Forest	Water
Mixed Subalpine Forest	
Mixed Xeric Forest	
Standing Burnt and Dead Timber	

was smoothed into distinct polygons of continuous land area delineation with unique spectral class.

Redmond and Prather (1996) mapped the SBWC Structural Stage layer as part of the Land Cover Database. These classes describe the vertical structure of the vegetation and were based on diameter and/or height classes (Table 6). The vertical structure was first characterized by Thomas (1979) and was used to describe sequential stand development following clearcutting in the Blue Mountains of Oregon and Washington (O'Hara et al. 1996).

Corrections similar to those made in the ICBEMP were made to the SBWC coverages when the three layers were cross referenced for errors. From these corrections, new layers of PVT, Cover Type and Structural Stage

TABLE 6. Selway-Bitterroot Wilderness Complex Project (SBWC) structural stages.

FOREST	OTHER
Seedlings/Sapling–< 5.0″ DBH	Agriculture
Pole–5.0-8.9″ DBH	Grasslands
Medium–9.0-21.0″ DBH	Rock/Barren/Alpine
Large/Very Large–> 21.0″ DBH	Urban
	Water
SHRUB	
Low Shrub–< 2.5′	
Medium Shrub–2.5-6.5′	
Tall Shrub–> 6.5′	

were generated by using the corrected list of unique combinations as a lookup table. A lookup table is used in the GIS to generate a new map based on a unique number. The unique number here would be based on the list of vegetative triplet (unique combinations of PVT, cover type, and structural stage categories) corrected for spatial agreement. These corrected layers are the final layers used in assigning the ecological attributes.

All five FARSITE vegetation and fuels data layers were created by assigning values to each PVT, cover type, structural stage combination and then linking these data to the mapped GIS polygons. Most values were quantified from over 1,000 ground-truth plots established within the SBWC, and from the Ottmar et al. (1997) fuels database. However, the fire behavior fuel models that describe how a fire will burn under normal conditions (Anderson 1982) were assigned to the vegetation triplets by fire managers and forest ecologists during a three-day workshop. A preliminary analysis of ground truth data produced a 60 to 80 percent accuracy in the FARSITE fuel layers using this method (Keane et al. 1997).

CONCLUSION

The mapping approach described in this paper allows the mapping of ecological attributes using the power of the GIS and three common vegetation classification systems. We have found the unique combinations of PVT, cover types, and structural stages (the vegetative triplets), created when developing the integrated classification, to be extremely well suited to derive and map ecological attributes. Furthermore, this integration into one classification is based on the ecological processes of plant succession that have been extensively documented in the development of hierarchal vegetation classification. By using the GIS, ecological attributes can be mapped to provide land managers additional information when addressing complex management questions. This approach has been successfully used at coarse- and mid-scale applications to map ecosystem and fuel characteristics. Hardy et al. (2000) describe another example of a database that could be used to map ecological attributes based on cover type and structural stage combinations.

REFERENCES

Anderson, H.E. 1982. Aids to determining fuel models for estimating fire behavior. Gen. Tech. Rep. INT-122. Ogden, UT: U.S. Department of Agriculture, Forest Service, Intermountain Forestry and Range Experimental Station. 22 p.

Arno, S.E., D.G. Simmerman, and R.E. Keane, 1985. Forest succession on four habitat types in western Montana. Gen. Tech. Rep. INT-177. Ogden, UT: U.S. Department of Agriculture, Forest Service, Intermountain Forest & Range Experiment Station. 74 p.

Bailey, R.G. 1995. [Revised]. Description of the ecoregions of the United States. Washington, DC: Misc. Publ. No. 1391. U.S. Department of Agriculture, Forest Service. 108 p.

Bailey R.G., M.E. Jensen, D.T. Clelend, and P.S. Bourgeron. 1996. Design and use of ecological mapping units. Pages 105-116. In: M.E. Jensen and P.S. Bourgeron editors, Eastside Forest Ecosystem Health Assessment, Volume II, Ecosystem Management: Principles and Applications. Gen. Tech. Rep. PNW-GTR-318. Portland OR: U.S. Department of Agriculture, Forest Service, Pacific Northwest Research Station. 376 p.

Cook, J.E. 1996. Implications of moderns successional theory for habitat typing: a review. Forest Science. 42(1):67-75.

Covington W.W., R.L. Everett, R. Steele, L.L. Irwin, T.A. Daer, and A.N.D. Auclair. 1994. Historical and anticipated changes in forest ecosystems of the inland west of the United States. Journal of Sustainable Forestry 2(1/2):13-63.

Daubenmire, R. 1952. Forest vegetation of northern Idaho and adjacent Washington, and its bearing on concepts of vegetation classification. Ecological Monographs. 22:301-330.

Daubenmire, R. and J.B. Daubenmire. 1968. Forest vegetation of eastern Washington and northern Idaho. Tech. Bull. 60. Pullman, WA: Washington Agriculture Experiment Station. 104 p.

ECOMAP 1993. National hierarchical framework of ecological units. Washington, DC: U.S. Department of Agriculture, Forest Service. 20 p.

Eyre, F.H., ed. 1980. Forest cover types of the United States and Canada. Washington, DC: Society of American Foresters. 147 p.

Finney, M.A. 1994. Modeling the spread and behavior of prescribed natural fires. In: 12th Conference on Fire and Forest Meteorology; October 26-28; Jekyll Island, GA. Washington, DC: Society of American Foresters. 6 p.

Fischer, W.C. and A.F. Bradley. 1985. Fire ecology of western Montana forest habitat types. Gen. Tech. Rep. INT-223. Ogden, UT: U.S. Department of Agriculture, Forest Service, Intermountain Forest & Range Experiment Station. 95 p.

Habeck, J.R. 1972. Fire ecology investigations in Selway-Bitterroot Wilderness–Historical considerations and current observations. University of Montana Publication No. R1-72-001.

Hann, W.J., M.E. Jensen, and R.E. Keane. 1988. Chapter 4: Ecosystem management handbook–ECODATA methods and field forms. USDA Forest Service Northern Region Handbook. On file at Northern Region, Missoula Montana.

Hardy, C.C., R.E. Burgan, and R.D. Ottmar. 2000. A database for spatial assessments of fire characteristics, fuels profiles, and PM10 emissions. In: Sampson, R.N., R.D. Atkinson, J.W. Lewis (Eds.), Mapping Wildfire Hazards and Risks. Papers from the American Forests scientific workshop, September 29-October 5, 1996, Pingree Park, CO. The Haworth Press, Inc., New York.

Hardy, C.C., R.E. Burgan, and J.P. Menakis. [In preparation]. Development of a current vegetation map for the Interior Columbia River Basin Scientific Assessment.

Hessburg, P.F., B.G. Smith, S.D. Kreiter, C.A. Miller, R.B. Salter, C.H. McNicoll, and W.J. Hann. 1999. Historical and current forest and range landscapes in the interior Columbia River basin and portions of the Klamath and Great Basins. Part I: Linking vegetation patterns and landscape vulnerability to potential insect and pathogen disturbances Gen. Tech. Rep. PNW-GTR-458.

Huschle G. and M. Hironaka. 1980. Classification and ordination of seral plant communities. Journal of Range Management 33(3): 179-182.

Jensen, M.E., W.J. Hann, R.E. Keane, J. Caratti, and P.S. Bourgeron. 1993. ECODATA–A multiresource database and analysis system for ecosystem description and evaluation. Pages 249-265. In: M.E. Jensen and P.S. Bourgeron editors, Eastside Forest Ecosystem Health Assessment, Volume II, Ecosystem Management: Principles and Applications. Gen. Tech. Rep. PNW-GTR-318. Portland OR: U.S. Department of Agriculture, Forest Service, Pacific Northwest Research Station. 376 p.

Keane, R.E., M.E. Jensen, and W.J. Hann. 1990. ECODATA and ECOPAC-analytical tools for integrated resource management. The Compiler 23:11-24.

Keane, R.E., J.L. Garner, K.M. Schmidt, D.G. Long, J.P. Menakis, and M.A. Finney. 1997. Development of input data layers for the FARSITE fire growth model for the Selway-Bitterroot Wilderness Complex, USA. GTR RMRS-GTR-3. Ogden, UT: USDA-FS, Rocky Mountain Research Station, 66 p.

Keane, R.E., D.G. Long, J.P. Menakis, W.J. Hann, and C.D. Bevins. 1996. Simulating coarse-scale vegetation dynamics using the Columbia River Basin SUccession Model–CRBSUM. Gen. Tech. Rep. INT-GTR-340. Ogden, UT: U.S. Department of Agriculture, Forest Service, Intermountain Research Station. 50 p.

Kessell, S.R., and W.C. Fischer. 1981. Predicting postfire plant succession on coniferous forest landscapes of the Northern Rocky Mountains. Res. Pap. INT-RP-484. Ogden, UT: U.S. Department of Agriculture, Forest Service, Intermountain Forest & Range Experiment Station. 122 p.

Kuchler, A.W. 1964. Potential natural vegetation of the conterminous United States. (Manual and map). New York: Am. Geog. Soc. Spec. Publ. 36, 1965 rev. 116 p.

Loveland, T.R., J.M. Merchant, D.O. Ohlen, and J.F. Brown. 1991. Development of a land-cover characteristics databases for the conterminous U.S. Photogrammetric Engineering and Remote Sensing. 57(11): 1453-1463.

Menakis, J.P., D.G. Long, R.E. Keane, and W.J. Hann 1996. The development of key broadscale layers and characterization files. In: Keane, Robert E., Jones, Jefferey L., Riley, Laurienne S., Hann, and J. Wendal, (Technical Editors). Compliation of administrative reports: multi-scale landscape dynamics in the Basin and portions of the Klamath and Great basins (Irregular pagination). On file with: USDA, Forest Service; DOI Bureau of Land Management; Interior Columbia Basin Eco-system Management Project, 112 E. Poplar, Walla, Walla, WA 99362.

O'Hara K.L., P.A. Latham, P. Hessburg, and B.G. Smith. 1996. A structural classifi-cation for inland northwest forest vegetation. Western Journal of Applied Forest-ry. 11(3): 97-102.

Oliver, C.D. 1981. Forest development in North America following major distur-bances. Forest Ecology and Management. 3: 151-168.

Oliver, C.D. and B.C. Larson. 1990. Forest Stand Dynamics. New York: McGraw-Hill. 467 p.

Ottmar, R.D., E. Alvarado, P.F. Hessburg, [and others]. 2000. Historical and current forest and range landscapes in the Interior Columbia River Basin and portions of the Klamath and Great Basins. Part II: Linking vegetation patterns and potential smoke production and fire behavior. Gen.Tech. Rep. PNW-GTR-XXX. Portland, OR: U.S. Department of Agriculture, Forest Service, Pacific Northwest Research Station. [in press]

Pfister, R.D., B.L. Kovalchik, S.F. Arno, and R.C. Presby. 1977. Forest habitat types of Montana. Gen. Tech. Rep. INT-34. Ogden, UT: U.S. Department of Agricul-ture, Forest Service, Intermountain Forest & Range Experiment Station. 174 p.

Pfister, R.D. 1991. Land classification in the western United States. Pages 9-17. In: D.L. Mengel and D.T. Tew, editors, Proceedings of a symposium. Ecological Land Classification: Applications to Identify the Productive Potential of Southern Forests. Gen. Tech. Rep. SE-68. Asheville, NC: U.S. Department of Agriculture, Forest Service, Southeastern Forest Experiment Station. 149 p.

Quigley, T.M., R.W. Haynes, and R.T. Graham (eds). 1996. Integrated scientific assessment for ecosystem management in the Interior Columbia Basin and por-tions of the Klamath and Great Basins. Gen. Tech. Rep. PNW-GTR-382. Portland OR: U.S. Department of Agriculture, Forest Service, Pacific Northwest Research Station. 303 p.

Redmond, R.L. and M.L. Prather. 1996. Mapping existing vegetation and land cover across western Montana and north Idaho. Report on file at: USDA Forest Service, Northern Region for completion of contract 53-0343-4-000012.

Reid, M., P. Bourgeron, H. Humphries, and M. Jensen. 1995. Documentation of the modeling of potential vegetation at the three spatial scales using biophysical settings in the Columbia River Basin. Report on file at: Northern Region Ecosys-tem Management Project for completion of contract 5304H1-6890.

Shao, G., G. Zhao, S. Zhao, H.H. Shurgart, S. Wang, and J. Schaller. 1996. Forest cover types derived from Landsat Thematic Mapper imagery for Changbai Moun-tain area of China. Canadian Journal Forest Research 26:206-216.

Shiflet, T.N., ed. 1994. Rangeland cover types of the United States. Denver CO: Society of Range Management. 151 p.

Steele, R. 1984. An approach to classifying seral vegetation within habitat types. Northwest Science. 58: 29-39.

Steele, R. and K. Geier-Hayes. 1987. The grand fir/blue huckleberry habitat type in central Idaho: succession and management. Gen. Tech. Rep. INT-228. Ogden, UT: U.S. Department of Agriculture, Forest Service, Intermountain Research Station. 66 p.

Steele, R. and K. Geier-Hayes. 1989. The Douglas-fir/ninebark habitat type in central Idaho: succession and management. Gen. Tech. Rep. INT-252. Ogden, UT: U.S. Department of Agriculture, Forest Service, Intermountain Research Station. 65 p.

Steele, R. and K. Geier-Hayes. 1993. The Douglas-fir/pinegrass habitat type in central Idaho: succession and management. Gen. Tech. Rep. INT-298. Ogden, UT: U.S. Department of Agriculture, Forest Service, Intermountain Research Station. 83 p.

Thomas, J.W. (Tech Ed.). 1979. Wildlife habitats in managed forests: The Blue Mountain of Oregon and Washington. USDA For. Serv. Agric. Hand. No. 553.

Willard, E. and B. Villnow. 1996. Development of coarse-scale rangeland structural stages for the Interior Columbia River Basin.

Chapter 11

Inherent Disturbance Regimes:
A Reference for Evaluating the Long-Term
Maintenance of Ecosystems

R. Everett
J. Townsley
D. Baumgartner

SUMMARY. A science-based ecosystem management approach requires valid reference points to assess the long-term maintenance of forest systems. Historical range of variability (HRV) in vegetation patterns has served as the initial reference point and has support in the coarse-filter approach to conserving biodiversity (Hunter 1991). The HRV becomes less useful as a reference with increasing human disturbance on the landscape, public aversion to treatments to restore historical states, and continued public expectations for non-historical conditions. The inherent disturbance regime (IDR) is put forward as an alternative or to be used in combination with HRV as a suitable refer-

R. Everett (Retired) was affiliated with USDA Forest Service, Forest Sciences Laboratory, Wenatchee, WA 98801. J. Townsley is affiliated with USDA Forest Service, Okanogan National Forest, Okanogan, WA 99840. D. Baumgartner is affiliated with Department of Natural Resource Sciences, Washington State University, Pullman, WA 99164.

The authors wish to thank Dr. Gary Daterman and Dr. Dave Sandberg, Pacific Northwest Research Station; Dr. Roger Chapman, Washington State University; and Dr. Jon Haber, Region 1, USFS for their review and comments on the manuscript.

[Haworth co-indexing entry note]: "Chapter 11. Inherent Disturbance Regimes: A Reference for Evaluating the Long-Term Maintenance of Ecosystems." Everett, R., J. Townsley, and D. Baumgartner. Co-published simultaneously in *Journal of Sustainable Forestry* (Food Products Press, an imprint of The Haworth Press, Inc.) Vol. 11, No. 1/2, 2000, pp. 265-288; and: *Mapping Wildfire Hazards and Risks* (ed: R. Neil Sampson, R. Dwight Atkinson, and Joe W. Lewis) Food Products Press, an imprint of The Haworth Press, Inc., 2000, pp. 265-288. Single or multiple copies of this article are available for a fee from The Haworth Document Delivery Service [1-800-342-9678, 9:00 a.m. - 5:00 p.m. (EST). E-mail address: getinfo@haworthpressinc.com].

ence point for evaluating long-term maintenance of human-altered ecosystems. The IDR is defined by the types of disturbance; their frequency, intensity, and extent in turn defines the vegetation composition and structure supported over time. Although public expectations for vegetation conditions may differ from historical, these altered states may also be supported by IDR and have long-term maintenance potential. The use of a disturbance reference (IDR) rather than a condition reference (HRV) appears more compatible with the dynamic nature of forests. As an accurate description of historical conditions and disturbance regimes is difficult we recommend that both references be used reiteratively to better refine the other. Increased discontinuity between public expectations for forest conditions and the ability of those conditions to be supported by disturbance regimes may lead to catastrophic disturbance events and a decline in long-term site nutrient capital and biodiversity. *[Article copies available for a fee from The Haworth Document Delivery Service: 1-800-342-9678. E-mail address: <getinfo@haworthpressinc.com> Website: <http://www.haworthpressinc.com>]*

KEYWORDS. Ecosystem management, disturbance regime, historical range of variability

INTRODUCTION

A series of legislative acts, including the Wilderness Act (1964), National Environmental Policy Act (1969), Forest and Rangeland Renewable Resources Planning Act (1974), National Forest Management Practices Act (1976), Endangered Species Act (1978), and Resource Planning Act (1990), require land management agencies to make assessments and manage public lands on a sustainable basis for resource conditions, species viability, and commodity production. A science-based ecosystem management approach was developed to meet this direction that requires valid reference points for defining sustainable ecosystem conditions (Overbay 1992; Bourgeron and Jensen 1994). The initial reference point was the "historical range of variability" (HRV) in vegetation pattern and composition prior to Eurosettlement (USDA Forest Service 1992; Morgan et al. 1994). Historical range of variability differs from previous reference points such as climatic climax in that HRV accommodates the occurrence and effects of disturbance in ecosystems, and suggests that an array of conditions is a more suitable reference than a single ecological state.

Historical range of variability is a useful starting point for defining vegetation characteristic of sustainable ecosystems and understanding cause and effects of ecosystem change over time (Morgan et al. 1994). However, the HRV concept focuses on past conditions, is not structured to integrate pre-

and post-Eurosettlement conditions, and if rigidly followed, restricts land manager decision space to create resource conditions required to meet emerging public expectations. Strict adherence to HRV standards may not allow landscapes to change in accordance with shifts in global climate. A broader, scientific-based reference point is needed that accommodates human expectations and interactions with ecosystem function and structure.

To accommodate human expectations, the reference point should be able to evaluate a wide spectrum of potential resource conditions for long-term maintenance under the probable disturbance regimes indigenous to the area. Each geographic area has an array of disturbance types (such as insects, pathogens, fire and/or mass wasting) that can manifest themselves with varying frequencies, intensities, and extents characteristic of the vegetation, biophysical environments, and climatic conditions. The set of disturbances than can affect vegetation is an inherent characteristic of the arrangement of vegetation patches, species composition, stand density, and representation of microsites across the landscape. This composite of disturbances and their combined effects at different scales broadly defines the inherent disturbance regimes (IDR) of the area. The IDR ultimately determines the vegetation structure and composition that is supportable in the long-term (Everett et al. 1996b). The usefulness of the IDR reference in evaluating the long-term maintenance of landscape and stand conditions, including those meeting historical or emerging public expectations, should be tested.

MANAGING FOR VEGETATION STRUCTURE OR DISTURBANCE IN CREATING DESIRED LANDSCAPES

Land management has traditionally been a process of managing vegetation pattern, density and composition to provide resource conditions, conserve sensitive species and habitats, and provide resource products. Managing landscape pattern within the historical range of variability (HRV) for vegetation conditions is one approach to integrate multiple-use while theoretically conserving biodiversity and site productivity. The stated assumption is that by maintaining vegetation conditions within the range of historical occurrence the species and processes that have developed with or at least existed in historical landscapes will be maintained over time (Hunter 1991).

Management of vegetation pattern and composition for specific historical states is made difficult because of the dynamic nature of ecosystems and ubiquitous unplanned disturbance events (Agee and Johnson 1988). This is further complicated by climatic flux that occurs from decade to decade (Oliver and Larson 1990). Disturbance is a normal and inevitable component of ecosystems (Cooper 1913; Sprugel 1991) with forest systems in a constant state of change (Botkin and Sobel 1975; Oliver et al. 1994). The "disturbance pulse" theory speaks to the specific need for disturbance in the long-term maintenance of some ecosystem components such as the "fire-dependent" California chaparral (Odum 1969). Numerous wildlife and plant species are dependent upon disturbance to continually create transitional habitat and the viability of those species becomes less certain when disturbance effects are reduced by management (Oliver et al. 1994).

Managing for disturbance is working with natural processes rather than managing for static vegetation conditions. Our understanding of disturbance regimes and resulting vegetation patterns is developing rapidly (Pickett and White 1985; Urban et al. 1987; Turner et al. 1993; Bourgeron and Jensen 1994; Baker 1995). Models are being developed that integrate multiple types of disturbances (Hagle et al. 1995) and disturbance regime assessments are being conducted across large landscapes in the western United States (Lehmkuhl et al. 1994; Hessburg et al., in press). Maintaining "natural" disturbance regimes (Morgan et al. 1994) and managing for changing landscape patterns within the HRV is one approach to manage for disturbance, but it has the added restrictive caveat of required historical similarity that may or may not meet public expectations.

Disturbance management is managing for the disturbance effects that create and maintain desired resource conditions (Agee and Johnson 1988) and commodity outflows (Oliver 1992). "It seems prudent . . . to regulate forest development and disturbance so that social benefits can be realized in an organized and predictable fashion" (Carlson et al. 1995, p. 34). Also, disturbance management is managing for specific changes and rates of change that meet public expectations and conserve species, processes, and site nutrient capital (Morgan et al. 1994). Disturbance regimes are used to characterize the spatial and temporal patterns of disturbance (Wargo 1995) that drive the dynamic patch mosaic of vegetation pattern (Watt 1947; Stone and Ezrati 1996). Within-stand disturbance profiles, a multivariate metric of stand structure, have been used to evaluate small scale disturbance patterns, the effects of management intervention on disturbance events, and subsequent ecosystem response (Lundquist 1995). An assumption is made that by conserving disturbance processes, ecosystem renewal is ongoing and the full array of vegetation stages and associated species are conserved (Oliver 1992). Renewal may not be possible on non-equilibrium disturbance landscapes where the frequency of disturbance is less than the recovery interval and a majority of the landscape is disturbed (Turner et al. 1993; Baker 1995).

Disturbance events alter site nutrient capital (Hungerford et al. 1991), but are assumed to be an ecosystem function that associated vegetation and other species have co-existed with previously. Also, although site nutrient capital may be depleted locally the disturbance event and nutrient translocation may be required for maintenance of larger scale systems. For example, rains that cause local soil erosion also provide for annual flood pulses that provide nutrient enrichment and wildlife habitat in floodplains downstream (Odum 1969; Sparks 1995). The hazard for unprecedented loss in site nutrient capital is more likely if disturbances are prevented until an uncontrollable and severe disturbance occurs (Oliver et al. 1994; Carlson et al. 1995; MacCleery 1995). The alteration of disturbance regimes and subsequent accumulation of greater aboveground forest biomass (Oliver et al. 1994) sets the stage for more severe fire disturbance events and greater off-site nutrient translocation through volatilization or erosion processes (Grier 1975). Direct N losses from underburning in ponderosa pine was estimated at 70 lb per acre, but up to 800 lb N per acre can be lost in severe wildfire events (Grier 1975; Covington and Sackett 1984; Hungerford et al. 1991).

The opportunity for disturbance to occur may be largely due to the biophysical characteristics of the forest ecosystem. We define inherent disturbance regime (IDR) as the probable scale, intensity, and periodicity of disturbance that derives from the accumulation and spatial arrangement of biomass as affected by abiotic factors such as topography and climatic, and biotic factors such as insects, pathogens, and human actions. The use of the inherent disturbance regimes (IDR) as a guide in the application of disturbance in natural resource management should reduce the uncertainty of ecosystem response. The IDR approach is prospective rather than retrospective and is more closely tied to disturbance management than historical or static structural preference. The IDR should be more flexible than HRV in evaluating potential forest outcomes, and at least as certain of long-term maintenance of larger scale systems.

THE HRV REFERENCE

The range of historical (natural) variability (HRV) is the array of compositions, structures, and processes that have occurred over time for a given area (USDA Forest Service 1992). The period just prior to significant Eurosettlement effects has often been chosen as the reference point for its consistency with current climatic conditions and because pre-Eurosettlement conditions are assumed to be sustainable over time. In landscape ecology the kinds and properties of vegetation patterns are commonly used as the reference criteria for HRV. The reference time frame varies according to resource information available such as 50 years for aerial photographs (Lehmkuhl et al. 1994) to a century or more from stand reconstruction (Bonnicksen and Stone 1982). The HRV concept has been used in land management to compare the disparity between current landscapes and those that occurred historically as a measure of uncertainty in long-term maintenance of ecosystems (Caraher and Knapp 1993). Long-term maintenance may be thought of in terms of the potential for multiple rotations of existing vegetation, or in terms of the conservation of site nutrient capital and biodiversity at appropriate scales.

Caraher and Knapp (1993) provided a clear use of the HRV reference in their evaluation of forest health in the Blue Mountains of eastern Oregon. Their results suggest that early seral plant communities are currently under-represented in fir forests in the Tucannon watershed, within HRV in the Lower Grande Ronde, and above HRV in the upper Grand Ronde (Figure 11.1). These watersheds were all under-represented in late seral, park-like stands and all within HRV for late seral multistory stands.

An estimation of the historical range of variability for a forest type is made difficult because of the absence of historical information on vegetation development over large areas (Christensen 1988; Swanson et al. 1994). Early survey reports (Plummer 1902), historical photo points (Gruell 1980; Skovlin and Thomas 1995) and records of early settler experiences (Robbins and Wolf 1994) provide insight into historical conditions. However, there remains limited scientific information on historical landscape conditions on which to base professional judgments. As we do not have continuous historical information for large landscapes, other alternative approaches have been sug-

FIGURE 11.1. Comparison of current stand types (*) with their historical range in variability (adapted from Caraher and Knapp 1993).

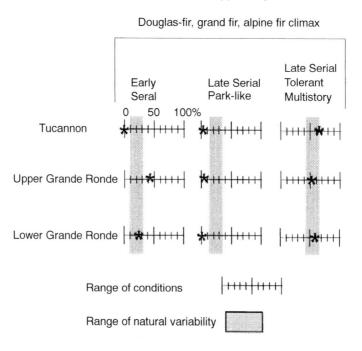

gested. Time-sequence of change estimates for a given vegetation type can be developed by observing its current state at multiple locations (Watt 1947; Stone and Ezrati 1996). Because vegetation initiation for a forest type does not occur simultaneously on all landscapes, the array of vegetation conditions existing in multiple watersheds at a single point in time has been used as a surrogate in the estimation of HRV in recent large-scale assessments (Hessburg et al. 1994; Lehmkuhl et al. 1994). In these assessments, the array of vegetation pattern, structure, and composition depicted in historical aerial photographs (1940s) of replicated watersheds was used to estimate HRV.

Although historical aerial photos provide a good characterization of previous landscapes, their limited time-frame (1940s and later) does not permit an accurate estimate of landscapes prior to Eurosettlement. Eurosettlement rapidly altered disturbance regimes; by the 1940s some forest stands were two to eight fire-free intervals out of historical conditions with associated changes in species composition and stand density (Figure 11.2, from data in Everett et al., 1997). The extended fire-free interval following Eurosettlement resulted in increased stand density for grand fir stands in eastern Washington. Figure 11.3 shows significant changes in species composition and stand structure from 1890 to 1940 to 1990 for a climax grand fir stand in eastern Washington. General changes in landscape conditions from 1940 to 1990 could be seen in a comparison of aerial photos, but without knowledge of pre-1940 conditions, an erroneous estimate of pre-Eurosettlement conditions and a faulty HRV would result.

FIGURE 11.2. Reduction in fire events and resulting increased stand density over time for the grand fir (*Abies grandis*) series on the east slope of the Washington Cascades (adapted from Everett et al., 1997).

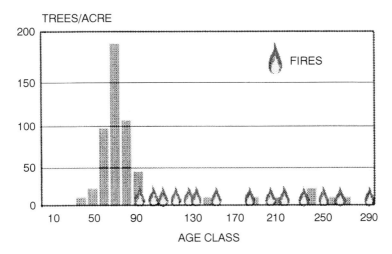

FIGURE 11.3. Comparison of stand structure and composition between 1892, 1942, and 1992 on the Porky basin spotted owl nest stand in eastern Washington (adapted from Everett et al., 1997). (ABGR, *Abies grandis*, grand fir; PSME, *Pseudotsuga menziesii*, Douglas-fir).

1892	1942	1992
ABGR–47%	ABGR–89%	ABGR–97%
PSME–53%	PSME–11%	PSME–3%

If HRV considers only those conditions that occurred in the past then the concept is restricted to that portion of possible vegetation conditions that were realized and not the larger population of events and outcomes that were possible and could be sustainable over time. Using HRV to describe static desired forest structure is working against rather than with ecosystem processes. Ecosystems are dynamic and moving forward, not backward, in time and extensive efforts to define and achieve exact historical vegetation conditions may not be warranted. Plant communities are unique in space and time and chance plays a major role in species co-occurrence (Gleason 1926; Watt 1947).

Several factors suggest that plant communities can be thought of as transient assemblages: the ebb and flow of whole biota along elevational and cardinal gradients in response to geomorphic and climatic changes; the continuous occurrence of fire, insect, and pathogen disturbance; the regional enrichment of local species pools (Ricklefs 1987); and the fact that species respond individually to disturbance events. The character of post-disturbance vegetation is highly dependent upon the type and severity of disturbance, the vegetation present at time of disturbance, and the immediate post-disturbance conditions. Given the low probability of pre-disturbance vegetation, disturbance extent and severity, and post-disturbance conditions always being the same, then vegetation change is inevitable and duplication of HRV conditions is difficult. Chaos theory suggests that plant community trajectories with similar, but not exact initial conditions could diverge exponentially over time (Stone and Ezrati 1996). Applications of the HRV reference that focus on the variation within HRV rather than achieving a single state appear more appropriate (Morgan et al. 1994).

The HRV concept rests upon the tenuous basic assumption that past and future climatic conditions will be similar. Forests and rangelands that existed prior to Eurosettlement originated during the Little Ice Age (Oliver and Larson 1990) and should future climatic conditions differ from the past, disturbance regimes and associated assemblages of plants would be expected to vary. Implicit in the HRV concept is the speculative assumption that historical vegetation conditions are sustainable under current altered disturbance regimes and resource conditions. Forest ecosystems have changed dramatically since Eurosettlement (MacCleary 1995). The potential for returning to historical conditions is severely limited by changes in site potential, species composition, and the requirement to meet public expectations. The usefulness of the HRV concept can be limited when historical conditions are no longer possible or required restoration treatments are unacceptable. Post-European settlement disturbance has been sufficiently severe on some sites to cause soil loss, a decline in site nutrient capital, loss of indigenous species, and the inability to re-establish previous vegetation (Perry et al. 1989). For example, the invasion of juniper and pinyon into the shrub steppe/grasslands is associated with increased erosion and reduction in indigenous species plant populations and soil seed reserves (Koniak and Everett 1982) leading to potential site conversions (Everett 1984).

The value of the HRV concept declines as the extent of non-historical species and processes increases over time. There are no historical reference points for exotic plants (i.e., knapweed, *Centaurea* sp.; cheatgrass, *Bromus tectorum*), exotic diseases (i.e., chestnut blight fungus, *Cryphonectria parasitica* and white pine blister rust, *Cronartium ribicola*) or exotic insects (i.e., Asian gypsy moth, *Lymantria dispar*) or their impacts on vegetation structure and composition (Bridges 1995). Similarly, there are no historical reference points for human habitation, roading, timber harvest, or livestock grazing and the vegetation patterns created by these activities.

Our ability to restore historical conditions is limited because of the reduced role indigenous people have in vegetation management. Numerous plant communities were created and maintained by indigenous people and one of the major contributing factors to post-Eurosettlement vegetation change has been the absence of the indigenous people's use of fire (Kay 1995). As the HRV includes the effects of indigenous people on the landscape (MacCleary 1995) any attempt to restore vegetation to within HRV assumes that the previous effects of indigenous people can be restored.

Restoration of HRV is in doubt when the public neither desires the return to historical conditions nor the treatments required for the restoration to occur. The livestock operators who depend upon the early spring forage from cheatgrass may not wish to return to historical conditions. Segments of the public who desire the return to historical open pine stands may find smoke from required prescribed burning unacceptable (Shelby and Speaker 1990), and every house that is built in the urban-wildland interface is a public statement for conditions other than historical.

THE INHERENT DISTURBANCE REGIME REFERENCE

The IDR reference is a general estimate of the frequency, extent, and intensity of major disturbances (insects, pathogens, fire, wind throw, floods and mass wasting) that shaped the characteristics of landscape and stand vegetation prior to Eurosettlement. Because of chance events in a somewhat chaotic environment, there is not one rigidly-defined disturbance regime for an area, but a general disturbance portfolio that operates within a swarm of varying probabilities of occurrence. The IDR reference of disturbance regimes prior to Eurosettlement is "coarsely" compared to current disturbance regimes for commonality, with the assumption that historical disturbance regimes provided long-term maintenance of ecosystems at one or more landscape scales. Although the IDR reference has an historical base, its output is an evaluation of current landscapes (that meet public expectations) for long-term maintenance capability. The inherent disturbance regime concept emphasizes the long-term maintenance of vegetation and associated species and processes through conservation of disturbances characteristic of the biophysical environment of an area (Everett and Lehmkuhl 1996). It is based on the view that ecosystems are continually evolving and that a "changing landscape, primarily in response to non-human processes and disturbance factors, is more realistic than an implied static landscape" based on some "vignette of primitive America" (Parsons et al. 1986 in Agee and Johnson 1988, p. 10).

The IDR concept considers the types and levels of disturbance required to achieve desired current, historical, or previously unachieved conditions. Disturbances considered are inherent to the biophysical characteristics of the area and the vegetation, associated species, and processes have previously interacted with these disturbances. Disturbance events caused by indigenous people are included in the IDR reference because it is assumed they played a significant role in defining the vegetation character of pre-Eurosettlement conditions (Kay 1995; MacCleary 1995) and that the resulting vegetation patterns were maintained over long periods of time (Uebelacker 1986).

The gaps in our knowledge of disturbance regimes requires integrating multiple sources of information to estimate the IDR. Although we may not know the individual disturbance events for large landscapes there is a wealth of information on disturbance types (insect, disease, fire, mass wasting), and their extent, intensity and frequently by vegetation type and biophysical environment (Odum 1969; Urban et al. 1987; Bourgeron and Jensen 1994; Carlson et al. 1995). There is both current and historical information on insect outbreaks (Swetnam and Lynch 1989; Veblen et al. 1991; Hessburg et al. 1994) and root diseases (Filip and Goheen 1984; Hessburg et al. 1994). Probabilities for insect or pathogen activities and the probability of transition from one successional stage to another can be defined for individual biophysical environments and their associate assemblages of plants and animals, which can be used to model change across large landscapes (Hagle et al. 1995). Fire regimes (intensity, frequency, and extent) and their effects have been described in general for vegetation types throughout the western United States (Agee 1993). Changes in fire regimes and resulting stand and landscape structure following Eurosettlement has been similarly discussed (Covington et al. 1995). For specific isolated sites there may be excellent fire history information, but at larger landscape scales the information is less precise or often missing. Large regional assessments for landscape structure and disturbance have done much to increase the information base in the Columbia River Basin (Hessburg et al. 1994; Lehmkuhl et al. 1994), but these remain statistical samples rather than continuous map themes.

For individual areas of interest the analysis of current landscape stand types, their age, and stand structure provides an estimate of previous disturbance regimes (frequency, extent, and severity) (Bourgeron and Jensen 1994). However, attempts to define historical disturbance regimes from current landscape and stand configurations is made difficult because the most recent disturbance may mask a preceding event (Swanson et al. 1994). Also, reduced disturbance effects since Eurosettlement has caused increasing homogeneity of landscape patterns in some areas and reduce apparent historical disturbance levels (Lehmkuhl et al. 1994).

Additional sources of disturbance data may improve the estimation of IDR for specific locations. Fire scar analyses provide detailed information on pre-Eurosettlement fire regimes (Agee 1993), and tree core comparisons between host and non-host tree species provide insight into previous insect epidemics (Swetnam and Lynch 1989). Knowledge of the current root disease pockets and rates of spread can allow estimates of historical levels of infestations and comparisons between current conditions (Hagle et al. 1995). Also, the dating of tree scars associated with mass wasting events provides data on the frequency and extent of this type of disturbance (Shuffling 1993).

We suggest that the general character of the IDR can be tentatively refined from an amorphous cloud of possibilities (Swanson et al. 1994) to a series of general response surfaces with frequency, intensity, and extent as the x, y, and z axes. Figure 11.4 depicts a hypothetical disturbance response curve based on the principle that smaller, less severe disturbance events occur more often than large-scale severe events in systems capable of long-term maintenance

FIGURE 11.4. Hypothetical fire disturbance regimes for historical and current conditions of ponderosa pine and spruce/fir forests. Response surface for ponderosa pine has shifted to the right as a result of reduced fire effects and the buildup of forest fuels and more closely resembles the fire regime of the spruce/fir forests.

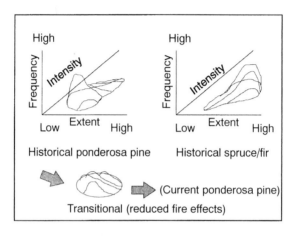

(Turner et al. 1993). This type of curve could represent the probability for different types of fire events given an ignition in the historical dry ponderosa pine forests of the inland west that were characterized by a high frequency-low severity fire regime. The majority of fire events were low intensity (underburns) and occurred with high frequency (2-16 yr. fire-return intervals, Agee 1993), but could vary in extent because of continuity in herbaceous and shrubby understory fuels (USDA Forest Service 1995). The curve has shifted to the right which is characteristic of the low frequency-high severity fire regimes of moist, high elevation spruce/fir forests. This shift is because of reduced fire effects and the subsequent occurrence of a dense conifer understory. The reader is cautioned that the response surfaces are hypothetical and the response surface is probably best thought of as "a fuzzy layer with variable thickness" reflecting different confidence intervals for the probable occurrence of an event of specific severity, extent, and frequency.

Multiple disturbance agents may operate in parallel or in a series to affect change in vegetation (Wargo 1995). Insect, pathogen, and fire disturbance regimes of an area are often interrelated and their response surfaces may be similar in form, albeit displaced in time. Similar response surfaces should occur for fire and insects with the co-occurrence of endemic levels of insects and frequent low intensity fires, and the infrequent occurrence of a large insect event followed by a large stand-replacement fire. Differences in response surfaces among disturbance types could be expected among different biophysical environments. Examples would be the relative shift in importance from insect/fire dominant disturbance to pathogens when the number of root rot pockets increase or an increase in importance of mass wasting disturbance with greater slope, elevation, and precipitation.

Inherent disturbance regimes are hierarchical in nature, with one disturbance type nested within another (Urban et al. 1987; Bourgeron and Jensen 1994). Because of nested disturbances at different scales the area required to manage for disturbances can be much larger than the specific site of interest. Everett and Lehmkuhl (1996) provide the example of managing for disturbance in rare plant habitat that is monitored at multiple scales. Disturbances that were recognized as management concerns included individual tree fall, in square meters; canopy openings, in tens of meters; changes in stand density and the effect on hydrologic responses over tens of hectares; and elk grazing and trampling over thousands of hectares.

THE EFFECTIVE DISTURBANCE REGIME

Vegetation conditions compatible with IDR should be sustainable over time, exclusive of human effects. However, humans and their actions are a significant part of ecosystems and ecosystem function. With increasing human influence on the landscape the IDR becomes less capable of describing the disturbance regimes for an area. Combined, the IDR and current human-induced disturbances define the effective disturbance regime (EDR) that is shaping the character of current vegetation. For example, we speculate that in the ponderosa pine type the initial reduction in fire effects from livestock removal of fine fuels, roading, selective tree harvesting, and active fire suppression (Covington et al. 1995) caused the response surface to flattened and collapse, reflecting reduced fire frequency, severity, and extent. Continued reduction in fire effects created dense conifer understories and increased the amounts and continuity of fuels which shifted the IDR fire response surface toward a low frequency-high severity fire regime (Figure 11.4).

The long-term maintenance of current or desired forest structure and composition is not automatically in question on sites where EDR differs significantly from IDR. "Ecosystems may evolve that are different from any that existed before and may stabilize at one of several levels" (Agee and Johnson 1988, p. 5). Different equilibrium species assemblages can be maintained under the same fire regime (Allen and Wyleto 1983; Urban et al. 1987) as has occurred following cheatgrass and pinyon-juniper invasion of adjacent plant communities. The uncertainty of long-term maintenance of current vegetation increases if the shape of the IDR probability curve shifts to more severe disturbance events than occurred historically and the potential for decline in site nutrient capital or biodiversity is evident. In the case of ponderosa pine, increased vegetation biomass may not be supportable under the EDR and the increased severity of disturbance may prolong the recovery period for nutrients and biodiversity (Grier 1975; Perry et al. 1989).

APPLICABILITY OF IDR

The IDR reference can evaluate a broader array of conditions than HRV, but is similarly restricted to a set of possible conditions that are both desired by the public and supportable under the disturbance regimes existing prior to Eurosettlement. The complete array of vegetation conditions supportable by pre-Eurosettlement disturbance regimes would include historical conditions as well as other conditions unrealized in the past that better meet current public expectations. There is no necessity to enumerate and evaluate every possible vegetation condition under IDR, but we can evaluate those conditions desired by the public. If the public desired only historical conditions then the IDR reference would be superfluous.

The IDR reference, like the HRV, has restrictions on its use. Difficulties in applying the IDR approach occur when disturbance required to achieve desired conditions is unacceptable, and if there are incompatibilities between the desired vegetation and the inherent disturbance regime of the area. The conflict between prescribed burning and air quality speaks to the public's aversion against required disturbance events, and the desire for dense pine/ Douglas-fir stands for spotted owl habitat speaks to the public's desire for forest conditions not supportable by the current (effective) high frequency-low severity disturbance regimes of the vegetation type (Everett et al., 1997).

Meeting species viability using the coarse-filter approach of required disturbance under IDR is no less risky than using historical conditions under HRV. If the public desires conditions that have no historical counterpart there are no grounds for assuming continued species viability even if the desired condition is supportable under IDR. Conversely, if a species is not in harmony with the IDR and rate of change for the area its continued viability is less certain.

USING HRV AND IDR IN COMBINATION

There are philosophical and methodological reasons for utilizing both HRV and IDR references. Agee and Johnson (1988) recommended managing for both vegetation (conditions) structure and disturbance (processes) to increase the long-term maintenance of ecosystems. Achieving desired conditions has been the prevalent management direction and is perhaps more readily visualized by the public than managing for disturbance regimes. However, the public's "condition view point" is usually static in nature and does not incorporate the inevitable changes that occur in dynamic systems. Complex descriptions of site-specific conditions and changes in conditions over time for large landscapes may overwhelm the general public and may be unrealistic in terms of commodity production given reoccurring unplanned disturbance events. The explanation of managing lands within the bounds of IDR may be more direct and would allow greater flexibility in resource condition outcomes.

A combined HRV/IDR reference could evaluate both conditions and rate and type of change in conditions for an area over a period of time. The HRV provides an historical condition basis and IDR provides an historical rate of change basis to evaluate species viability (Hunter 1991). The IDR provides the ability to evaluate emerging non-historical conditions for long-term maintenance under historical and current disturbance regimes. Also, IDR could be used in project planning as a tool for determining the means and time frame to produce desired conditions (Jon Haber, Planning Staff, USFS R1, personal communication, 1996).

As knowledge of vegetation patterns and disturbance regimes of an area provides complementary information, the optimum approach would be a reiterative process that uses each in turn to better define the other (Swanson et al. 1994; Wargo 1995). As an example, a first approximation of IDR can be garnered from our general knowledge of disturbance regimes by biophysical environments, which can in turn be used to estimate historical vegetation conditions that could be supported and their spatial locations on the landscape. Gathering local disturbance/stand information at these locations would further refine estimates of landscape and stand historical conditions (Bonnicksen and Stone 1982) and provide a refined approximation of IDR for individual locations and the larger area.

The combined HRV/IDR approach was recently used in defining the desired future forest on the 140,000 acre Tyee Burn in eastern Washington. The identified desired forest condition was one that met public expectations as expressed in land-use allocations and had reduced hazard for catastrophic fire events. Historical vegetation types and their locations were estimated from analysis of stand exam information and previous research plots (Figure 11.5). Fire, insect, pathogen, and mass wasting disturbance regimes were estimated for each vegetation type using scientific opinion supported by existing knowledge of disturbance in the forest types present and the collection of local disturbance information (USDA Forest Service 1995).

Public expectations for resource conditions were expressed by land-use designations which defined the desired vegetation characteristics (Figure 11.6). The required vegetation structure for the emphasized use was evaluated for long-term maintenance by IDR. In the historical ponderosa pine type the desired vegetation for livestock grazing and deer winter range was compat-

FIGURE 11.5. Historical vegetation defined by stand exam data. The type and extent of historical vegetation was used to define the types and area of occurrence for disturbance regimes (adapted from Everett et al. 1996b). (ABLA2, *Abies lasocarpa*, subalpine fir; PIPO, *Pinus ponderosa*, ponderosa pine; PICO, *Pinus contorta*, lodgepole pine; PSME, *Pseudotsuga menziesii*, Douglas-fir).

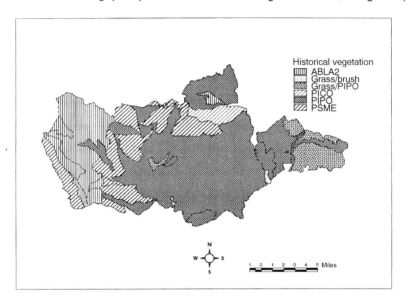

ible with the high frequency-low severity fire regime of the area. Denser, higher elevation spruce/fir forests were compatible with public expectations for late successional reserves, scenic viewsheds, and general matrix forest with an IDR characterized by a low frequency-high severity fire regime. Uncertainty in long-term maintenance of desired landscape and stand characteristics increased in lower elevation historical ponderosa pine areas where the public expectations for denser forest stands in late successional reserves, riparian protection zones, or scenic viewsheds is counter to the vegetation supported by the IDR. The previous discontinuity in IDR and the desired vegetation was a significant factor in the occurrence, severity, and extent of the previous burn (Everett et al. 1996a).

REFERENCE POINT EVALUATION AS PART OF THE ADAPTIVE MANAGEMENT PROCESS

Both HRV and IDR should only be considered as general reference guides in land management. Because detailed information on historical landscape

FIGURE 11.6. Land-use allocations on the Tyee burn and the vegetation structure needed to provide for the emphasized uses of the allocations (adapted from Everett et al. 1996b). Vegetation structure for scenic allocations varies with the dominant forest type and the structure of matrix (general) forest reflects variable emphasized uses ascribed to specific locations.

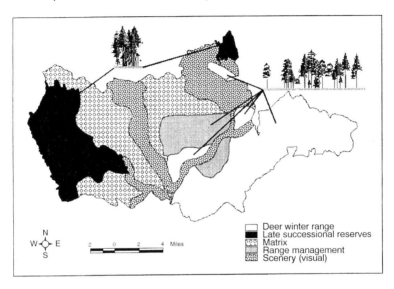

conditions and disturbance regimes for large scale landscapes is scant, HRV and IDR should be used under an adaptive management process with continual revision as more information is developed. Changing climatic conditions and associated change in vegetation composition and structure and disturbance regimes would require a re-evaluation of the applicability of both HRV and IDR references. As IDR is supported by scientific information on disturbance regimes by vegetation type and biophysical environment, it is less dependent on the availability of local historical information than HRV. The stated generality of the IDR approach, its tentative use in the absence of specific site information, and its evaluation of a wide array of post disturbance vegetation states for long-term maintenance may make this reference point more applicable than HRV across large landscapes until more detailed site-specific information on historical conditions can be obtained.

POTENTIAL MISUSE OF REFERENCE POINTS: REQUIRED RATE OF CHANGE

The HRV reference is subject to misuse if landscapes are managed for any one historical condition. An array of changing conditions describe an ecosystem and no one "snap-shot" in time reflects the conditions required for long-term maintenance of the ecosystem structure or function. Attempts to manage for the mean or the extremes of historical conditions ignores the important components of the entire HRV in the long-term maintenance of nutrient cycling and biodiversity. Managing for a dynamic steady state of the vegetation mosaic (Oliver 1992) within the HRV does not speak directly to required change nor rate of change as described by Samson (1992) or Morgan et al. (1994), but would meet this requirement.

The IDR could similarly be misused if a single portion of the disturbance response surface was emphasized. The nature of the response surface implicitly states that different intensities and extents of disturbance occur with different frequencies and it is the integration of the whole (the surface) that defines the IDR of the system. Dominant disturbance events in the IDR define the rate of change in vegetation conditions previously tolerated (or required) by ecosystem species, components, and processes. Morgan et al. (1994) ascribed the same significance to rate of change as to magnitude of change in defining historical range of variability. To reduce uncertainty in the management of disturbance the initiation of a disturbance (rate of change) may need to be tied to its probability of occurrence as defined by the IDR response surface. Until we know more about the tolerance of species and processes to altered rates of change, the most conservative approach may be to mimic historical rates of change where possible.

CONCLUSION

The historical range of variability (HRV) concept has been used as a reference in the management of landscape pattern and composition. The HRV concept provides a benchmark reference for changing resource management focus from an emphasis on product outputs to achieving long-term resource conditions. The HRV remains our best reference for restoration of ecosystems to historical levels. Historical conditions are not always desired, can be difficult to restore, may no longer be supportable under altered disturbance regimes, and may not be possible should future climatic conditions differ from the past. The value of the HRV reference will decline with increasing public demands for resource conditions that have no historical homologues and with increasing human effects that alter the disturbance regimes that created and maintained historical stand and landscape characteristics.

The inherent disturbance regime (IDR) concept is used as a reference benchmark in the management of disturbance events. The IDR reference provides increased flexibility over HRV by evaluating the long-term maintenance of desired (historical and otherwise) stand and landscape characteristics under the disturbance regimes of the area. The reference for long-term maintenance of desired conditions is whether the disparity between the IDR and the current effective disturbance regime (EDR = IDR + human effects) increases the hazard for more severe disturbance events and subsequent decline in site nutrient capital or biodiversity. Adhering to the IDR reference for the probability of occurrence of a disturbance of specific magnitude and extent should provide for change and rate of change in stand and landscape conditions similar to that which species and processes coexisted with in the past. We believe that IDR can evaluate a wider array of conditions for long-term maintenance than HRV and will increase land manager decision space to meet emerging public expectations.

The use of HRV and IDR to recreate or maintain historical conditions involves evaluating the disparity between current conditions/disturbance regimes with their historical estimates and defining required actions to readjust. In the evaluation of desired non-historical conditions, the activity is to assess long-term maintenance potential of the desired condition under the EDR and evaluate whether the disparity with IDR and HRV represents a significant hazard to conserving biodiversity and nutrient capital at the appropriate scales.

There is no one unique natural vegetation pattern or disturbance regime (Sprugel 1991), therefore both HRV or IDR references should only be viewed as general guides and tentative in most instances. Our knowledge of historical conditions and disturbance regimes is developing rapidly but remains incomplete for much of the landscape. Historical landscape information based on scientific opinion and aerial photographs may need to be redefined as more information is developed. Also, current large scale assessments (Everett et al. 1994; Quigley et al. 1996) that provide estimates of historical conditions are sample based and lack continuous map themes over large areas. Similarly, we know much about disturbance regime characteristics for vegetation types, vegetation structures, and biophysical areas, but that information is not continuous across the landscape. More information by specific locality will increase our confidence in estimating both historical conditions and disturbance regimes (Morgan et al. 1994; Swanson et al. 1994). This process of discovery will proceed most rapidly when HRV and IDR are used in combination to reciprocally define historical conditions and IDR for specific areas.

The IDR reference is an evolving experimental concept that should be applied to land management only under an adaptive management process which continually re-evaluates developing knowledge of disturbance regimes, their interactions, and resulting resource conditions. Both IDR and HRV reference points can only be a portion of a multivariate approach in evaluating long-term maintenance of ecosystems; both references focus on the biological characteristics of the ecosystem and need to be integrated with the public's socioeconomic expectations. Public expectations may include landscape conditions that did not exist historically and disturbance regimes that differ from those inherent to the area. Increased discontinuity between public expectations and vegetation supported by IDR may increase the hazard for more severe disturbance events and a failure to maintain long-term site potential or biodiversity. Public awareness of this disconnect is critical in creating achievable "desired conditions" and refinement of their expectations (Jensen et al. 1996).

REFERENCES

Agee, J.K. 1993. *Fire Ecology of Pacific Northwest Forests.* Island Press. Washington, DC. 493 p.

Agee, J.K. and D.R. Johnson (comps). 1988. Ecosystem management for parks and wilderness: Workshop synthesis. Institute of Forest Resources. Contribution No. 62. 39 p.

Allen, T.F.H. and E.P. Wyleto. 1983. A hierarchical model for the complexity of plant communities. *Journal of Theoretical Biology* 101:529-540.

Baker, W.L. 1995. Long-term response of disturbance landscapes to human intervention and global change. *Landscape Ecology* 10:143-159.

Bonnicksen, T.M. and E.C. Stone. 1982. Reconstruction of presettlement giant sequoia-mixed conifer forest community using the aggregation approach. *Ecology* 63:1134-1148.

Botkin, D.B. and M.T. Sobel. 1975. Stability in time-varying ecosystems. *The American Naturalist* 109:625-646.

Bourgeron, P.S. and M.E. Jensen. 1994. An overview of ecological principles for ecosystem management. In: pp. 45-57. *Ecosystem Management: Principles and Applications.* PNW-GTR-318. USDA, Forest Service, Pacific Northwest Research Station, Portland, OR.

Bridges, J.R. 1995. Exotic pests: Major threats to forest health. In: pp. 105-116. *Forest Health Through Silviculture; Proceedings of the 1995 National Silviculture Workshop.* Mescalero, New Mexico. May 8-11, 1995. L.G. Eskew comp. RM-GTR-267. USDA Forest Service, Rocky Mountain Forest and Range Experiment Station, Fort Collins, CO.

Caraher, D. and W.H. Knapp. 1993. Assessing ecosystem health in the Blue Mountains. In: *Silviculture: from the Cradle of Forestry to Ecosystem Management: Proceedings of the National Silviculture Workshop*, November 1-4, 1993, Henderson, NC. GTR-SE-88. USDA Forest Service, Southeastern Forest Experiment Station, Asheville, NC. 258 p.

Carlson, C.E, S.F. Arno, J. Chew, and C.A. Stewart. 1995. Forest development leading to disturbances. In: pp. 26-36. L.G. Eskew comp., *Forest Health Through Silviculture; Proceedings of the 1995 National Silviculture Workshop.* May 8-11, 1995, Mescalero, New Mexico. RM-GTR-267. USDA Forest Service, Rocky Mountain Forest and Range Experiment Station, Fort Collins, CO.

Christensen, N.L. 1988. Succession and natural disturbance paradigms, problems and preservation of natural ecosystems. In: pp. 62-86. J. Agee and D.R. Johnson, eds., *Ecosystem Management for Parks and Wilderness*. University of Washington Press, Seattle.

Cooper, W.S. 1913. The climatic forest of Isle Royal, Lake Superior and its development. I, II, and III. *Botanical Gazette* 55:1-44;115-140,189-235.

Covington, W.W., R.L. Everett, R. Steele, L.L. Irwin, T.A. Daer, and A.N.D. Auclair. 1995. Historical and anticipated changes in forest ecosystems of the inland west of the United States. *Journal of Sustainable Forestry* 2:13-63.

Covington, W.W. and S.S. Sackett. 1984. The effect of a prescribed burn in southwestern ponderosa pine on organic matter and nutrients in woody debris and forest floor. *Forest Science* 30:183-192.

Everett, R.L. 1984. Great Basin pinyon and juniper communities and their response to management. In: pp. 53-61. *Proceedings of the Symposium on the Cultural, Physical and Biological Characteristics of Range Livestock Industry in the Great Basin*. February 11-14, 1994. Salt Lake City, UT. Society for Range Management.

Everett, R.L., P.H. Hessburg, M.E. Jensen, and B.T. Bormann. 1994. Eastside forest ecosystem health assessment. Vol. 1. Executive SUMMARY. PNW-GTR-317. USDA Forest Service, Pacific Northwest Research Station. 57 p.

Everett, R.L. and J.F. Lehmkuhl. 1996. An emphasis-use approach to conserving biodiversity. *Wildlife Society Bulletin* 24:192-199.

Everett, R.L., A.E. Camp, and R. Schellhaas. 1996a. Building a new forest with fire protection in mind. In: pp. 253-258. Proceedings of the Society of American Foresters National Convention. Oct. 28-Nov. 1, 1995, Portland MA., Society of American Foresters, Bethesda, MD.

Everett, R.L., R. Schellhaas, T. Anderson, J. Lehmkuhl, and A. Camp. 1996b. Restoration of ecosystem integrity and lands use allocation objectives in altered watersheds. In: pp. 271-280. Proceedings of the 1996 Watershed Restoration Management annual conference, July 14-17, 1996, Syracuse, NY. Herndon, VA.

Everett, R.R. Schellhaas, D. Spurbeck, P. Ohlson, D. Keenum, and T. Anderson. 1997. Forest structure of current spotted owl nest stands and their historical conditions on the east slope of the Pacific Northwest Cascades, USA. *Forest Ecology and Management* 94:1-14.

Filip, G.M. and D.J. Goheen. 1984. Root diseases cause severe mortality in white and grand fir stands of the Pacific Northwest. *Forest Science* 30:138-142.

Gleason, H.A. 1926. The individualistic concept of the plant association. *Bulletin of the Torrey Botanical Club* 53:7-26.

Grier, C.C. 1975. Wildfire effect on nutrient distribution and leaching in a coniferous ecosystem. *Canadian Journal of Forest Research* 5:599-607.

Gruell, G.E. 1980. Fire's influence on wildlife habitat on the Bridger-Teton National Forest, Wyoming. Volume 1: Photographic record and analysis. Res. Pap. INT-235. USDA Forest Service, Intermountain Forest and Range Experiment Station, Ogden, UT. 207 p.

Hagle, S.K., S. Kegley and S.B. Williams. 1995. Assessing pathogen and insect succession functions in forest ecosystems. In: pp. 117-127. L.G. Eskew comp., *Forest Health Through Silviculture; Proceedings of the 1995 National Silviculture Workshop*. May 8-11, 1995, Mescalero, New Mexico. RM-GTR-267. USDA Forest Service, Rocky Mountain Forest and Range Experiment Station, Fort Collins, CO.

Hessburg, P.F., R.G. Mitchell, and G.M. Filip. 1994. Historical and current roles of insects and pathogens in eastern Oregon and Washington forest landscapes. PNW-GTR-327. USDA Forest Service, Pacific Northwest Research Station, Portland, OR. 72 p.

Hessburg, P.F., B.G. Smith, C.A. Miller, S.D. Kreiter, and R.B. Salter. 1999. Modeling change in potential landscape vulnerability to forest insect and pathogen disturbances: methods for forested subwatersheds sampled in the midscale interior Columbia River Basin assessment. PNW-GTR-454. Portland, OR: USDA Forest Service, Pacific Northwest Research Station. 56 p.

Hungerford, R.D., M.G. Harrington, W.H. Frandsen, K.C. Ryan, and G.J. Niehalf. 1991. Influence of fire on factors that affect site productivity. In: pp. 32-50. *Proceedings: Management and Productivity of Western-Montane Forest Soils.* April 10-12, 1990, Boise ID. GTR-INT-280. USDA Forest Service, Intermountain Research Station, Ogden, UT.

Hunter, M. 1991. Coping with ignorance: The coarse-filter strategy for maintaining biodiversity. In: pp. 256-281. K.A. Kohn, ed., *Balancing on the Brink of Extinction–The Endangered Species Act and Lessons for the Future.* Island Press, Washington, DC.

Jensen, M.E., P.S. Bourgeron, R.L. Everett, and I. Goodman. 1996. Ecosystem management: a landscape ecology perspective. *Water Resources Bulletin* 32:1-14.

Kay, C.E. 1995. Aboriginal overkill and native burning: implications for modern ecosystem management. *Western Journal of Applied Forestry* 10(4):121-126.

Koniak, S. and R.L. Everett. 1982. Seed reserves in soils of successional stages of pinyon woodlands. *American Midland Naturalist* 108:295-303.

Lehmkuhl, J.F., P.F. Hessburg, R.D. Ottmar, M.H. Huff, and R.L. Everett. 1994. Historic and current forest landscapes in eastern Oregon and Washington, Part I: Vegetation pattern and insect and disease hazards. PNW-GTR-328. USDA Forest Service, Pacific Northwest Research Station, Portland, OR. 88p.

Lundquist, J.E. 1995. Disturbance profile–a measure of small-scale disturbance patterns in ponderosa pine stands. *Forest Ecology and Management* 74:49-59.

MacCleary, D.W. 1995. The way to a healthy future for National Forest ecosystems in the West: What role can silviculture and prescribed fire play? In: pp. 37-45. L. G. Eskew comp., *Forest Health Through Silviculture; Proceedings of the 1995 National Silviculture Workshop.* May 8-11, 1995, Mescalero, New Mexico. RM-GTR-267. USDA Forest Service, Rocky Mountain Forest and Range Experiment Station, Fort Collins, CO.

Morgan, P., G.H. Aplet, J.B. Haufler, H.C. Humphries, M.M. Moore, and W.D. Wilson. 1994. Historical range of variability: a useful tool for evaluating ecosystem change. *Journal of Sustainable Forestry* 2:87-111.

Odum, E.P. 1969. The strategy of ecosystem development. *Science* 164:262-270.

Oliver, C.D. 1992. A landscape approach: achieving biodiversity and economic productivity. *Journal of Forestry* 90(2):20-25.

Oliver, C.D., D.E. Ferguson, A.E. Harvey, H.S. Malany, J.M. Mandzak, and R.W.

Mutch. 1994. Managing ecosystems for forest health: An approach and the effects on uses and values. *Journal of Sustainable Forestry* 2:113-133.

Oliver, C.D. and B.C. Larson. 1990. *Forest Stand Dynamics*. McGraw Hill, New York, NY. 467 p.

Overbay, J.C. 1992. Ecosystem management. In: pp. 3-15. *Proceeding of the National Workshop: Taking an Ecological Approach to Management*. April, 27-30, 1992, Salt Lake City, UT. WO-WSA-3. USDA Forest Service, Watershed and Air Management, Washington, DC.

Parsons, D.J., D.M. Graber, J.K. Agee, and J.W. van Wagtendonk. 1986. Natural fire management in national parks. *Environmental Management* 10:21-24.

Perry, D.A., M.P. Amaranthus, J.G. Borchers, and others. 1989. Bootstrapping in ecosystems. *Bioscience* 39:230-237.

Pickett, S.T.A. and P.S. White. 1985. *The Ecology of Natural Disturbance and Patch Dynamics*. Academic Press, London. 472 p.

Plummer, F.G. 1902. Forest conditions in the Cascade Range, Washington. Professional paper No. 6. USDI, US Geological Survey. 39 p.

Quigley, T.M., R.W. Haynes and R.T. Graham (tech. eds.). 1996. An integrated scientific assessment for ecosystem management in the Interior Columbia Basin and portions of the Klamath and Great Basins. PNW-GTR-382. USDA Forest Service, Pacific Northwest Research Station, Portland, OR.

Ricklefs, R.E. 1987. Community diversity: Relative roles of local and regional processes. *Science* 235:167-171.

Robbins, W.G. and D.W. Wolf. 1994. Landscapes and the intermontane northwest: an environmental history. PNW-GTR-319. USDA Forest Service, Pacific Northwest Research Station, Portland, OR. 32 p.

Samson, F.B. 1992. Conserving biological diversity in sustainable ecological systems. *Transaction of the 57th North American Wildland and Natural Resources Conference*. 57:308-320.

Shelby, B. and R.W. Speaker. 1990. Public attitudes and perceptions about prescribed burning. In: pp. 253-260. J. Alstad, S. Radosevich, and D. Sandberg. eds., *Natural and Prescribed Fire in Pacific Northwest*. Oregon State University Press, Corvallis.

Shuffling, R. 1993. Induction of vertical zones in sub-alpine valley forest by avalanche-formed fuel breaks. *Landscape Ecology* 8:127-138.

Skovlin, J.M. and J.W. Thomas. 1995. Interpreting long-term trends in Blue Mountains ecosystems from repeat photography. PNW-GTR-315. USDA Forest Service, Pacific Northwest Research Station. Portland, OR. 102 p.

Sparks, R.E. 1995. Need for ecosystem management of large rivers and their floodplains. *Bioscience* 45:168-182.

Sprugel, D.G. 1991. Disturbance, equilibrium and environmental variability: What is "natural" vegetation in a changing environment. *Biological Conservation* 58:1-18.

Stone, L. and Ezrati, S. 1996. Chaos, cycles and spatiotemporal dynamics in plant ecology. *Journal of Ecology* 84:279-291.

Swanson, F.J., J.A. Jones, D.O. Wallin, and J.H. Cissel. 1994. Natural variability-implications for ecosystem management. In: pp. 80-94. Ecosystem Management:

Principles and Applications. PNW-GTR-318. USDA Forest Service, Pacific Northwest Research Station, Portland, OR.

Swetnam, T.W. and A. Lynch. 1989. A tree-ring reconstruction of western spruce budworm history in the southern Rocky Mountains. *Forest Science* 35:962-986.

Turner, M.G., W.H. Romme, R.H. Gardner, R.V. O'Neill, and T.K. Kratz. 1993. A revised concept of landscape equilibrium: Disturbance and stability on scaled landscapes. *Landscape Ecology* 8:213-227.

Uebelacker, M.L. 1986. Geographic explorations in the southern Cascades of eastern Washington: changing land, people and resources. Ph.D. dissertation. University of Oregon, Eugene. 216 p.

Urban, D.L., R.V. O'Neill, and H.H. Shugart, Jr. 1987. Landscape ecology: A hierarchical perspective can help scientists understand spatial patterns. *Bioscience* 119-127.

U.S. Department of Agriculture. 1992. Our approach to sustaining ecological systems. Forest Service, Northern Region, Missoula, MT. 15 p.

U.S. Department of Agriculture. 1995. Preliminary Report: Tyee/Hatchery Creek Fire Restoration. Forest Service, Pacific Northwest Research Station, Wenatchee Forestry Sciences Lab., Wenatchee, WA. 21 p.

Veblen, T.T., K.S. Hadley, M.S. Reid, and A.J. Rebertus. 1991. The response of subalpine forests to spruce beetle outbreak in Colorado. *Ecology* 72:213-231.

Wargo, P.M. 1995. Disturbance in forest ecosystems caused by pathogens and insects. In: pp. 20-25. L.G. Eskew comp., *Forest Health Through Silviculture; Proceedings of the 1995 National Silviculture Workshop*. May 8-11, 1995, Mescalero, New Mexico. RM-GTR-267. USDA Forest Service, Rocky Mountain Forest and Range Experiment Station, Fort Collins, CO.

Watt, A.S. 1947. Pattern and process in the plant community. *Journal of Ecology* 35:1-22.

Chapter 12

A Wildfire and Emissions Policy Model for the Boise National Forest

Leon F. Neuenschwander
R. Neil Sampson

SUMMARY. We utilized the Boise National Forest's Hazard/Risk model, along with fire history records and fire behavior models, to estimate the current and anticipated levels of large wildfires and associated greenhouse gas and particulate emissions based on the forest condition and wildfire regime on the BNF. The model indicated that the forests at greatest risk of large, intense wildfires are the dense ponderosa pine-Douglas-fir forests that make up over 1.1 million acres on the forest. We conclude that without an aggressive treatment program to reduce large areas of contiguous heavy fuel loadings the forest will be burned at an annual average rate of about 7.5% of the remaining at-risk forest. Using recent fire data to develop average patterns of intensity in wildfires within this forest type, we estimate that emissions will average around 1 million tons of carbon (C) per year over the next 20 years as the bulk of the ponderosa pine forests are burned. An aggressive treatment program featuring the removal of fuels where necessary, and prescribed fire as a means of re-introducing fire to these ecosystems, would result in a 30-50 percent reduction in the average annual wildfire

Leon F. Neuenschwander is affiliated with College of Forestry, Wildlife and Range Sciences, University of Idaho, Moscow, ID 83844. R. Neil Sampson is affiliated with American Forests, Washington, DC, and The Sampson Group, Inc., Alexandria, VA 22310.

[Haworth co-indexing entry note]: "Chapter 12. A Wildfire and Emissions Policy Model for the Boise National Forest." Neuenschwander, Leon F., and R. Neil Sampson. Co-published simultaneously in *Journal of Sustainable Forestry* (Food Products Press, an imprint of The Haworth Press, Inc.) Vol. 11, No. 1/2, 2000, pp. 289-309; and: *Mapping Wildfire Hazards and Risks* (ed: R. Neil Sampson, R. Dwight Atkinson, and Joe W. Lewis) Food Products Press, an imprint of The Haworth Press, Inc., 2000, pp. 289-309. Single or multiple copies of this article are available for a fee from The Haworth Document Delivery Service [1-800-342-9678, 9:00 a.m. - 5:00 p.m. (EST). E-mail address: getinfo@haworthpressinc.com].

experienced in the dense ponderosa pine forests, a 14-35% decrease in the average annual C emissions, and a 10-31% decrease in particulate emissions. We argue that the most effective way to curb emissions is with an aggressive treatment program linked to a landscape-based ecosystem management plan. This would have the effect of breaking up large contiguous landscape patterns so that fires become more patchy and diverse in their environmental impact, resulting in significantly reduced emissions as well as improved landscape diversity. *[Article copies available for a fee from The Haworth Document Delivery Service: 1-800-342-9678. E-mail address: <getinfo@haworthpressinc.com> Website: <http://www.haworthpressinc.com>]*

KEYWORDS. Land treatment, fuel treatment, wildfire risk, carbon emissions, particulates

INTRODUCTION

The Boise National Forest (BNF) covers 2.6 million acres in southwestern Idaho. About half of the forest (1.2 million acres) consists of dry ponderosa pine (*Pinus ponderosa*) forests that were historically characterized by frequent, low-intensity wildfires (Steele 1988; Covington et al. 1994; Neuenschwander and Dether 1995). Since 1986, concern for the health and future of these forests has mounted as large, high-intensity wildfires have affected almost 500,000 acres of ponderosa pine (BNF 1996). Ponderosa pine, which currently is estimated to occupy around 31.5 million acres across the Inland Western region of the United States (Powell et al. 1993) has also been identified as one of the more threatened ecosystems in the country (Noss et al. 1995).

The forest health concerns in the ponderosa pine forest type center around the changes in forest structure, species composition, nutrient cycling, and fire regime that have resulted from decades of fire suppression, grazing and logging (BNF 1996; Neuenschwander and Dether 1995; Covington et al. 1994; Steele et al. 1986; Covington and Sackett 1986). These forests, which once were characterized by large, widely-spaced ponderosa pines that survived the frequent, low-intensity wildfires, are now dominated by dense stands of small pines, inland Douglas-fir (*Psuedotsuga menziesii* var *glauca*), true firs (*Abies grandis, Abies concolor,* etc.) and lodgepole pine *(Pinus contorta)* (Covington et al. 1994). As a result of the altered stand conditions, current wildfires are characterized by high-intensity stand-replacing crown fires covering large areas that may, if subsequent weather conditions, stand regeneration, and wildfire patterns are not favorable, convert from pine forests to grass and shrubland (Neuenschwander and Dether 1995).

These conditions are not limited to the BNF, but characterize large regions of ponderosa pine and mixed conifer forests of the Inland West (Sampson et al. 1994; Covington et al. 1994; Arno and Brown 1991). Developing useful strategic planning tools to guide ecological restoration efforts over such an enormous area seems essential in view of the limited capability and short time frames facing land managers. These tools must provide a practical approach to the restoration of non-lethal fire regimes in these forests, as well as restoring fire's role in maintaining grasslands threatened by the encroachment of woody vegetation (Arno 1996; Brown 1995).

The BNF offers an excellent laboratory in which to develop and test such a tool, both because it is representative of a regional situation, but also because it has been conducting an intensive effort to study the forest health situation in the BNF, and develop management strategies to restore the ponderosa pine forest to a more sustainable condition. As part of this effort, the BNF managers have been involved in a cooperative forest health research project with American Forests, University of Idaho, Intermountain Research Station, Idaho Department of Lands, and Boise Cascade Corporation. One product of that effort has been the development of a state-of-the-art Geographic Information System (GIS) which was used as the basis for development of a hazard-risk assessment model in cooperation with the University of Idaho (BNF 1996).

We utilized the BNF hazard-risk model as the basis for developing a method of estimating the likely future effect of wildfires on the remaining ponderosa pine forests on the BNF, and quantifying the potential impact of management treatment options designed to restore these forests to a more fire-tolerant, sustainable condition through combinations of mechanical thinning, prescribed fire, and associated restoration practices.

In addition to estimating potential effects on the forest ecosystems, we also looked at the impacts on the carbon (C) cycle to estimate whether or not forest treatment could affect the buildup of atmospheric greenhouse gases (GHG) which have been identified as primary contributors to global climate change (IPCC 1996) and the production of smoke emissions (PM) that are increasingly cited as a contributor to both public health hazards and visibility impairment (Rigg et al. 1999).

METHODS

We utilized the BNF hazard-risk model to identify the amount of ponderosa pine forest that appears to be at high risk to large, intense wildfires (as of early 1992, when the vegetative condition data was gathered). An estimate of the likelihood and extent of future wildfire ignitions was developed using the fire history records available from the BNF. The probability of those ignitions growing into large, unmanageable fires was estimated from the records of recent wildfire behavior. Estimates of the probable impact of wildfires on forest vegetation (mortality) under a variety of treatment options was developed by utilizing existing plot data from the BNF in a fire behavior model (FIRESUM) across a range of weather conditions characteristic of the highest-incidence wildfire period. This allowed the selection of treatment options that provide the highest probability of survival of the large ponderosa pines under future fire conditions.

The current wildfire data and the selected treatment options were simulated in an adapted version of the CONSUME model to estimate the biomass that would be consumed under both wildfire and prescribed fire conditions. Biomass consumption was then converted into C and PM emissions per acre of forest burned or treated. A 20-year step model was then developed to indicate how future treatment options might affect wildfire extent, forest structural diversity, and emissions for the coming two decades.

Since each step of the analysis produced results that were inputs for future steps, the following sections detail the methods and results of those intermediate steps.

The Forest Area at Risk

The BNF hazard-risk model divides the landscape by watershed and sub-watershed, and uses five data layers (vegetation, slope, elevation, aspect, and historical fire occurrence by location) to identify those areas at highest relative risk for the occurrence of large, intense wildfires that burn outside the historical range for such events (BNF 1996). The areas at highest risk were identified from 1992 AVHRR satellite imagery as moderate- or high-density image classes where forests containing ponderosa pine forests as a major seral species occupy more than 25% of a subwatershed and where moderate to high numbers of fires have occurred in recent years. The model returned an estimate that up to 152 out of 378 subwatersheds (1,196,781 acres) are at risk to these fires (Figure 12.1).

Determining the Probability of Large Fires

Large forest fires in the ponderosa pine forests have been uncommon since settlement, and intense wildfires that killed most or all of the large trees were historically rare, but both have become increasingly common in the last decade. Since 1956, only 4 wildfires in the BNF ponderosa pine forests have exceeded 50,000 acres. All of those have occurred since 1989. We used the historical records, which indicate the final size for each ignition, to quantify the size development of past wildfires in these forests. The data indicate that out of 1087 total ignitions over 5 years, 11 (or about 1%) grew larger than 5,000 acres in size (Table 1).

FIGURE 12.1 The Boise National Forest, with subwatersheds at highest risk of wildfire shown as darker areas in inset. (Source: BNF 1996)

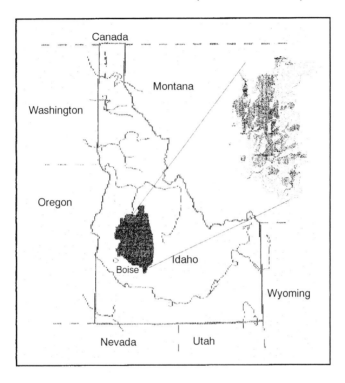

TABLE 1. Wildfire size on the Boise National Forest (1986-1990) by final size, with the probability that a wildfire will expand into the next size class.

Final Fire Size	Number of Fires	Probability of Increasing Size	Annual Average Number
Total Ignitions	1087	0.52	217
> 0.25 acre	564	0.18	113
> 10 acres	100	0.49	20
> 100 acres	49	0.67	10
> 1,000 acres	33	0.84	6.6
> 3,000 acres	28	0.40	5.6
> 5,000 acres	11		2.2

As Table 1 shows, half of all fire ignitions are suppressed or go out naturally at a very small size. More important for our considerations, however, are the indications that if a wildfire achieves a size of 100 acres, it then has a 22 percent probability (11 out of 49) of growing to 5,000 acres or larger. Once they attain that size, attempts at suppression or control are of little effect, and the fire will continue to grow until a weather change stops its advance or it burns into an area with little or no available fuel. As seen in 1992 and 1994 with the Foothills and Rabbit Creek fires, these large fires can impact 150,000 acres or more.

Table 1 suggests that the BNF will experience an average of at least 2 wildfires of 5,000 acres or larger each year. Large fires in recent years have been in the 150,000 to 200,000 acre range. The ultimate size and behavior of any large wildfire is, of course, highly dependent on weather and fuel conditions. On the basis of the large size of the continuous areas of high-hazard fuel conditions (Figure 12.1), and the experience of recent years, we used professional judgment to estimate that the average annual wildfire acres over the Boise National Forest will be in the range of 7.5 percent of the forested area identified as being at risk (1,196,781 acres in 1992). This is consistent with the experience since the late 1980s.

This approach depends on several assumptions, including:

- In forests where vegetative (fuel) conditions are outside HRV, large, intense wildfires will continue to occur at the same general frequency as they did in the recent past;
- Ponderosa pine forests with high-risk fuel conditions will continue to exhibit those conditions until they are treated through silvicultural management (including prescribed fire) or experience a wildfire;
- Areas currently exhibiting low-risk fuel conditions and ignition histories will not convert into high-risk condition within the near term;
- The number and distribution of ignitions will continue to follow the historical pattern;
- Wildfires that ignite will continue to grow from very small to 100 acres, and larger, at the same general rates as in the recent past; and,
- Wildfire suppression policies and capabilities will remain roughly similar to those of the recent past. (There are new federal policies for wildland fire [USDI/USDA 1995], but we had no basis upon which to estimate the practical effects of those policy changes in the near future.)

These assumptions provide a conservative base for estimating future conditions, for at least three reasons. Areas other than the dense ponderosa pine forests also burn in these wildfires, often with high intensity, but we did not estimate that likelihood. We also did not estimate the probability of areas re-burning within the 20-year time frame of the model. As human settlements continue to expand into wildland systems, it increases the future likelihood of anthropogenic ignitions.

On the BNF, wildfires of 1,000 acres or more typically occur during a 29-day window centered around August 22, the date of highest occurrence. Wildfires of 100 acres or more occur over a 135-day window. These dates provide a means of selecting the most probable weather conditions within which to estimate behavior of future wildfires (assuming future fires will fall within those time windows, with weather conditions similar to those contained in the historical record for those dates).

Achieving Fire-Tolerant Forests

We adapted the FIRESUM (Ver. 1.2) model to evaluate the impact of different fire regimes on ponderosa pine forests with the different stand and fuel characteristics that would be found in untreated and treated conditions (Keane 1989). FIRESUM, when provided with detailed stand and fire weather information, returns estimates of fire behavior and tree mortality. Our goal was to determine forest conditions that would return a reasonably high probability that the larger ponderosa pine trees ($10''$ or 25 cm DBH) would survive a fire event typical of the BNF ponderosa pine zone. Forest data from sample plots on the BNF were utilized to represent five vegetative conditions that are commonly found in untreated (1-3) and treated (4-5) areas within these forests:

1. High density   Basal area = 57 m^2/ha (250 ft^2/ac)
2. Medium density   Basal area = 35 m^2/ha (150 ft^2/ac)
3. Low density   Basal area = 18 m^2/ha (80 ft^2/ac)
4. 100 trees/acre   Basal area = 40 m^2/ha (175 ft^2/ac)
5. 50 trees/acre   Basal area = 33 m^2/ha (144 ft^2/ac)

Wildfire weather conditions were simulated by using data from nearby stations, adjusted to the elevation of the plot data. Medians were calculated for the daily high temperature and windspeed in the most common wildfire season (July 29-Sept. 11), then the 25th and 75th percentiles of this data were used as ranges for daily high temperatures and high wind speeds. Relative humidities were set at ranges of 10-20 percent based on professional judgment. The weather ranges developed for wildfire conditions were:

Daily high temperatures–81.7 to 86.6°F;
Daily high wind speeds–8.7 to 11.7 mph; and,
Relative humidities–10 to 20 percent.

Weather data ranges for prescribed fire on these sites were developed from preferred prescription parameters and were:

Daily high temperatures–60 to 75°F;
Daily high wind speeds–0 to 10 mph; and,
Relative humidities–25 to 35 percent.

Seven fire scenarios were developed to illustrate a range of possible future conditions, based on professional experience in the region, as follows:

1. Wildfires in years 5, 21, 37, 53, 69, 85, and 101;
2. Wildfires in years 5, 37, 69, and 101;
3. Wildfires in years 5, 55, and 101;
4. Wildfires in years 5 and 101;
5. Wildfire in year 101 only;
6. Prescribed fires in years 5, 21, 37, 53, 69, 85, and 101; and,
7. Prescribed fires in years 5, 21, 37, and 53 followed by wildfire in year 57.

The fire weather ranges were entered into a random number generator to obtain 2,000 random sets of wildfire weather variables and 500 sets of prescribed fire weather variables. With these variable sets, we completed 2,500 FIRESUM runs using fire scenarios 1-6 on each of the five sample plot data sets and using fire scenario 7 on plot 3 only.

Simulation results and graphs generated from these model runs allowed us to identify stand densities and fire regimes that provide the highest stability and fire tolerance in these stands. They suggest that management prescriptions using prescribed fire can provide a sustainable, diverse ponderosa pine ecosystem if properly designed and executed. Figure 12.2 illustrates the combined effect of using the various fire scenarios and weather conditions on the five stand conditions, and the Rx fire on plot 3. The FIRESUM output indicates that the survival probability of the larger trees remains at 50% or less without thinning to maintain tree densities of 100 per acre or less, or the use of prescribed fire. After prescribed fire treatment on plot 3 (the low-density plot), the wildfire survival probabilities of 10″ (25 cm) DBH trees increase to over 50%, indicating the likelihood of a more sustainable forest condition.

Estimating Fuel Consumption and C Emissions

The BNF plot data and fire mortality results from FIRESUM were then input into the CONSUME (Ver. 1.0) model to estimate the biomass that would be consumed under wildfire and prescribed fire conditions. We added estimates of crown fuel consumption to the CONSUME results to get total fuel consumed, since the model considers only surface fuels. These estimates were derived by using an average of the mortality statistics generated by 100 FIRESUM runs to estimate the crown fuel consumption for all trees that did

FIGURE 12.2. Average probability of survival by tree size (DBH), for 5 selected ponderosa pine stands under a range of wildfire regimes, compared to a prescribed (Rx burn) treatment.

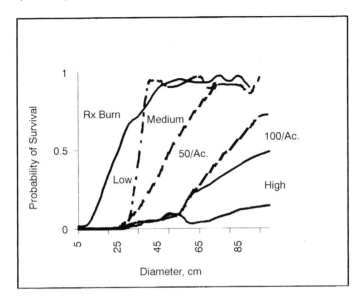

not survive the first fire (at year 5) in the high and low density plots. Crown weights were estimated from Snell and Brown (1980). In calculating total fuel consumption, we assumed that:

- Under high-intensity wildfire conditions, approximately 90% of all duff fuels are consumed regardless of stand density or duff depth; and,
- For the average number of trees that die in 100 FIRESUM runs, all crown fuels $\leq 3''$ in diameter are completely consumed regardless of pre-burn treatment or fire scenario.

Fuel consumed was converted to C emissions by multiplying total fuel by 0.5 to convert to C content, and multiplying by .9 to account for charcoal and ash formation (Agee 1993). The resulting estimates are shown in Table 2.

We utilized these figures to estimate the average fuel consumption of two recent fires in the ponderosa pine area on the BNF, based on the intensity patterns estimated by BNF field staff (BNF 1995). This resulted in an estimated average of 47.2 tons of fuel consumed per acre of wildfire (Table 3).

This average was used to estimate the fuel consumption of modern wildfire events, based on the following assumptions:

TABLE 2. Total fuel consumption and carbon emissions under various fire intensities.

Fire Regime	Total Fuel Consumption	C Emission
	(tons/acre)	
High intensity wildfire (100% tree mortality)	79.5	36
Moderate intensity wildfire in high density stand	74	33
Low intensity wildfire in low density stand	20	9
Prescribed fire in low density stand	11	5

TABLE 3. Fuel consumption and carbon emission estimates developed from fire intensity data for two 1994 wildfire events by the Boise National Forest.

Fire/Intensity	Acres Affected	Fuel Consumed (tons)	C Emissions (tons)	Average Fuel (tons/acre)
Bannock Creek				
Low	800	16,000	7,200	
Medium	350	25,900	11,655	
High	700	55,650	25,043	
Unburned	0	0	0	
TOTAL	1,850	97,550	43,898	52.7
Rabbit Creek				
Low	60,700	1,214,000	546,300	
Medium	25,750	1,905,500	857,475	
High	31,100	2,472,450	1,112,603	
Unburned	1,100	0	0	
TOTAL	118,650	5,591,950	2,516,378	47.1
TOTAL BOTH	119,400	5,689,500	2,560,275	47.2

- All treated areas would be burned as part of the treatment, with some requiring mechanical fuel reduction prior to burning and others able to be burned within prescription under existing fuel conditions; and,
- In burning treated areas with Rx fire, all fuels more than 3″ in diameter would either be removed as part of the fuel treatment program or would not burn in the fire due to its low intensity.

These estimates are believed to under-estimate the total C emissions associated with fire events in this forest area because:

- There are no estimates of soil organic matter loss in CONSUME. Field mapping of these fires indicated a significant area affected by severe soil heating, which means that organic matter was oxidized and C emitted, but these emissions were not quantified;

- The amount and size of the material removed in pre-burn thinning treatments was estimated, but its C fate was not estimated. Part of it might be converted to pulp or solid wood products which would have an additional C storage life, and some would be used to replace fossil fuels in energy production, but from 20 to 30 percent is emitted as a result of harvesting, transporting, and processing of wood products (Row and Phelps 1996); and,
- Following wildfire extinction, C emissions will continue to occur on burned sites due to increased heat and moisture, elevated decomposition levels, and the probability of re-burns.

Estimating the Effects of Forest Treatment

A spreadsheet model was developed to incorporate the estimates developed above into an indicator of the likely future for the ponderosa pine forests on the BNF. The model tested the effects of three possible treatment options:

- A No-Treatment option (NT), where little, if any, of the dense ponderosa pine forests are treated, but where they continue to experience wildfire at the average annual rate of about 7.5 percent of the remaining unburned forest.
- A Target Goals option (TG): An aggressive treatment option with 30,000 acres of prescribed fire (with or without preliminary thinning or fuel treatment) accomplished each year. In this option, the acres to be treated are picked on the basis of the agency's ability to meet annual goals, not on any strategic landscape location value.
- A Strategic Treatment option (ST): The same amount of treatment as TG (30,000 acres per year), tied to an ecosystem management program that considers the landscape impact of the treated acres, seeking out those areas that, after treatment, will be most effective in breaking up large continuous at-risk areas so that resulting wildfires become smaller and less-intense in their effects.

In the strategic treatment option, the areas likely to be chosen for high-priority treatment are those adjacent to anthropogenic ignition sources (power lines, roads, railroads, urban intermix, recreational sites), areas subject to a high frequency of lightning strikes, and those that, by their position on the landscape, can be effective in breaking up the continuity of large areas of heavy fuels. For example, south and west-facing slopes may be priority-treated because they are more naturally open stands that, when treated, provide fire managers with effective fire management options.

We used professional judgment to estimate the "multiplier effect" likely to be achieved by good strategic selection of treatment sites, and created a decay function in the model so that the multiplier effect diminished quickly as the best sites were treated first. This resulted in an estimate that, in the first year of treatment, good strategic selection would result in 5 acres being removed from the "at risk" category for each acre treated. We then reduced that "multiplier" over a 5-year period, on the assumption that, if the most advantageous acres were chosen first, the second year's selection would be less effective, until after 5 years, the "multiplier effect" would be exhausted.

RESULTS

Effect of Forest Treatment on Fire Extent

Sample model results are shown in Table 4. On the basis of our estimate that 7.5% of the acres at risk in the ponderosa pine forest will burn annually, the model indicated that large wildfires will burn around 80% of the dense ponderosa pine-Douglas fir forests within a 20-year period–a prediction that has been made previously (Neuenschwander and Dether 1995). Under this scenario, the damage to other resources at risk was not estimated, but would be considerable. This assumption does not estimate the potential for re-burning of previously-burned watersheds, or the burning of grasslands, brush fields, or other forest types that are intermixed with the dense ponderosa pine-Douglas fir forests.

Wildfire extent would be reduced 25 to 40 percent by the two treatment alternatives, according to the model. This is due to the assumption that the treated areas, if they burn in a subsequent natural event, could be allowed to burn as a "prescribed natural fire" since the fire's effects should be largely nonlethal in the more fire-tolerant forest condition.

The total of fire of all types on the forest would, however, increase from 10 to over 33 percent under the treatment alternatives. As Table 4 indicates, an annual average somewhere in the range of 50,000 to 60,000 acres is predicted by the model. An average 20-year fire return interval, which is consistent with some estimates of historical fire intervals in ponderosa pine, would mean an average of 60,000 acres per year in a 1.2 million-acre forest.

Table 4 under-estimates future wildfire effects on the BNF for at least two reasons. First, it does not consider the likelihood of re-burns affecting many areas, particularly those where fire kills thick stands of small trees, which are then left to blow down and build major fuel loads for a subsequent fire. More importantly, it does not consider the other vegetation types and areas on the forest that are also subject to ignition and periodic wildfire. By considering the ponderosa pine forests that are in dense vegetative condition, we have

TABLE 4. Results of two treatment alternatives (compared to no treatment) on wildfire extent, carbon emissions, and air quality on the Boise National Forest.

	Model Alternative			Advantage (% Change)	
Attribute	No Treatment	Target Goals	Strategic Treatment	TG/NT	ST/NT
Wildfire (Average Acres per Year)					
At Year 0	89,759	89,759	89,759	0.00	0.00
At Year 5	60,783	52,691	33,380	− 13.31	− 45.08
At Year 10	41,162	25,997	12,920	− 36.84	− 68.61
At Year 15	27,874	7,921	0	− 71.58	
At Year 20	18,876	0	0		
Total in 20 Years (ac)	963,977	685,627	488,247	− 28.88	− 49.35
Annual Average (ac)	45,904	32,649	23,250	− 28.88	− 49.35
Total Fire Average Acres (wildfire + Rx) Per Year					
At Year 0	89,759	90,759	90,759	1.11	1.11
At Year 5	60,783	82,691	63,380	36.04	4.27
At Year 10	41,162	55,997	42,920	36.04	4.27
At Year 15	27,874	37,921	30,000	36.04	7.63
At Year 20	18,876	30,000	30,000	58.93	58.93
Total in 20 Years (ac)	963,977	1,286,627	1,089,247	33.47	13.00
Annual Average (ac)	45,904	61,268	51,869	33.47	13.00
Greenhouse Gas Emissions (1,000 tC/yr)					
At Year 0	1,907	1,912	1,912	0.26	0.26
At Year 5	1,292	1,268	858	− 1.82	− 33.59
At Year 10	875	701	423	− 19.86	− 51.63
At Year 15	592	317	149	− 46.51	− 74.93
At Year 20	401	149	149	− 62.98	− 62.98
Total in 20 Years (ac)	20,484	17,544	13,350	− 14.35	− 34.83
Annual Average (ac)	978	835	636	− 14.35	− 34.83
Air Pollution Emissions (tPM/yr)					
At Year 0	67,497	67,716	67,716	0.32	0.32
At Year 5	45,708	46,201	31,680	1.08	− 30.69
At Year 10	30,953	26,128	16,294	− 15.59	− 47.36
At Year 15	20,961	12,535	6,579	− 40.20	− 68.61
At Year 20	14,194	6,579	6,579	− 53.65	− 53.65
Total in 20 Years (ac)	724,890	647,367	498,942	− 10.69	− 31.17
Annual Average (ac)	34,519	30,827	23,759	− 10.69	− 31.17

focused on some of the highest-risk areas, but other areas are susceptible to wildfires, as well. To the extent that these other areas burn and produce both C and PM emissions, they are not included in Table 4.

Effect of Forest Treatment on C Emissions

We calculated C emissions on the basis of fuel consumed, using a factor of 0.45 tons C per ton of dry biomass burned. Average fuel consumption in wildfires was 47.2 tons/acre (Table 3) and fuel consumption for Rx fire was

11 tons/acre (Table 2). Since our fuel consumption model (CONSUME) dealt primarily with above-ground biomass and duff consumption, the effects of fire severity in oxidizing and emitting soil C are not included.

The model indicates that average annual C emissions from wildfires on the BNF would be around 1.9 million tons per year at the beginning of the treatment program (Table 4). If a treatment program including prescribed fire were initiated at the level suggested, total emissions would rise to about 2 million tons per year in year 3 due to the addition of prescribed fire emissions to the existing wildfire regime, then decline rapidly once wildfire acres began to be reduced. The strategic treatment option produced the most rapid decline, cutting emissions by over half in 10 years of treatment, and by one-third over the 20-year projection. Target goal treatment reduces emissions to the same general levels by year 18, but with higher annual averages, as shown in Figure 12.3.

Effect of Forest Treatment on Particulate Matter (PM) Emissions

These were also estimated on the basis of fuel consumed, as estimated by CONSUME. Factors for total PM emissions were developed from Brown and

FIGURE 12.3. Model estimates of average annual carbon emissions from combined wildfire and prescribed fire, Boise National Forest.

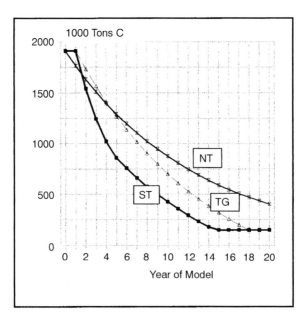

Bradshaw's (1994) estimates of emission production for ponderosa pine forests in the Selway-Bitterroot Wilderness. We used their estimates of emissions from crown fires in ponderosa pine (31.85 lb PM/ton of fuel consumed) for the wildfire estimates and their estimates of surface fires in the same forest type (39.87 lb PM/ton of fuel consumed) for the Rx fire. These suggest that the higher level of smoldering fire during prescribed fires results in a higher level of PM production per ton of fuel consumed. If this is the case, PM production will not fall as fast as C emissions if forest managers increasingly utilize Rx fire as a management tool. The reductions could still be significant, however, as indicated in Figure 12.4 and Table 4, due to the major differences in fuel consumption between wildfire and Rx fire (Table 2). Other data suggest that the differences between wildfire and Rx fire particulate emissions may be less than we used, or even reversed. Hardy (pers. comm.) cites estimates for ponderosa pine broadcast-burned slash as 25 lbs PM_{10} per ton of fuel consumed (Ward et al. 1989) and 30 lbs PM_{10} per ton of fuel burned in forest wildfires (Hardy et al. 1992). Had we used these estimates, the relative reduction due to strategic treatment would have been even greater.

FIGURE 12.4. Average annual PM emissions over 20-year model simulation, by treatment option.

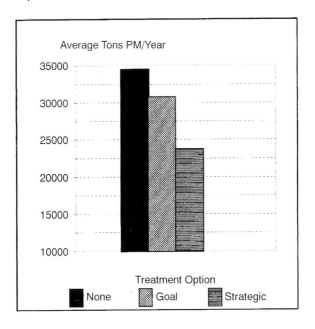

The Landscape Implications of Treatment

The landscape implications of the treatment program, and its effectiveness in helping to prevent a near-total loss of the ponderosa pine forests in the Boise National Forest over the next 20 years is shown in Figure 12.5. The implications are significant. If no treatment program is established, the model results suggest that, after 20 years, over 950,000 acres (some 80%) of the 1992 inventory of dense ponderosa pine forest will have been affected by a lethal wildfire. This does not count any area that will have been re-burned during that period and converted to brush cover, although it is difficult to imagine that level of intense wildfire occurring without a significant impact in terms of re-burning.

That suggests a forest in which a large proportion is in early successional stages, with little or no significant proportion in large seral-stage or "old growth" ponderosa pine. That may favor species adapted to early succession-al habitats, but bodes ill for those needing later successional stages.

A 30,000 acre per year treatment program (TG) would reduce wildfire acres significantly. Under this option, the model indicates that 665,500 acres would burn in lethal wildfires while 530,000 acres were treated over the

FIGURE 12.5. Model estimate of landscape diversity after 20 years under three management options, ponderosa pine forest type, Boise National Forest.

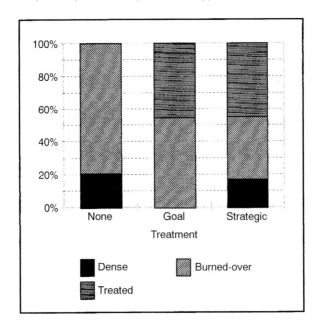

20-year period. That would reduce the amount of lethal wildfire by about 1/3, which would improve diversity across the landscape and lower wildfire's economic impact.

Landscape diversity, however, would be even further improved under the strategic treatment option (ST), primarily because the current forest condition–ponderosa pine with heavy understory and fuel buildup–could remain in patches across the landscape, favoring the wildlife and plant species associated with that forest condition. Under this option, around 500,000 acres of wildfire would be expected (almost a 50% reduction from the untreated estimate) and 457,000 acres would be treated in 20 years. This would leave about 20% of the current "acres at risk" on the landscape at the end of 20 years. These areas would still be susceptible to lethal wildfires, but if the patch size had been reduced through strategic treatment, and the landscape-level fuel continuity reduced, the risks of their ignition by intense wildfires from other areas would decrease significantly. If this can be achieved on the landscape through an intensive program of strategically-positioned treatment projects, the result could be a more diverse forested landscape containing a broad array of forest conditions and habitats, reasonably tolerant of a fire regime based on periodic low- and mixed-intensity wildfires, and, hopefully, sustainable (Figure 12.5).

CONCLUSIONS

An aggressive treatment program increases the total amount of fire-affected acres on the BNF, as the prescribed (Rx) fire is added to the average annual wildfire. The very different nature of Rx fire, however, results in significant reductions in total C and PM emissions, as well as an increase in the amount of ponderosa pine forest placed in a more sustainable, fire-tolerant condition.

Forests in this condition (after treatment) may still suffer high-intensity wildfire under extreme conditions, so treatment is not fool-proof, nor should it be inferred that treatment will return these forests to a pre-settlement condition, or effectively restore a "natural" fire regime. What can be said, with confidence, is that treatment will significantly increase the probability that these ponderosa pine forests will tolerate most wildfire events; that the wildfires that do occur will be smaller, less intense, and less destructive as a result; and that forest managers will have more available fire management options as a result.

We can also state with confidence that a continuing treatment regime will be needed on these forests; a one-time treatment will not suffice. With the historical Native American influence removed, land use ownership and settlement patterns changing the western landscape, and long-term atmospheric and climate changes that we do not yet understand, these forests cannot be "self-sustaining" without continued management, in our view.

That treatment regime, in addition to making ponderosa pine forests more fire-tolerant, also faces an enormous challenge in restoring fire-tolerant eco-systems on those areas that have already been hit by large high-intensity wildfires on the BNF, and areas that will inevitably burn in spite of aggressive treatment areas. Those areas, many of which have suffered high-severity soil impacts as well, are at the mercy of weather conditions for at least two decades before they can begin to have pine regeneration large enough to withstand either Rx treatment or wildfire events (Figure 12.2 indicates the high levels of mortality of small trees in any fire event).

This makes the risk of re-burning in these recovering forests a serious consideration. If the regeneration is killed before the onset of sexual maturity, the lack of a seed source and the altered soil and micro-climate may preclude further natural regeneration and make successful reforestation difficult, if not impossible. Such a situation will most likely result in conversion of large areas of former ponderosa pine forest into brush cover that will likely persist for decades, if not centuries. To the extent that the public expresses a desire to keep these formerly-forested lands in forest, a continued effort at restoration management will be required.

Under the most favorable future circumstances, the need to keep fire as an essential part of ecosystem functions in these forests means an ongoing, aggressive prescribed fire program. In the ponderosa pine forests alone, considering that they occupy over 1.2 million acres on the BNF, a 20-year average fire return interval would suggest that something on the order of 60,000 acres would burn in the average year. We have, in the interests of practicality, established a goal of one-half this amount for our treatment options. Whether that is adequate or not remains to be seen. What seems certain, however, is that the people living in and near the BNF need to accept fire as a normal part of these forests, and where they have established expec-tations of no fire—or no air pollution—they will be disappointed.

Those statements apply not just to the citizens who live near the BNF, but to thousands of communities across the West, who face increasingly danger-ous wildland conditions and the risk of wildfire events that, in addition to their damage to the land, create major air pollution events that, while brief, can be extremely hazardous to people who have respiratory ailments. Heavi-ly-impacted areas may be at risk of serious watershed damage for as long as a decade or more following an intense wildfire, particularly on steep, unstable slopes where tree roots which once stabilized the soil and helped prevent mass soil movement in extreme weather events have died and decomposed.

The climatic conditions affecting these forests, and the random nature of events, guarantees a continued occurrence of high-intensity, lethal wildfires—

even in those forest types where such fire was historically infrequent. The goal of ecosystem management can not be absolute control. Instead, ecosystem management seeks to improve the likelihood of desired things happening, and decrease the probability of undesirable things (from either a human or ecological standpoint). We believe that to be a responsible goal, one that can be attained by a strategic management system that effectively utilizes the ecological science, monitoring capacity, and geographically-based decision models available to managers as they enter the 21st Century.

AUTHORS NOTE

This work was supported in part by the Environmental Protection Agency, under a cooperative agreement between American Forests and the Office of Economy and Environment. Steven Winnett was EPA's Project Manager, and the authors appreciate his guidance and patience through a lengthy research period. EPA has not officially reviewed or approved any of the conclusions presented in this report, and the conclusions are the sole responsibility of the authors.

The authors would also like to thank University of Idaho graduate students Lorri Ondricek, Calvin Farris, and Kobe Hankins, whose research and analysis were an important part of the final report.

REFERENCES

Agee, J.K. 1993. *Fire Ecology of Pacific Northwest Forests*. Washington, DC: Island Press.

Arno, Stephen F. 1996. The seminal importance of fire in ecosystem management–impetus for this publication. In Hardy, C.C. and S.F. Arno (eds) *The Use of Fire in Forest Restoration* (Gen. Tech. Rep. INT-GTR-341). Ogden, UT: USDA Forest Service, Intermountain Research Station.

Arno, S.F. and J.K. Brown. 1991. Overcoming the paradox in managing wildland fire. *Western Wildlands* 17(1):40-46.

Boise National Forest. 1996. *Resources at Risk: A Fire-Based Hazard/Risk Assessment for the Boise National Forest*, Boise, ID: Boise National Forest. 34 pp. plus Appendices.

Boise National Forest. 1995. Boise River Wildfire Recovery: Final Environmental Assessment. III:38-83.

Brown, J.K. 1995. Fire regimes and their relevance to ecosystem management. In Proceedings: 1994 Society of American Foresters/Canadian Institute of Forestry Convention, September 18-22, 1994, Anchorage, Alaska. Bethesda, MD: Society of American Foresters. pp. 171-178.

Brown, J.K. and L.S. Bradshaw. 1994. Comparisons of particulate emissions and smoke impacts from presettlement, full suppression, and prescribed natural fire periods in the Selway-Bitterroot Wilderness. *Int. J. Wildland Fire* 4(3):143-155.

Covington, W.W., R.L. Everett, R. Steele, L.L. Irwin, T.A. Daer, and A.N.D. Auclair. 1994. Historical and anticipated changes in forest ecosystems of the Inland West of the United States. *Journal of Sustainable Forestry*, 2(1/2):13-64.

Covington, W.W. and S.S. Sackett. 1986. Effect of periodic burning on soil nitrogen concentrations in ponderosa pine. *Soil Sci. Soc. Am. J.* 50:452-457.

Hardy, C.C., D.E. Ward, and W. Einfeld. 1992. $PM_{2.5}$ emissions from a major wildfire using a GIS: rectification of airborne measurements. In Proceedings of the 29th Annual Meeting of the Pacific Northwest International Section, Air and Waste Management Association, Nov. 11-22, Bellevue, WA. Pittsburgh, PA: Air and Waste Management Association.

IPCC. 1996. *Climate Change 1995: Impacts, Adaptations and Mitigation of Climate Change: Scientific-Technical Analyses*. Contribution of Working Group II to the Second Assessment Report of the Intergovernmental Panel on Climate Change. Watson, R.T., Zinyowera, M.C., and R.H. Moss (Eds). Cambridge: Cambridge University Press. 879 pp.

Keane, R.E., S.F. Arno, and J.K. Brown. 1989. *FIRESUM–An Ecological Process Model for Fire Succession in Western Conifer Forests*. Gen Tech Rep INT-266. USDA Forest Service, Intermountain Research Station.

Neuenschwander, Leon and Diedra Dether. 1995. Ponderosa pine: An Idaho ecosystem at risk. *Idaho Research* (Summer, 1995). Moscow, ID: University of Idaho. 9-12.

Noss, R.F., E.T. Laroe, III, and J.M. Scott. 1995. Endangered Ecosystems of the United States: A preliminary assessment of loss and degradation. Washington: USDI National Biological Service, Biological Report 28. p. 12, App. A.

Powell, D.S., J.L. Faulkner, D.R. Darr, Z. Zhu, and D.W. MacCleery. 1993. Forest Resources of the United States, 1992. USDA Forest Service, Rocky Mountain Forest and Range Experiment Station, General Technical Report RM-234. 132 p.

Rigg, H.G., R. Stocker, C. Campbell, B. Polkowsky, T. Woodruff, and P. Lahm. 1999. A screening method for identifying potential air quality risks from extreme wildfires. In Sampson, R.N., R.D. Atkinson, J.W. Lewis (Eds.), *Mapping Wildfire Hazards and Risks*. Papers from the American Forests scientific workshop, September 29-October 5, 1996, Pingree Park, CO. The Haworth Press, Inc., New York.

Row, C. and R.B. Phelps. 1996. Wood carbon flows and storage after timber harvest. In Sampson, R.N. and D. Hair (eds) Forests and Global Change, Volume 2: Forest Management Opportunities for Mitigating Carbon Emissions. Washington: American Forests. pp.59-90.

Sampson, R.N., D.L. Adams, S.S. Hamilton, S.P. Mealey, R. Steele, and D. Van De Graaff. 1994. Assessing forest ecosystem health in the Inland West. *Journal of Sustainable Forestry* 2(1/2):3-12.

Snell, J.A.K. and J.K. Brown. 1980. Handbook for predicting residue weights of Pacific Northwest conifers. General Technical Report PNW-103. USDA Forest Service, Pacific Northwest Forest and Range Experiment Station. February 1980. p. 9.

Steele, R. 1988. Ecological relationships of ponderosa pine. In: Baurmgarner, D.A. and J.E. Lotan (Eds.) *Ponderosa Pine, the Species and Its Management*. Pullman, WA: Cooperative Extension Service. pp. 71-76.

Steele, R., S.F. Arno, and K. Geier-Hayes. 1986. Wildfire patterns change in central Idaho's ponderosa pine-Douglas-fir forest. *Western Journal of Applied Forestry* 1(1):16-18.

USDI/USDA. 1995. *Federal Wildland Fire Management: Policy & Program Review* (Final Report). Washington, DC: U.S. Department of the Interior; U.S. Department of Agriculture. 45 pp.

Ward, D.E., C.C. Hardy, D.V. Sandburg, and T.E. Reinhardt. 1989. Part III–emissions characterization. In: Sandberg, D.V., Ward, D.E. and R.D. Ottmer (comps) *Mitigation of Prescribed Fire Atmospheric Pollution Through Increased Utilization of Hardwoods, Piled Residues, and Long-Needled Conifers.* (Final Report; U.S. DOE.BPA. Seattle, WA: USDA Forest Service.

Chapter 13

Methodology for Determining Wildfire and Prescribed Fire Air Quality Impacts on Areas in the Western United States

Roger A. Stocker

SUMMARY. Colorado has long been known for its scenic vistas and majestic mountains. Wildfires which are common in the summer months can significantly impact scenic environments close to the wildfire location. There are also concern as to the health impact to people in towns and cities located on the urban/wildland interfaces where the impact can also include loss of life and property. For these and other reasons, a concerted effort is underway to better understand the implications of these wildfires on areas potentially impacted by smoke.

This study created a methodology for evaluating air quality impacts on both a statewide and regional basis using a state-of-the-art mesoscale meteorological model to simulate the meteorological conditions which correlate with prescribed fire and wildfire activity. These meteorological fields are input into an air quality model which simulates transport and secondary aerosol formation for certain pollutants. This modeling effort makes use of climatological analyses derived from

Roger A. Stocker is affiliated with WESTAR, Portland, OR 97204.

This work was sponsored by EPA OPPE (grant #A000679-93). Special thanks to Dr. Roger Ottmar of the USDA Forest Service who provided the needed heat release and emissions data for use in the Colorado Pilot Program and agency personnel at the National Interagency Fire Center in Boise, ID who provided all available historical information regarding prescribed fire and wildfire activity.

[Haworth co-indexing entry note]: "Chapter 13. Methodology for Determining Wildfire and Prescribed Fire Air Quality Impacts on Areas in the Western United States." Stocker, Roger A. Co-published simultaneously in *Journal of Sustainable Forestry* (Food Products Press, an imprint of The Haworth Press, Inc.) Vol. 11, No. 1/2, 2000, pp. 311-328; and: *Mapping Wildfire Hazards and Risks* (ed: R. Neil Sampson, R. Dwight Atkinson, and Joe W. Lewis) Food Products Press, an imprint of The Haworth Press, Inc., 2000, pp. 311-328. Single or multiple copies of this article are available for a fee from The Haworth Document Delivery Service [1-800-342-9678, 9:00 a.m. - 5:00 p.m. (EST). E-mail address: getinfo@haworthpressinc.com].

observational data provided by the Grand Canyon Visibility Transport Commission (GCVTC). Locally, the model examines impacts within the urban/wildland interface to the extent possible. The regional analysis evaluates the impact of long range transport on neighboring states and the relative impacts on states downwind from areas with the potential for extreme wildfire events. *[Article copies available for a fee from The Haworth Document Delivery Service: 1-800-342-9678. E-mail address: <getinfo@haworthpressinc.com> Website: <http://www.haworthpressinc.com>]*

KEYWORDS. Wildfire, air quality, prescribed fire, particulates

INTRODUCTION

The idea that wildland fires in the United States are important to ecosystem management has received a lot of recent attention. The management practices in the early 20th century, which reflected a philosophy that all wildland fires were destructive and needed to be suppressed, has led to many changes in wildland ecosystems. From an air quality perspective, the primary change has been the increase in fuel loading which has led to increases in catastrophic wildfires in the area.

To combat this problem, the Federal Land Managers (FLMs) have proposed that more prescribed fire needs to be introduced into land management practices. Recent information from the Grand Canyon Visibility Transport Commission (GCVTC) has suggested that in some areas, as much as a 16 fold increase in prescribed fire activity will be needed in the western United States over the next 50 years to combat the problem (Table 1). This has led to questions regarding the potential smoke impacts to air quality of these proposed prescribed fires.

A recent study conducted in the interior Columbia River Basin (ICRB) has examined the potential smoke impact from prescribed fire and wildfire activity in a large area encompassing portions of Washington, Idaho and Oregon. One major finding suggests that the potential air quality impacts from prescribed fire is much less (order of magnitude differences) when compared to wildfires even when current levels of prescribed fire activity are increased by 1500% (Scire and Tino, 1996).

This study was conducted to determine the impact of wildland fires from a regional perspective. Air quality impacts local to the fires (< 50 km) were not examined and therefore, findings do not extend to local impacts. Also, emissions from simulated fires in this study were arbitrarily capped by the mixing layer height. This procedure leads to the assumption that all emissions from prescribed and wildfires are contained entirely within the mixed layer. This is not true for most large wildfires, where a significant portion of the emissions

TABLE 1. Preliminary PM$_{2.5}$ emissions estimates for 1995 and 2040 based on work from the GCVTC for the represented states. Numbers include optimum smoke management reduction using biomass utilization and mechanical treatments.

State	1995 PM$_{2.5}$ emissions (tons)	2040 PM$_{2.5}$ emissions (tons)
AZ	23536	64496
CA	17152	100745
CO	2485	26953
ID	24638	54039
NM	23312	65226
NV	664	8350
OR	10707	45062
UT	3558	57185
WA	4324	21918
WY	1140	17394

can extend well beyond this height due to the extreme temperatures generated. This can lead to overestimation of pollution concentrations at the surface. The results from this study were not verified with field observations, so the full effect of these assumptions is not known.

Recent and proposed EPA actions (Natural Events Policy and the new NAAQS standards for PM and regional haze under the direction of the EPA) will have a direct impact on the ability to manage prescribed fire programs. Under the new Natural Events Policy, wildfire emissions can be excluded in determining violations to the NAAQS while prescribed fire emissions can not. This policy presents a potential road block to the proposed increases in prescribed fire suggested by the FLMs. The other major development comes in the new standards for PM$_{2.5}$ and regional haze being developed by the EPA. These new health and visibility standards will directly influence prescribed fire activity in the future. Even the relatively small impact from prescribed fire ramp-ups suggested by the ICRB study (~10 μg/m^3 from PM$_{2.5}$ over a 24 hour period) would represent a large single-source percentage of a 50 μg/m^3 PM$_{2.5}$ standard. This raises the question of equity since point source emitters are being asked to reduce their emissions, while FLMs are proposing increases.

The modeling study discussed here will identify the potential air quality impact on a West-wide basis from both prescribed fire and wildfire. To do this, a sophisticated meteorological model coupled with an air quality model will be exercised over the western United States for "typical" meteorological conditions conducive for both prescribed fires and wildfires. This paper outlines a methodology for examining the problem on a West-wide basis while presenting preliminary results from a pilot program performed for Colorado wildfires.

METHODOLOGY

The challenge is to estimate the air quality impacts of both prescribed fire and wildfire activity (current and future) on downwind environments (receptors) in order to estimate the health risks and visibility impairment associated with each type of activity. Knowing that both wildfire and prescribed fire smoke impacts to downwind receptors are episodic in nature, it is important that the evaluation to health standards (NAAQS) be performed on the shortest possible time interval. The 24 hour time interval represents the shortest duration available for comparison with the PM NAAQS but might not be sufficient for evaluating acute impacts in receptor locations when high levels of smoke are present.

Wildfire and prescribed fire can not be treated alike in a modeling framework due to differences in emission factors, temporal occurrence, burn area size, and ecological damage. The different methodology for these two types of fires has led to two separate phases within this project. Phase I estimates impacts from wildfire; Phase II considers prescribed fire. The current work focuses on analysis of a worst case scenario utilizing the methodology outlined in Phase I.

Phase I

The impacts from wildfires are restricted to fires of 25,000 acres or larger. This represents an area 38.6 square miles in size, indicating a large but not unrealistic wildfire. Several wildfires of 100,000 acres or larger have occurred in the West in recent years. Historically, wildfire activity peaks in the summer months when solar insolation is strongest and moisture conditions are low. Figure 13.1 shows historical wildfire activity for areas managed by the Fish and Wildlife Service in selected Western states.

Phase I will use "typical" meteorological conditions for August to estimate concentrations of $PM_{2.5}$ on receptors in the West. Wildfire activity does not peak in the same month throughout the West, starting early in the southern States and migrating northward as summer progresses (Figure 13.1). However, due to limitations in computer power, this study simulated conditions for only one month, August, 1992, on the basis of the data in Figure 13.1. The year was chosen due to extensive climatological work in the Grand Canyon Visibility Transport Commission (GCVTC) study which determined that the 1992 conditions represented "typical" meteorology for the region (Stocker et al. 1996 and Cover et al. 1990).

FIGURE 13.1. Fish and Wildlife Service wildfire frequency histogram for selected States from an 11-year historical record.

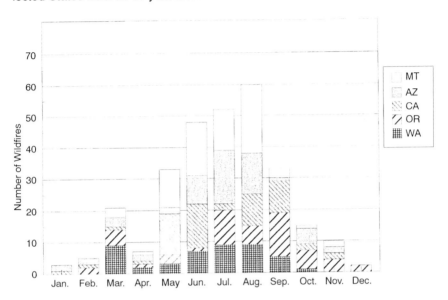

Once the event time is chosen, it is necessary to predict where wildfires are most likely to occur. After these locations are determined, the air quality impacts of these fires at a downwind receptor can be estimated. A number of studies have been conducted which searched for synoptic scale weather patterns or values of atmospheric variables which could be used as predictors of wildfire outbreaks. In one such study, a synoptic weather pattern analysis breaks the country into zones to account for the migrational shift mentioned previously and suggests common weather patterns associated with increased wildfire activity (Heilman 1995 and Takle et al. 1994). The atmospheric weather variables that positively correlate with wildfire activity are such things as wind speed, wind shear, and elevation-adjusted dewpoint temperature (Potter 1995). Also, the National Fire Danger Rating System can provide information into the likelihood of a wildfire in a region. The combined impact of these tests can help predict wildfire activity during the simulated meteorological conditions in the selected one-month period. The evaluation technique tests whether some large area might experience a 25,000 acre wildfire; with the specific location placed stochastically within a vegetation class selected on the likelihood that it would experience a large wildfire (i.e., arid grasslands would have a higher probability of occurrence than high alpine timber classes).

The meteorology will be provided by the mesoscale meteorological input. This information will then be input into an air quality model along with the emissions information linked to the dominant vegetation type in an area and time of year. The impacts from the air quality model will then be used to determine both health and visibility impacts to downwind receptors where appropriate.

Phase II

Most of the methodology outlined above is the same for the prescribed fire case studies. However, the implementation of this methodology will be quite different.

Prescribed fires, on average, do not consume the area that large wildfires do. Typical prescribed fires of timber harvest activity fuels are on the order of 50 acres in size. Thus, modeling techniques appropriate for large wildfire activity are not appropriate for smaller burns, as will be discussed in the next section. Also, prescribed fire activity takes place at different seasons of the year, under different meteorological conditions. Figure 13.2 shows the prescribed fire activity for the Fish and Wildlife Service (FWS) corresponding to the same 11 year period as illustrated in Figure 13.1 Figure 13.2 illustrates that the spring and fall seasons are the most common burn seasons, with May being highest in the spring and September in the fall. There are large regional differences in the prescribed fire activity, but in most western states, prescribed fire occurs in both spring and fall. This suggests that two one-month segments will need to be simulated for prescribed fire. By selecting spring

FIGURE 13.2. Fish and Wildlife Service prescribed fire frequency for selected States from an 11-year historical record.

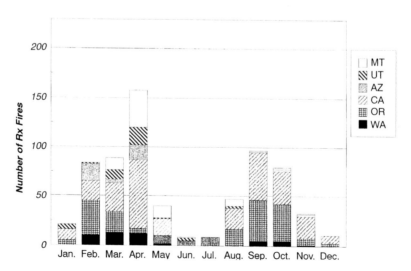

and fall periods for prescribed fire, and summer for wildfire does not mean that wildfires will only be evaluated during the summer and prescribed fire only during the spring and fall. The combined impact from prescribed fire and wildfire will need to be evaluated together where appropriate to estimate the potential for smoke impact from both sources.

The testing criteria for prescribed fire is more difficult than for wildfire, with many more factors to consider. One of the important factors in the West is the complex terrain which produces complex meteorological and transport wind patterns. Where certain locations, such as urban and Class I areas, need to be protected from smoke intrusion, it is critical to be able to accurately predict transport winds between sources and receptors, but the complex terrain and wind patterns makes this difficult. These factors suggest that local information is needed to determine whether an area will be in prescription on a given day. While meteorological variables can be used to suggest when prescribed fires should not be allowed, they do not provide sufficient information to determine if a fire should occur in an area. Thus, meteorological variables (see Table 2) will only be used for sideboard testing to eliminate the possibility of fire starts on days which fall outside the sideboards. In the model, a prescribed fire will be located stochastically within a vegetation type when the meteorological criteria are within the sideboard. Therefore, this scheme will utilize a stochastic element not only to place a fire randomly within a vegetation type, as in the wildfire case, but also to determine when a fire will occur. This could lead to the placement of a test fire in a location with transport winds blowing towards a sensitive area, but since this can happen in reality and this project is being performed to examine worst case situations, this seems reasonable.

MODEL DESCRIPTIONS

The evaluation of long range transport in the western United States deals with two important issues not easily addressed by the models used in the

TABLE 2. Meteorological variables and values used in the determination of the prescribed fire sideboard. These values represent "in prescription" limits.

Meteorological variable	Value
wind speed (30 feet)	5-20 mph
mixing height	2500-3000 feet
10 hr fuel moisture	12-16%
wind direction	variable (dependent on locations of sensitive receptors)

environmental sciences. The first is meteorological flow patterns in complex terrain and the second is secondary aerosol production associated with pollution transport over large distances. Both of these issues are of critical concern in administering and evaluating impacts associated with the NAAQS.

The following sections briefly describe the meteorological model and air quality models used in this study to evaluate smoke impacts. Figure 13.3 shows a schematic of the linkages between the inventory, meteorological and air quality models for both the prescribed fire and the wildfire portions of this study.

MM5

MM5 is a primitive equation mesoscale grid model capable of predicting the meteorological conditions of an area based on coarse temporal and spatial observations of 3D wind speed and direction, temperature, pressure, and humidity and 2D information on vegetation, land use, and terrain. The coarse 3D meteorological information is provided to the model at a 12 hour frequency. Only the initial time corresponds to a pure synoptic observation. Subsequent times are determined by a combination of derived conditions based on the physics in the model, with a "nudging" term that accounts for changes in synoptic conditions over the simulation run. The meteorological conditions derived from running the model reflect conditions for whatever grid spacing

FIGURE 13.3. Flowchart of the wildland fire modeling system.

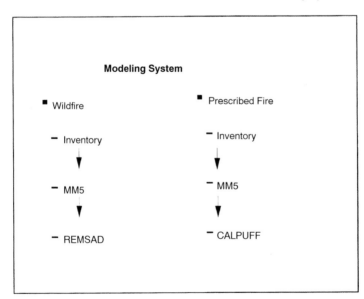

is chosen for use. Horizontal grid spacing can range from very localized meteorology on the order of 1 km to large scale meteorology on the order of the input fields used to initialize and update the model. The MM5 also has the capability of using one- and two-way nesting to enable the user to simulate large scale meteorology on one grid while telescoping down to a finer resolution for more localized meteorology on a finer grid. The two-way nesting option allows for communication of information both ways between the large and small scales. The one-way option allows only for communication from the large down to the small scale.

For the purposes of this study, it is important that local meteorology be simulated where possible due to the complex terrain that produces well defined mesoscale flow patterns when the large scale winds are not strong ($|V|<10$ m/s). This is often the case with prescribed burning. Figure 13.4 shows the domain size and horizontal grid spacing that will be used for this work with the largest grid extending from 140°W longitude to capture regional storm activity that might influence fire activity within the smaller scale grids. The small scale grids will be the primary information grids used by the air quality models to account for transport patterns and meteorological variables influencing chemical reactions along the path. A one-way interactive grid is used, which allows both grids to be run at different times with the coarse grid information being fed into the small scale grid.

FIGURE 13.4. Meteorological domain for synoptic and local scale meteorological information. Grid 1 represents a 50 km grid spacing while grids 2 and 3 show 10 km grid spacing.

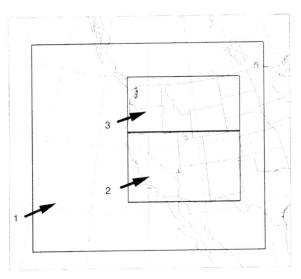

The simulation of a complete month will be done by merging four eight-day simulations. This is done to make sure that the synoptic scale model fields do not deviate significantly from what the observations suggest, as they might if run continuously for one full month. This procedure has been successfully used in the GCVTC work to simulate an entire year of meteorological conditions (Stocker et al. 1996).

Air Quality Models

There are two models being run in this study. The Regulatory Modeling System for Aerosols and Deposition (REMSAD) is a grid based model designed to simulate large area wildfire sources and large area receptors at some downwind distance from the source. The other model (CALPUFF) is a puff model, commonly used for simulating point sources, which is used for simulating pollution dispersion/transformation/deposition of a prescribed fire plume. CALPUFF has undergone extensive review by a national committee, the Interagency Workgroup on Air Quality Models (IWAQM), in hopes that it will someday be available as an EPA-approved tool for air quality work. Recent changes to this model include a direct linkage with the Emissions Production Model developed by the USDA Forest Service to calculate smoke emissions from wildland fire activity.

REMSAD is a public domain Eulerian-based grid model developed at Systems Application International based on code developed for the UAM-V proprietary ozone model. REMSAD has two modes of operation: (1) PM and (2) Toxics. Only the PM mode will be used for this study. The users guide for REMSAD contains further information regarding the toxics mode (Systems Application International, 1996).

The PM mode of this model is efficient enough so that large domains may be used to simulate long range transport and chemical transformations over long transport times. The chemistry accounted for by this model is outlined in Table 3 with much of the input information being obtained from the meteorological model outlined in the previous section. This grid model has the capability of performing calculations involving two-way nesting as defined in the last section. This allows for the direct linkage between the MM5 results and REMSAD with little modification. Due to REMSAD's inherent Eulerian grid structure, sub-grid scale information is not available to the user. Therefore, with a 25,000 acre fire, the smallest receptor size is also restricted to 25,000 acres or an area of 100 km^2. This means that no wildland/urban interface questions can be answered and the impact of emissions on small cities downwind will not be addressed.

CALPUFF is a Lagrangian-based puff model developed by Earthtech which has the ability to treat point sources effectively, but has only a basic chemistry option. The advantage of this scheme over a Eulerian approach is that it has the

TABLE 3. Input and output variables used by the REMSAD model (adapted from Guthrie et al. 1995).

Inputs	Outputs
Humidity	Hydroxyl Radical
Cloud Cover	NO_x
Solar Radiation	SO_2
NO_x	NH_3
SO_2	Total Sulfate
NH_3	Ammonium Sulfate
Ozone	Ammonium Nitrate
	Nitric Acid

ability of examining impacts where sources and receptors are in close proximity to one another (i.e., urban/wildland interface) and can tag emissions from any source to determine where pollution at a receptor originated.

A major drawback of this approach is that obtaining an accurate concentration estimate at a receptor requires a large number of puffs from a source. This is due to the way that concentrations are calculated in this framework. Basically, the receptor is defined as a given volume and puffs are counted up within this volume to obtain a concentration. Minor variations on this approach such as "slug" calculations and kernel density estimators have served to reduce the number of particles needed to obtain accurate concentration estimates by stretching out the influence of a puff over a defined area. However, this does not solve the problem and when long range transport with its associated dispersion is taken into account, puffs from sources can number into the 10's of thousands in order to obtain accurate concentration numbers.

One other deficiency is its treatment of chemistry. Linear chemistry can be handled easily within the Lagrangian framework; however, secondary nonlinear chemistry is not handled well since there is no simple way to treat the interaction of a puff and the regional background concentration of precursors, which is necessary when looking at nonlinear chemistry.

INVENTORY

The starting point of any modeling system is the inventory. If the inventory does not accurately estimate the emissions or background concentrations of interest, the results from the most accurate modeling system are doomed to failure. For this reason, a strong emphasis is made on obtaining an inventory with the same level of accuracy and resolution as produced by the meteorological and air quality models used.

There are two current inventories and one proposed effort which could be used in a project of this nature. The first is a seasonal smoke emission inventory developed on a 50 km grid which estimates current and future prescribed fire and wildfire activities in the western United States. In this inventory, the lands in the West, excluding Montana, were classified according to ownership, vegetation cover type, and land allocation class (LAC). The LAC was used to describe general management goals determining where and when fire would be used and the feasibility of using mechanical treatment. Information in this database included as vegetation type, natural fire rotation, and current and future prescribed burn activity. This information was then processed through an emissions model which calculated seasonal emissions from both prescribed fires and wildfires over the western United States for the years 1995, 2015 and 2040. The major drawback to using this inventory in the current modeling effort is that it includes emissions at a 50 km grid spacing, which is not compatible with the meteorological input at 10 km. Averaging the meteorology to the 50 km grid spacing would eliminate most of the important small scale transport information.

Another inventory was used by the Interior Columbia River Basin (ICRB) study and involved the computation of emissions from this smaller area. The primary information from this study is the heat release data generated by the Emission Production Model (EPM) developed by the USDA Forest Service. The EPM model produced heat release information for 228 prescribed and wildfires. Within the prescribed fires, 126 simulations were performed. These simulations varied by vegetation type (shrubs, grass, mixed conifer, and ponderosa pine), burn type (underburn, broadcast, and pile), and size (5-220 acres). The remaining simulations accounted for wildfires and varied only by the vegetation types previously mentioned and size (50-50,000 acres). This information is useful for determining how high in the atmosphere a fire plume will rise before the plume becomes nonbuoyant (Ottmar et al. 1997).

The inventory that seems most promising for use with this study has not been completed at this time. It will use information from both the GCVTC and ICRB data sets. A vegetation map from the ICRB study based on the Land Cover Characteristics (LCC) defined in Loveland et al. (1991) will identify 159 LCC classes with 1 km grid spacing. A biophysical map will be based on the potential vegetation types (PVT) as delineated by Kuchler (1964) and others. This will be further refined using a Digital Elevation Map (DEM) to obtain information on slope, elevation, and aspect. A panel of experts will be assembled to refine the LCC based on the biophysical map and produce a vegetation map with the same vegetation classes as used in the GCVTC inventory. Once the information from this inventory has been linked to the vegetation classes in the GCVTC, the potential exists for using GCVTC information regarding future projections for the current inventory. However, the 50 km information from the GCVTC will need to be linked in some way to the 1 km information in this inventory. This will be done by aggregating the GCVTC information via spatial analysis onto an ecoregion as defined in Bailey (1995). This will then provide an estimation for a pre-scribed fire treatment action assuming that the calculated variance within each ecoregion meets some acceptable tolerance. The vegetation and treat-ment will then be calculated using an emissions model described in Hardy et al. (1996a) to calculate the emissions (Hardy, 1996b).

SPECIFIC APPLICATION–COLORADO PILOT PROGRAM

The Colorado workshop's goal was to develop a methodology for risk assessment of wildfire by integrating information from potentially-impacted resources and values, including habitat protection, endangered species, soils, and air quality. Experts from each of these fields were assembled in Colorado and asked to develop a methodology for targeting "high risk" areas with these areas being determined by using specific information pertaining to Colorado, while keeping in mind that this scheme should be robust enough to work for other areas in the West.

A crucial element in the air quality section involved the determination of the air quality downwind from a potential source area. The Phase I modeling methodology described in the last few sections was used to evaluate impacts from wildfire on a statewide basis. To do this, MM5 was run for a seven day event in 1994 starting on July 2 and ending on July 8–a period corresponding to a severe fire on the west slope of the Rocky Mountains at Storm King Mountain. The transport winds at the surface for this time period suggested uniform southwesterly winds on the west side of the Rockies, variable winds in the mountains and southeasterly winds east of the Rockies. This pattern persisted for the first four days of the simulation, followed by the approach of a cold front which passed through the area on day 5 and 6 of the simulation.

The emissions inventory data were taken directly from the ICRB study which produced emissions and heat release information based on a 1,000 acre wildfire in mixed conifer and ponderosa pine forest types. (A major problem with the EPM model run for the ICRB was that it did not handle the smolder-ing consumption of a fire well, with the EPM-modeled fire producing a smoldering phase that was too short.) In order to circumvent this problem and obtain a more realistic 25,000 acre fire profile, five 1,000 acre fires were initialized over a span of five days to create a 25,000 acre fire profile lasting five days. (One 1,000 acre fire was started at hour 1, 5, 10, 15 and 20 of each day with each fire lasting approximately 13 hours.) The combined heat re-lease profile for this fire can be seen in Figure 13.5. In this Figure, the

FIGURE 13.5. Heat release profile for the aggregated 25,000 acre simulated Program wildfire. (Total emissions of PM were also mapped to this profile.) (Units are in btu/sec × 10^{-7}.)

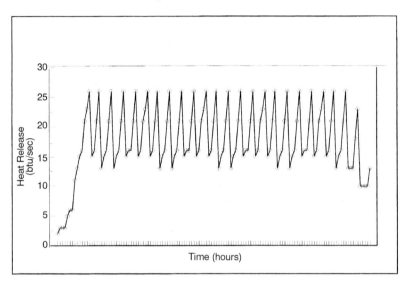

emissions over the time of the fire were seen to exhibit peaks and valleys in fire intensity as one to three fires were active in differing stages of a burn. Heat release information is used to calculate effective nonbuoyant plume heights in the air quality models. As will be shown in a later Figure, areas close to the source of a wildfire do not always experience the maximum impact due to the elevated release height of this nonbuoyant plume.

Figure 13.6 shows an elevation map used to establish the area burned by the simulated fires in the model. The range of 5500-8000 ft MSL was chosen to represent the elevation ranges of a mixed conifer and ponderosa pine forest. All of the highlighted cells were burned in this simulation to test worst case impacts downwind from any potential source location in Colorado. The exact time of the wildfire start in any of the burned cells was determined randomly within the first 24 hours of the simulation. This simulation was not conducted to represent some possible real case but to examine potential worst case impacts from the combined influence of wildfire emissions. One of the tasks assigned to the fire emissions group at the meeting was to estimate the most likely locations in Colorado to experience wildfires. This information in turn can be used to more accurately place fires in the grid to represent a possible "real" impact from multiple prescribed fire occurrences.

Figure 13.7 shows the maximum concentration estimates of 24 hr average

FIGURE 13.6. DEM map identifying locations for the wildfire simulated for the Colorado Pilot Programs. (Scale represents elevation in meters.)

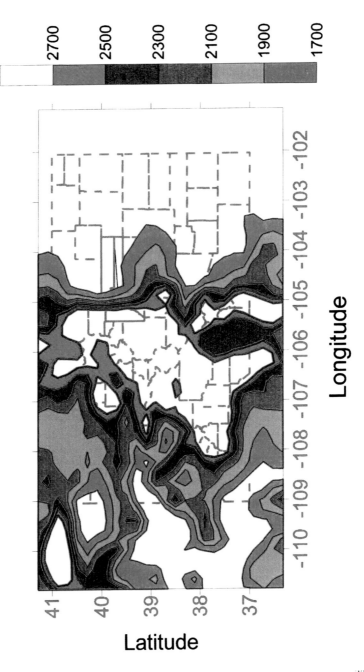

FIGURE 13.7. Maximum 24-hour PM$_{2.5}$ concentrations (μg/m^3) for the Colorado Pilot Program, July 2 (7 pm MST) to July 8 (12 noon), 1992.

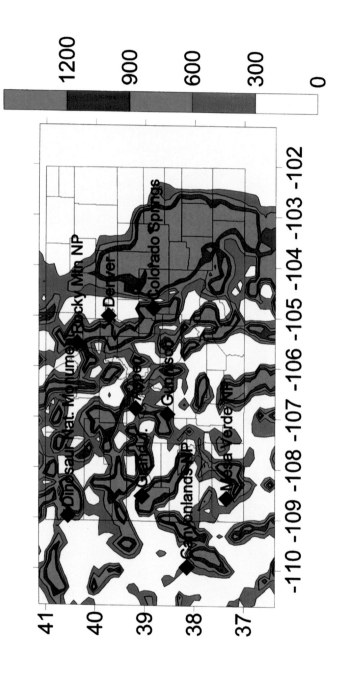

$PM_{2.5}$ for the source areas shown in Figure 13.6. Estimates of impact range from none to 1500 $\mu g/m^3$ for a 24 hour average. The current 24 hour standard for PM_{10} ($PM_{2.5}$ is a subset of PM_{10}) is 150 $\mu g/m^3$. Thus, the simulated concentration values are an order of magnitude over the health standard. This represents a worst-case estimate of the potential impact from burning most of the Colorado forested areas and only represent some possible upper impact estimate. This suggests, however, that if many forested areas were to experience a wildfire, most of Colorado would be impacted. The next step in this process is to use emissions information specific to Colorado linked with a probability map to run another simulation accounting for the most likely number of fires, and determine the impact to large urban areas and Class I regions.

CONCLUSIONS

This work presents a methodology by which air quality impacts from both wildfire and prescribed fire can be evaluated over distances covering entire states or multi-state regions. The methodology recognizes the unique problems associated with the complex terrain in the western United States and couples an emissions inventory with an advanced meteorological and air quality model to determine the impacts of both prescribed fire and wildfire. A specific example of an application for Colorado demonstrates how this methodology might be applied on a state-by-state basis to evaluate intrastate transport issues. This demonstration predicts that most of the state of Colorado has the potential to be significantly impacted by wildfire activity in the forested areas of the State.

REFERENCES

Bailey, R.G. 1995. Description of the ecoregions of the United States. 2nd ed. rev. and expanded (1st ed. 1980). Misc. Publ, No. 1391 (rev.), Washington, DC: USDA Forest Service. 108p. with separate map at 1:7,500,000.

Cover, D.E., P.S. Mitchell, M.D. Zeldon, and R.J. Farber. 1990. A computer aided meteorological classification scheme for the desert southwest. In: *Visibility and fine particles*, C.V. Mathai, Ed., TR-17, AWMA, Pittsburgh.

Guthrie, P.D., C.A. Emery, M.P. Ligocki, G.E. Mansell, A.M. Kuklin, and D. Gao. 1995. Development and preliminary testing of the Regulatory Modeling System for Aerosols and Deposition (REMSAD). Technical Memorandum, Systems Applications International, December, 1995.

Hardy, C.C., R.E. Burgan, R.D. Ottmar, and J.E. Deeming. 1996a. A database for spatial assessments of fire characteristics, fuel profiles, and PM10 emissions. In: USDA Forest Service, ICRB Scientific Assessment Landscape Ecology staff area Report; chapter 19B. USDA Forest Service, Intermountain Research Station, Report on file at Intermountain Fire Sciences Laboratory, Missoula, MT.

Hardy, C.C., R.E. Keane, and J. Menakis. 1996b. Development of coarse scale vegetation and fuels data in support of an air quality prediction system for the western United States. Proposal submitted to WESTAR Council for funding.

Heilman, W.E. 1995. Synoptic circulation and temperature patterns during severe wildland fires. In: *Preprint volume I of the ninth conference on applied climatology.* 15-20 January 1995. Dallas, TX.

Kuchler, A.W. 1964. Potential natural vegetation of the conterminous United States. (Spec. Pub. 36.) New York: American Geographical Society. 116 p. with separate map at 1:3,168,000.

Loveland, T.R., J.W. Merchant, D.O. Ohlen, and J.F. Brown. 1991. Development of a land-cover characteristics database for the conterminous U.S. *Photogrammetric engineering and remote sensing. 57*(11), 1453-1463.

Ottmar, R.D., E. Alvarado, and P.F. Hessburg. 2000. Historical and current forest and range landscapes in the Interior Columbia River Basin and portions of the Klamath and Great Basins. Part II: Linking vegetation patterns to potential smoke production and fire behavior. Gen. Tech. Rep. PNW-GTR-xxx. Portland, OR: U.S. Department of Agriculture, Forest Service, Pacific Northwest Research Station. (In Press).

Potter, B.E. 1995. Atmospheric stability, moisture and winds as indicators of wildfire risk. In: Preprint volume I of the ninth conference on applied climatology. 15-20 January 1995. Dallas, TX.

Systems Application International. 1996. *Users guide for the Regulatory Modeling System for Aerosols and Deposition (REMSAD).* SYSAPP-96/42. September, 1996.

Scire J.S. and V.R. Tino. 1996. Modeling of wildfire and prescribed burn scenarios in the Columbia River Basin. Final report. USDA Forest Service report no. 1459-01. March 1996.

Stocker, R.A. 1996. Numerical examination of long-range dispersion using RAMS with a climatological analysis for the MOHAVE field study. Masters thesis, Colorado State University, Fort Collins, CO, 128pp.

Takle, E.S., D.J. Bramer, W.E. Heilman, and M.R. Thompson. 1994. A synoptic climatology for forest fires in the NE US and future implications from GCM simulations. *J. Wildland Fire 4*(4): 217-224.

Index

329